LabVIEW™ for Data Acquisition

ISBN 0-13-015362-1

90000
9 780130 153623

NATIONAL INSTRUMENTS | VIRTUAL INSTRUMENTATION SERIES

Bruce Mihura
- LabVIEW for Data Acquisition

Lisa K. Wells • Jeffery Travis
- LabVIEW for Everyone

Mahesh L. Chugani, Abhay R. Samant, Michael Cerra
- LabVIEW Signal Processing

Barry Paton
- Sensors, Transducers, and LabVIEW

Rahman Jamal • Herbert Pichlik
- LabVIEW Applications and Solutions

Jeffrey Travis
- Internet Applications in LabVIEW

Shahid F. Khalid
- LabWindows/CVI Programming for Beginners

Hall T. Martin • Meg L. Martin
- LabVIEW for Automotive, Telecommunications, Semiconductor, Biomedical, and Other Applications

Jeffrey Y. Beyon
- Hands-On Exercise Manual for LabVIEW Programming, Data Acquisition, and Analysis

Jeffrey Y. Beyon
- LabVIEW Programming, Data Acquisition, and Analysis

LabVIEW™ for Data Acquisition

Bruce Mihura

Prentice Hall PTR
Upper Saddle River, NJ 07458
www.phptr.com

Library of Congress Cataloging-in-Publication Data
Mihura, Bruce.
LabVIEW for data acquisition / Bruce Mihura.
p. cm. -- (National Instruments virtual instrumentation series)
Includes index.
ISBN 0-13-015362-1
1. LabVIEW. 2. Physical measurements--Automation. 3. Scientific apparatus and instruments--Computer simulation. 4. Computer graphics. I. Series.

Q185 .M473 2001
530.8'1'028566633--dc21 2001021681

Editorial/production supervision: *Vincent Janoski*
Acquisitions editor: *Bernard Goodwin*
Editorial assistant: *Michelle Vincenti*
Marketing manager: *Dan DePasquale*
Manufacturing manager: *Alexis Heydt*
Cover design: *Nina Scuderi*
Cover design director: *Jerry Votta*

Published by Prentice Hall PTR
Prentice-Hall, Inc.
Upper Saddle River, NJ 07458

Prentice Hall books are widely used by corporations and government agencies for training, marketing, and resale. The publisher offers discounts on this book when ordered in bulk quantities. For more information, contact: Corporate Sales Department, Phone: 800-382-3419; Fax: 201-236-7141; E-mail: corpsales@prenhall.com; or write: Prentice Hall PTR, Corp. Sales Dept., One Lake Street, Upper Saddle River, NJ 07458.

Printed in the United States of America
10 9 8 7 6 5 4 3 2 1

ISBN 0-13-015362-1

Pearson Education LTD.
Pearson Education Australia PTY, Limited
Pearson Education Singapore, Pte. Ltd.
Pearson Education North Asia Ltd.
Pearson Education Canada, Ltd.
Pearson Educación de Mexico, S.A. de C.V.
Pearson Education—Japan
Pearson Education Malaysia, Pte. Ltd.
Pearson Education, Upper Saddle River, New Jersey

Contents

10 DAQ at a Distance—Networked and Distributed Systems 385

11 Alternate Software for DAQ 393

12 Finalizing Your LabVIEW Software 403

Preface

I've been interested in gadgets and computers since grade school, particularly gadgets connected to computers. My first real gadget-to-computer project, at age 16, was an alarm for my car made from an Atari computer with 16 Kbytes of RAM and a photoresistor. I caught no car thieves, but I did catch a cat trying to steal a nap on my car! Six years later, I was more than a little happy to get my first job out of college at NI (National Instruments), where I worked for years as a LabVIEW developer, actually getting paid to connect gadgets to computers! I quickly learned that *data acquisition* is the professional term for connecting certain types of gadgets to computers, like the photoresistor in my first car alarm.

If you already have an NI data acquisition board with analog input connected to a computer with LabVIEW 6i, and you are dying to watch it work, skip right ahead to start on Chapter 3, up to and including Section 3.1.5, then come back here.

LabVIEW and Data Acquisition

This book is written for people who intend to use National Instruments' LabVIEW (**Lab**oratory **V**irtual **I**nstrument **E**ngineering **W**orkstation) for data acquisition. LabVIEW has become very popular as the programming language of choice in the context of industrial, scientific, academic, and laboratory environments. LabVIEW is a graphical programming language in which you build the programs with pictures, not words. Data acquisition involves connecting computers to a wide variety of gadgets via electronic signals; the computers then control these gadgets or read data from these gadgets. The term *DAQ* will be used throughout the book instead of data acquisition.

In general, any place that "scientific measurements" must be taken is an appropriate place for LabVIEW. Following are a few examples of LabVIEW applications:

- Measuring pressure, temperature, and vibration in an airplane during the course of its flight
- Monitoring the pH of a chemical solution during processing
- Analyzing sound waves in an acoustics laboratory
- Monitoring and recording flow rates of liquids or gasses

To use LabVIEW for any of these four example applications, you need the following four components:

1. LabVIEW
2. A computer on which LabVIEW can run
3. A *data acquisition* device (changes an electrical signal into something the computer can read)
4. A *transducer* (changes a wide variety of real-world phenomena, like pressure, temperature, pH, sound, etc., into an electrical signal for the data acquisition device)

National Instruments currently makes LabVIEW and a wide variety of data acquisition devices. If you are going to use LabVIEW, I recommend buying your data acquisition device from National Instruments, as it generally simplifies its integration with LabVIEW.

Because LabVIEW is a graphical programming language, it is often quicker to develop than using a text-based language, and its programs are often much more robust.

LabVIEW not only runs on all Microsoft operating systems starting with the word Windows, but it runs on Apple Macintosh O/S, Sun Microsystem's Solaris, and certain Hewlett-Packard workstations as well.

Organization

If you have never written a program before in any computer language, you will likely find this book difficult to follow—if this is the case, consider starting with the book *LabVIEW for Everyone,* described in Chapter 1, Section 1.1. Nobody learns how to program for the first time, in any language, without spending much time—usually more time than they expect. Be encouraged to know that LabVIEW, like Microsoft Visual Basic, is one of the least painful languages to learn.

Data acquisition is fundamental to many LabVIEW applications. For this reason, this book is written not quite as a "for dummies" book, but more from the "LabVIEW newcomer" point of view. Many people using LabVIEW for the first time want to do data acquisition—so if you are completely new to LabVIEW, Chapter 1 is designed to teach you just enough LabVIEW to perform some meaningful data acquisition.

Chapter 1: Learning LabVIEW for the First Time. This chapter is designed to teach a LabVIEW newcomer just enough to perform useful data acquisition with LabVIEW. It is meant to be the quickest LabVIEW tutorial ever, but as a result, it's a bit like taking a drink from a fire hose. This chapter is not explicitly DAQ-specific, but it subtly focuses on DAQ-related issues.

Chapter 2: Signals and DAQ. Learn or review the fundamentals of data acquisition that you will find relevant with LabVIEW—or with any programming language.

Chapter 3: Basic DAQ Programming Using LabVIEW. Combine LabVIEW with data acquisition at a very fundamental level. Use real hardware and real wires, and manipulate real signals in this chapter.

Chapter 4: Simulation Techniques. Most of your developing can be done without real hardware, right in the comfort of your home or office.

Chapter 5: DAQ Debugging Techniques. This chapter tells you how to track down bugs (with a focus on data acquisition programming), should you ever make a programming mistake.

Chapter 6: Real-World DAQ Programming Techniques. This chapter is a version of Chapter 3. In order to focus on data acquisition issues without being hindered by limitations of your specific device, all hardware is simulated in this chapter. The most common real-world scenarios are covered in detail in this chapter, and these scenarios often require the advanced techniques covered herein.

Chapter 7: Transducers. Exactly what device do you need to convert your temperature (or pressure, humidity, etc.) into a signal compatible with you data acquisition device? This chapter points you in the right direction.

Chapter 8: Non-NI Hardware Alternatives. Do you already have a data acquisition device that you want to use with LabVIEW? Or do you just want to save some money on your data acquisition device and have time to spare? Read this chapter.

Chapter 9: Real-Time Issues. Suppose you must collect data at a rate of exactly 10 Hz, or 10 times per second—it is unacceptable to wait 0.11 seconds between data samples. Such issues are not at all obvious, but they are covered in this chapter.

Chapter 10: DAQ at a Distance—Networked and Distributed Systems. Suppose your computer must be placed hundreds of feet from the data you're collecting. Or, suppose you have many data sites, widely separated from one another and your computer. This chapter is for you.

Chapter 11: Alternate Software for DAQ. Suppose you want to perform data acquisition, but you want to use some software other than LabVIEW. For example, maybe you're already very familiar with another programming environment—see this chapter.

Chapter 12: Finalizing Your LabVIEW Software. Your LabVIEW software now works perfectly, and you don't want *anybody* changing it, not one bit! Here's how to accomplish that.

Appendix A: Fundamentals: Bits, Bytes, Files, and Data. If you have little programming experience or less, this appendix covers programming fundamentals that are needed in *any* programming language. Lab-

VIEW is no exception! It is not enough to know how to draw pretty pictures; you must understand what's going on underneath.

Appendix B: Top Ten DAQ Problems and Their Solutions. You'll find answers to some common problems in this appendix.

Appendix C: Saving LabVIEW's VIs. It is not obvious that you can sometimes cause permanent damage to LabVIEW itself, not just to your application, if you make a common blunder described herein.

Appendix D: Example Applications. Here is a collection of real-world LabVIEW applications I've personally done, presented so that you can be aware of the abilities of which LabVIEW is capable.

Appendix E: LabVIEW/DAQ Tips and Tricks. Review this list at some point while learning LabVIEW—these are time-saving tips and tricks.

Requirements

This book assumes you have version 6i of LabVIEW; you will gain the most from this book if you work through its examples with version 6i. Its most relevant and powerful features will be illustrated in this book. Ideally, you also have one of NI's multifunction DAQ devices. To work perfectly with this book, your DAQ device should have at least two analog inputs, two digital I/O ports, a counter/timer, and an analog output. However, if your DAQ device doesn't have all of these features, then let's hope you won't need them, so you can just skip the parts of the book that use them.

If you're a programmer, but have never programmed in anything but text-based languages, get ready for a surprise. Figure P–1 shows a "simple" example program that ships with LabVIEW 6i, where the user interface is not shown—just the guts of the program.

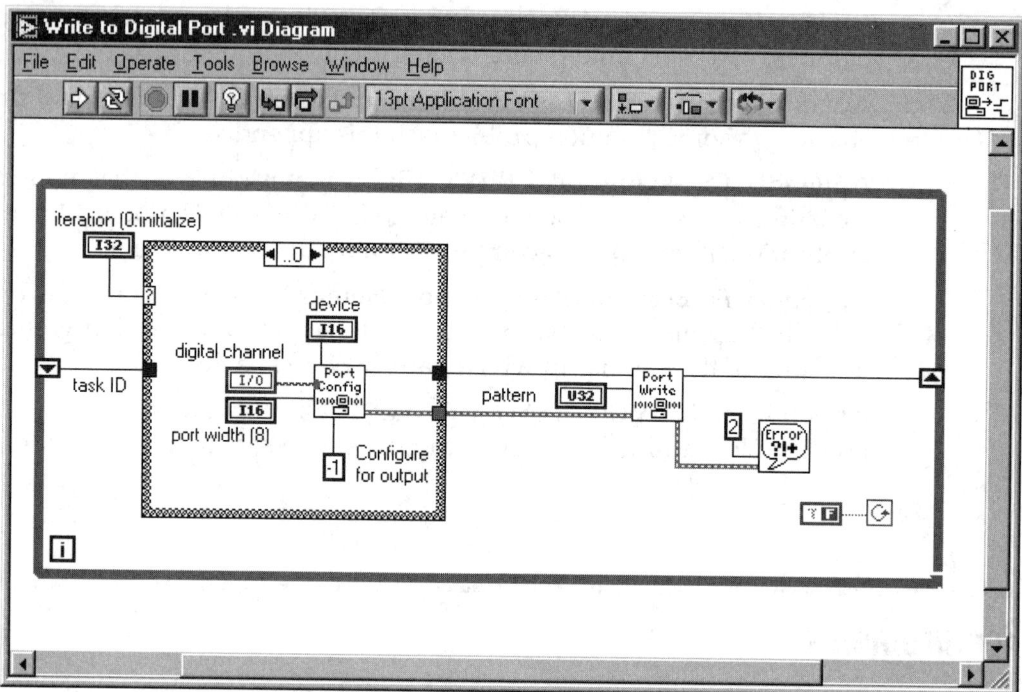

Figure P–1
An example LabVIEW code taken directly from National Instrument's data acquisition libraries.

DAQ, in the context of this book, means monitoring or controlling physical phenomena with a computer via electrical signals. These electrical signals are defined by their voltage or current levels, and are usually attached to some sort of scientific or industrial equipment by means of transducers that can convert physical values like pressure, temperature, position, flow rate, and so on to electricity (or vice versa). Most types of computers can interface to these transducers by means of DAQ devices. Companies like Agilent and Tektronix make a variety of high-quality, often specialized scientific instruments that perform very accurate and precise measurements, often in a laboratory environment. These instruments are generally connected to the computer via a special interface cable (GPIB, RS-232, etc.), but they can run without the computer connection, unlike the DAQ devices discussed in this book. These instruments will not be covered in this book to any great extent, even though they fit the definition as DAQ. They really *are* DAQ, in my opinion, but instead, we'll mostly use this book to describe DAQ as NI does—as generic analog and digital signals, because these signals are flexible enough to be used with an extremely wide variety of transducers. The usual example of the DAQ equipment discussed in this book is a device connected to a computer, which is also connected to a terminal block outside of the computer, reading or writing electrical signals connected to whatever you're interested in. Figure P–2 is a generic diagram of a typical DAQ system.

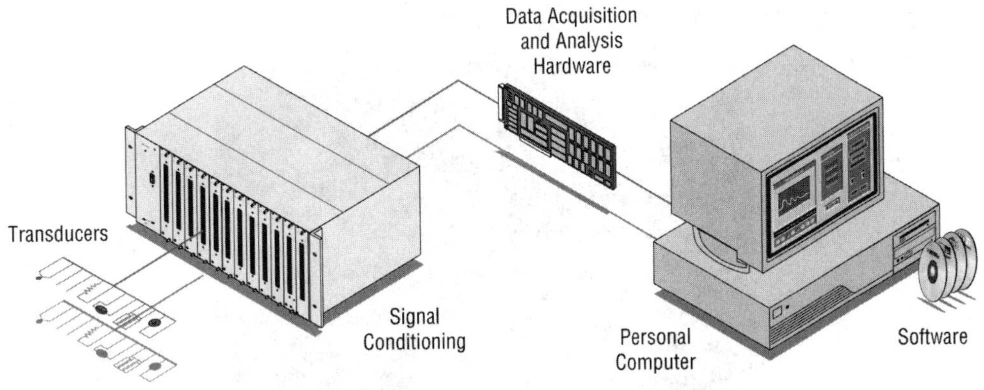

Figure P–2
A generic DAQ system.

For many DAQ systems, the Signal Conditioning unit shown in Figure P–2 may not be present—that unit might just be a simple connector block with no signal-conditioning electronics. In this case, it would be screw terminals or some sort of similar electrical/mechanical connection.

Going into a bit more detail, Figure P–3 shows more of the possible components of a DAQ system.

Figure P–4 shows what a DAQ system might really look like, using NI's SCXI hardware and built by someone who knows how to wire much more neatly than I.

A DAQ device may be a board inside the computer, or it might be some sort of external box connected to one of the computer's ports, such as the serial port (RS-232), a parallel port, USB, or IEEE 1394 (FireWire).

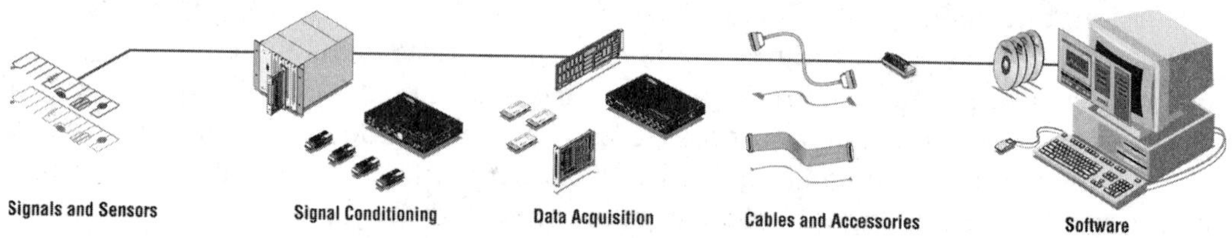

Figure P–3
A DAQ system showing more details and possibilities.

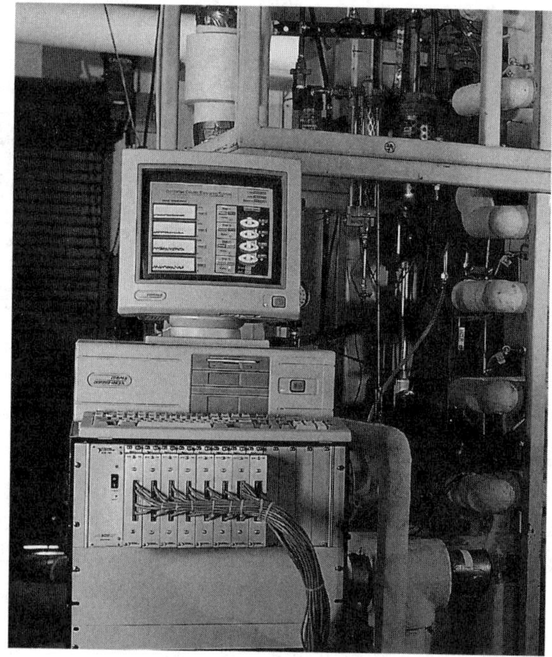

Figure P–4
A neatly configured DAQ system.

National Instruments categorizes LabVIEW's functionality into three parts:

1. Acquire
2. Analyze
3. Present

LabVIEW is designed to seamlessly integrate these basic components of data acquisition, and it does a great job, in my opinion. The acquisition part of LabVIEW makes it very easy to quickly collect data from NI's DAQ devices, as you have discovered if you've already done the first section of Chapter 3 through Section 3.1.5. The analysis libraries available with the full development system have an amazing list of functions involving FFTs, (**F**ast **F**ourier **T**ranforms) power spectrums, RMS (**R**oot **M**ean **S**quare) calculations, filters, curve-fitting, and many other mathematical functions. The presentation part of LabVIEW is perfectly tuned to very quickly present a wide variety data in a meaningful fashion.

One of NI's catch phrases is "The Software Is The Instrument." Or at least it used to be—I haven't heard that one in a while. At any rate, the comparison is that a modern lab can operate with a computer controlling everything, rather than with numerous separate instruments like oscilloscopes and function generators. This approach not only makes it inherently easier to analyze, manipulate, and control your data, since the data source is always connected to the computer by default, but it also makes it easier to upgrade or modify your instrument configuration.

This book is written as if you have a PC running a Windows operating system—not a Macintosh, UNIX platform, or other type of setup. However, if you do not have a PC, most of the information presented in this book will still be useful, provided your platform can run LabVIEW. You will need to mentally substitute the PC's <Ctrl> and <Alt> keys for whatever the corresponding keys are for your computer's LabVIEW interface, and also try to recognize the PC-specific issues that differ on your computer.

Help! I'm Stuck!

You have quite a few options when you get stuck in LabVIEW. Here they are, in the order you should normally try.

1. *Ask a live person who knows LabVIEW.* Ideally, there is somebody you work with, or know, who can answer some of your LabVIEW questions.
2. *Use LabVIEW's Help.* See Chapter 1, Section 1.3 for details on LabVIEW's internal help.

3. *Internet help*. Knowing how often Internet links die, I'm a bit reluctant to put any such links in this book, so I'll only put in a couple that I suspect will be around for a long time.

   First, try NI's Web site, *www.ni.com*. You can hunt around for help on the site, looking for links named "support" or "technical support" to begin with. Currently, the most helpful public forums seem to be *info-labview*, a busy LabVIEW mailing list, and the newsgroup *comp.lang.labview*; you will need to find these yourself, though, lest I give you broken links.

   Try a link to other links at my own Web site, whose availability is at the mercy of my ISP, at *www.LCtechnology.com/lvhelp.htm*. I will keep this current with your best Internet LabVIEW help options, at least for as long as I'm alive.
4. *Call NI*. Currently, the number is 1-800-IEEE-488.
5. *Hire a consultant*. NI has the National Instruments Alliance Program, which is a group of certifiable, uh, certified consultants like me, who can likely solve your technical problems relating to NI products. These companies range from one-man operations (such as mine at *www.LCtechnology.com/consult.htm*) to very large companies with better Web sites and hundreds of people.
6. *Give up*. Let's hope it won't come to this!

Contacting Me

I may be a moron for giving out my email address, but I'm at *bruce@LCtechnology.com*, and you may need to include the code word `lvdb` in the subject line should I need to implement a spam filter someday. Please follow these rules:

1. If you get stuck or find an error in the book, first try *www.LCtechnology.com/lvhelp.htm* to see if I've addressed the tough part of the book or the error. If it's not there, please email me!
2. If you're a LabVIEW pro, and you think I've left something out of this book, consider that this book is also designed for beginners before saying that I should have included such-and-such feature of LabVIEW. Must keep book short.
3. Please email me only if you have suggestions for improving the book.

Acknowledgments

First, a special thanks goes to Lisa and Ravi:

Lisa Well—Lisa K. Wells authored some of the earliest published LabVIEW books. She is the author of the *LabVIEW Student Edition User's Guide* (Prentice Hall, 1995), **LabVIEW Instructor's Guide** (Prentice Hall, 1995), and lead author of *LabVIEW for Everyone* (Prentice Hall, 1997).

Ravi Marawar—Ravi is Academic Program Manager at National Instruments and his primary responsibility is promotion of National Instruments tools in university teaching and instruction worldwide

Lisa was my connection to the publisher (you would not be reading this sentence without her), while Ravi was my National Instruments contact, who did a fantastic job supplying me with hardware, software, and internal technical support.

While writing the book, I posted a request to the LabVIEW-related newsgroups for people to proofread the original drafts to this book. Here are the people who really came through for me....

Dave Wayne, Analytical Chemist (inorganic mass spectrometry), Nuclear Materials & Technologies Division (NMT-15), Los Alamos National Laboratory

Brian Powell, LabVIEW R&D Senior Group Manager

Deborah Bryant, LabVIEW senior software engineer

Chip Henkel

Venu P. Nair

Michael Shinder is a biomedical research scientist constructively utilizing data acquisition and analysis to study neuroscience

Dr. Ali Ashayer Soltani, Research Associate at Penn State University

Learning LabVIEW for the First Time

1.1 Introduction

This chapter is written for version 6i of LabVIEW. If you do not have LabVIEW, and if the evaluation version of LabVIEW is not still with this book, you can get the free evaluation version (good for 30 days; programs cannot run for more than five minutes) or the real thing from National Instruments (*www.ni.com* or 1-800-IEEE-488).

If you know LabVIEW already and want to dive into DAQ now, skip this chapter except for Section 1.6, and go directly to Chapter 2 (for an overview of non-LabVIEW-related DAQ fundamentals) or Chapter 3, where working through Section 3.1.5 will hopefully get you a LabVIEW-DAQ connection.

I've intentionally made this chapter the most densely compacted, interactive introduction of LabVIEW possible—it will be like taking a drink from a fire hose. So unless you're already proficient in LabVIEW (and a speed-reader), don't expect to make it through this chapter in one day! Although compact, this chapter will introduce every aspect of LabVIEW needed to write the majority of useful DAQ programs.

If you have never programmed in any language before, you will not only need to learn LabVIEW, but you will need to learn basic programming concepts as well. First, become familiar with the information in Appendix A—preferably with an experienced software nerd around to explain any difficult concepts. (Note: Do not refer to your nerd as such when asking for help—try "professional.") If you have some programming experience but are new to LabVIEW, carry on with this chapter. Even if you are quite familiar with LabVIEW, at least skim this chapter to make sure you understand the concepts herein. But if this chapter seems too difficult, try the well-written *LabVIEW for Everyone*, by Lisa K. Wells and Jeffrey Travis, which serves a slower-paced, more thorough LabVIEW tutorial "for everyone." Whether Lisa bribed me to say this will never be publicly known.

During reviews, we considered omitting the next paragraph; however, I intend to tell things the way they are, and that means pointing out the negative along with the positive. Based upon the title, you may expect me to imply throughout this book that LabVIEW is *always* better than other languages for DAQ.

I have built DAQ applications in every style I can think of (three graphical languages, numerous text-based languages, assembly language on multiple computer platforms, microcontrollers in text *and* assembly, hardware-only, and more)—and many times over in each style. *Yet LabVIEW is* usually *my first choice for most DAQ jobs.* That's why I feel it worthwhile to write this book. But read on for alternatives. The truth is this: If you are already very good with C, or if your application is going to be extremely complex, you may be better off using NI's LabWindows/CVI for DAQ applications instead of LabVIEW. If you are good with Microsoft Visual Basic or Microsoft Visual C++, NI's Measurement Studio has excellent DAQ tools for you to use instead of LabVIEW. Or you may be able to get away without buying any NI software if you have a very simple application, by interfacing directly to NI-DAQ from another programming environment. See Chapter 11 for details on these options. Also, unless you follow the guidelines in Chapter 6, Section 6.1.1, LabVIEW can be tough to handle when your program gets very large or complex. Although DAQ is the focus of this book as a whole, it is not the focus of this chapter. This chapter is for LabVIEW in general, yet is subtly optimized to give you the necessary prerequisites for most DAQ applications.

I highly recommend that you work through the examples in this book with LabVIEW and a real DAQ device. If you don't have a real DAQ device, do what you can in LabVIEW. Like a math book, it is important that you read everything *sequentially* in this chapter and that you understand this chapter before proceeding to the rest of the book. Unlike a math book, this book is designed to make sense to most people. Check that—most people with any engineering or programming experience. In order to keep this chapter as short as possible, yet thorough enough to get you going with DAQ, each important topic is covered only once as the tutorial progresses, and any given topic may rely upon previous work; hence the need for sequential reading.

You should already know how to operate a computer on a very fundamental level. You should be familiar with

1. files and folders (folders are also called directories);
2. how to move, copy, and delete files and folders;
3. floppy disks and hard drives;
4. how to operate a text editor (like Microsoft Word, WordPad, or Notepad);
5. memory (also called RAM); and
6. the term *"operating system."*

Throughout this book, you will occasionally be instructed to *pop up* on an object. To pop up means to move your cursor so that it is over that object, and right-click (click with the right mouse button, not the left). Only one point on every cursor, called the *hot point,* actually "works" when you click; you may need to experiment to find exactly where the hot point is. An object on which you can pop up is said to have a *pop-up* menu. Finally, you will often need the Positioning tool or the Operating tool in order to get the proper pop-up menu—these tools will be introduced in Section 1.2.2.

As you're working through this book, you may close all VIs after each section, or your screen will become quite cluttered.

1.2 VI Basics

"What is a VI?," you might ask. VI (pronounced vee-eye) means Virtual Instrument, and that is what LabVIEW was originally geared for—making software versions of instruments that you might find in a laboratory. In a more general context, a VI is any program module written in LabVIEW.

1.2.1 The Front Panel, the Block Diagram, and Saving Your VIs

LabVIEW is a graphical programming language in which the programs are written using pictures, not words. Instead of having text, like

```
sumOfSquares = a * a + b * b;
```

LabVIEW instead has a graphical representation of its operations, as in the VI shown in Figure 1–1, saved as `Sum of Squares.vi`.

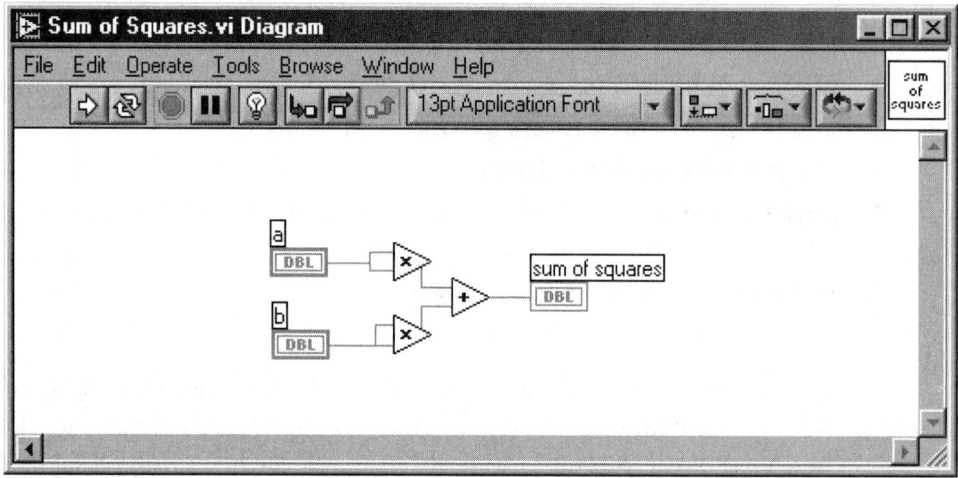

Figure 1–1
LabVIEW code (a block diagram) does not look like the usual text code found in most programming languages.

A VI consists of three major parts: (1) a front panel, (2) a block diagram, and (3) an icon/connector (consisting of an icon and a connector pane). We will discuss the first two of these three parts in this section.

The window in Figure 1–1 is called a block diagram, and it is always associated with a front panel, which is another window. Every front panel has at most one block diagram, and every block diagram has exactly one front panel. The front panel is what the user sees (sometimes called a GUI, or graphical user interface), and the block diagram is the *code,* or the heart of the program. A reasonable front panel for the block diagram in Figure 1–1 would be the example shown in Figure 1–2.

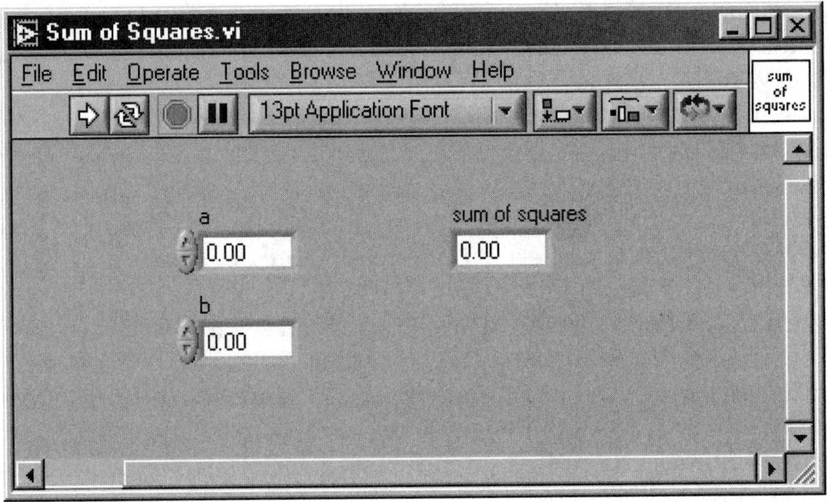

Figure 1–2
Example of a LabVIEW front panel.

The front panel is gray by default, and the block diagram is white by default. You may find it useful to keep them colored this way so you can instantly distinguish front panels from block diagrams during development. If you must change their colors, at least keep the block diagrams white and the front panels whatever subtle color you like; this helps both you and your users.

Initially, one of the most confusing parts of LabVIEW is the relationship between the front panel and the block diagram. Referring to Figures 1–1 and 1–2, you can see how the three rectangles on the block diagram have a correlation to those on the front panel, since their labels are the same. On the front panel, a user can type any number into the boxes labeled `a` or `b`. The sum of the squares of these numbers is calculated and shown on the front panel in the `sum of squares` box whenever the **Run** button ⇨ is pushed (the **Run** button can be found near the upper left of the front panel).

We will now use LabVIEW to create an empty front panel and block diagram (if you don't have LabVIEW, the evaluation version included with this book, or available free from NI, can get you started). Later in this lengthy chapter, we will build the VI shown in Figure 1–1.

First, create a folder on your hard drive where you should store all VIs produced in this book. Since the upcoming chapters will use VIs from previous chapters, this will make more sense than creating a separate folder per chapter. In this book, this folder will be `C:\Bruce\Projects\LV DAQ Book\VIs`.

In Windows, find the **Start** button in one of the corners of your screen, typically the lower left corner. Select **Start»Programs»National Instruments LabVIEW 6i** from the Windows task bar (this menu path may vary on your computer). If LabVIEW asks you to log in and you don't want to log in every time LabVIEW launches, see Appendix E, item 15. If LabVIEW was not running, you might see an introductory LabVIEW window asking if you want to create a new VI or open an existing one; if you see this window, create a new VI; otherwise, one will have been created for you. You will see, as shown in Figure 1–3, a new front panel come up, called **Untitled 1** (or, if LabVIEW was already running, the untitled window will be numbered according to the number of times an untitled window has been created—Untitled 2, Untitled 3, and so on).

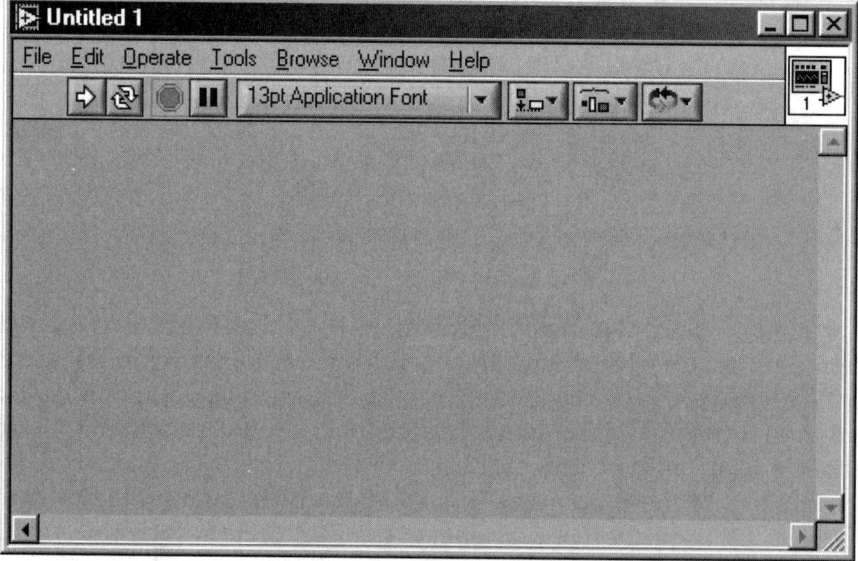

Figure 1–3
A VI's front panel, newly created.

This is a blank front panel upon which you will later add the boxes from `Sum of Squares.vi` seen in Figure 1–2. Instructions for adding these boxes will be in Section 1.2.4, but first, a few more LabVIEW basics will be introduced. To see the block diagram, which is currently hidden, select the **Window»Show Diagram** menu item. Alternatively, <Ctrl-E> is a keyboard shortcut for switching between front panel and block diagram views.

At this point, save the VI: Select the **File»Save** menu item, and save it as `Sum of Squares.vi` in the folder you created earlier. Quit LabVIEW now with the **File»Exit** menu item. Launch LabVIEW again, and this time, use the **File»Open...** menu item to open this same VI. You now know how to save a VI and open it again later. You could have also double-clicked the VI from your graphical operating system, because the `.vi` extension on the file name tells your operating system to use LabVIEW to open the VI. Another useful LabVIEW feature is the **File»Recently Opened Files** menu item.

Suppose you want to save all of your LabVIEW work into a single file, which might help you transfer your work from computer to computer. An LLB is a **L**ab**V**IEW **L**i**B**rary, which is a special LabVIEW file (ending with `.llb`) in which you can save your VIs into a single file as if it were a folder from LabVIEW's point of view. I prefer to not use LLBs in general, but to simply place all my VIs in a folder. When I want to save my VIs later as a single file, I use a compression utility like WinZip (available at *www.winzip.com*).

1.2.2 Controls, Indicators, and Tools

In Figure 1–2, the boxes on the front panel are called *controls* or *indicators*, depending on whether the data moves into or out of them. The boxes labeled `a` and `b` are controls, and the box labeled `sum of squares` is an indicator.

From the point of view of the block diagram, controls are sources of data, and indicators are destinations of data. If this is not clear, go back to your `Sum of Squares.vi` *block diagram, and see that* `a` *and* `b` *are sources of data, meaning that data comes out of them and goes elsewhere in the block diagram. Similarly, indicators on the block diagram, such as* `sum of squares`*, are destinations of data, meaning that data goes into them. From the point of view of the front panel, this logic is reversed: Controls are destinations of data (the source is the user or possibly another VI, as we'll see later) and indicators are sources of data (the destination is the user or another VI).*

When the **Run** button is pushed, *dataflow* begins. Dataflow occurs on the block diagram, not the front panel. In dataflow terms, a *node* is a "stopping point" for data on a dataflow diagram. In LabVIEW, wires connect nodes on a block diagram. A dataflow diagram consists of nodes, wires, and data packets. In a block diagram (such as the one in Figure 1–5), the nodes might look like those in Figure 1–4.

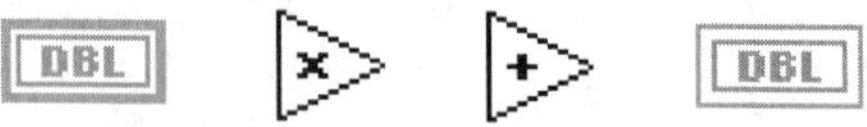

Figure 1–4
Examples of nodes on the block diagram.

The wires are the lines connecting these nodes, and the data packets are the two little circles on the wires.

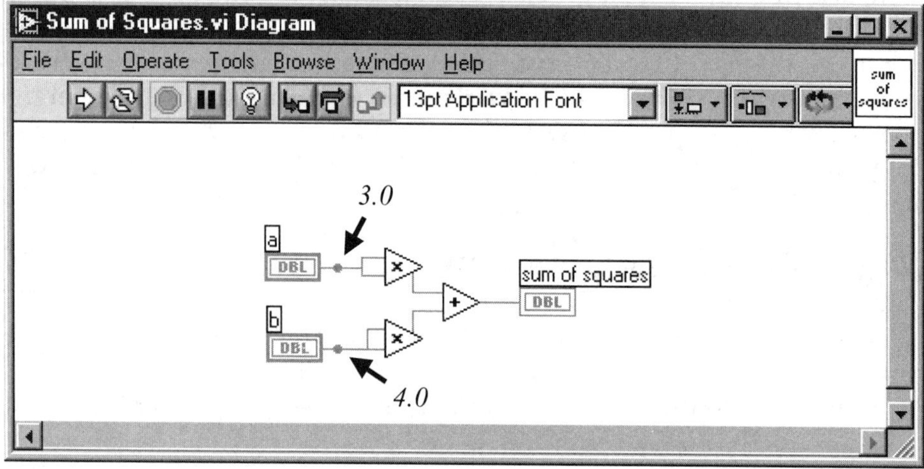

Figure 1–5
Data packets, drawn as little circles on a block diagram.

I used a drawing program to draw the data packets on the wires and the numbers with their lines on the block diagram in Figure 1–5; none of these are really a visible part of LabVIEW.

Data packets are usually invisible, but when writing your LabVIEW program, it is helpful to imagine them flowing from controls to indicators, which in Figure 1–5 would be from left to right. If they were *really* visible whenever a LabVIEW program ran, it would slow the programs down to a crawl. Later, we'll see a debugging tool called *execution highlighting* that allows you to see data packets similar to the ones shown above. The rules of dataflow are follows:

1. Data packets can flow in only one direction on any given wire.
2. Data must flow from one node to another.
3. A node can release data to its output wires only after it has received data on all of its input wires.

In LabVIEW, a *terminal* is an area to which a wire may be connected. The nodes shown in Figure 1–6 each have one or three terminals. In this case, all the nodes' terminals are wired, but some terminals on some nodes may remain unwired. Later, we will see nodes with many more terminals. Since all front panel objects have nodes with only one terminal, their block diagram connection points are usually called terminals rather than nodes, though either name is technically correct.

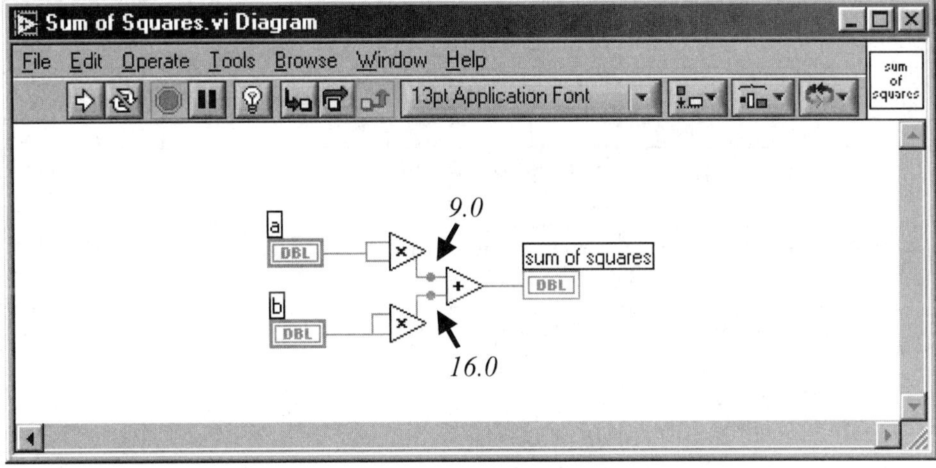

Figure 1–6
The same data packets as in Figure 1–5, a moment later.

Imagine the data packets (the little circles I've drawn) shown in Figure 1–5 coming out of the a and b terminals and moving to the right along the orange lines (orange on your screen, not in your book), called *wires*. These circles move towards the two Multiply functions ×. Just before the circles reach the Multiply functions, they split in two where the wires fork, so that the top Multiply function produces a $3 \times 3 = 9$ on the orange wire to its right. Similarly, the bottom Multiply function produces a $4 \times 4 = 16$ on the orange wire to its right, as shown in Figure 1–6.

Finally, when all data packets have hit the Add function, it sums the 9 and 16 to produce 25, which is then sent to the `sum of squares` node, thus showing up on the front panel. As users running the final program, we usually never see the block diagram in action; we see only front panels.

In this example, our data type is strictly numbers. Later, we will see other data types.

On the front panel and block diagram, the mouse cursor can serve several different functions. These basic tools are summarized in Table 1.1. To toggle between these different modes, use the <Tab> key.

Table 1.1 *Basic VI Editing Tools*

Icon	Tool	Purpose
	Operating	Changes the data value of a control on the front panel or of a constant on the block diagram.
	Positioning	Changes the position, size, or shape of an object.
	Labeling	Allows you to edit any editable text or to create a "free label."
	Coloring	Sets the color of an object to the "current color."
	Color Copying	Reads the current color of an object as the "current color."
	Wiring	**Front panel:** Used only in the connector pane (to be described later). **Block diagram:** Wires nodes together via their terminals.

1.2.3 Data Types

Make sure you're familiar with the concepts in Appendix A, if you're not already.

In the last section, we saw little data packets flowing along wires. As shown there, they had the data type of floating-point (numeric) data, which can have a fractional part, such as 2.5, 0.0000323, or 3.14159. For now, let's discuss the three most basic data types in LabVIEW: (1) numeric, (2) Boolean, and (3) string.

Numeric data types represent numbers, as the name suggests, and they can come in many flavors. First, we will describe all numeric subcategories and their limitations, then we will show you how to create and use them in LabVIEW.

Starting with the three major numeric categories, Table 1.2 lists the numeric data types. Table 1.3 lists the different types of integers.

How is this relevant to your LabVIEW/DAQ programming? Suppose you have a DAQ board and are acquiring analog data. At the hardware level, your data comes in as an integer, then is usually transparently converted to a floating-point number for you. Depending on the precision of your DAQ device, you might need a certain range of integers to be able to distinguish all the different values that could come from the board. This will be discussed in more detail in the "Resolution" section of Chapter 2, Section 2.2.1, where 12-bit and 16-bit integers are discussed. Appendix A offers even more fundamental information about numeric types.

Be careful when changing from one data type to the other on the block diagram, and realize that numbers could be changed without your knowledge, unless you notice the gray circle. Generally, a little gray splotch will appear, warning you of such a change in data type, as seen on the left side of the `i16` indicator in Figure 1–7.

Table 1.2 *Numeric Data Types*

Major Numeric Categories	Description
Integer	These numbers cannot have a fractional part; examples of integers are -2, 0, or 123.
Floating-Point	These numbers may have a fractional part, like 0.5, -12.345, or 0.0000001, although a fractional part is not required.
Complex	These numbers have both a real and an imaginary part, and are composed of two floating-point numbers. For most DAQ work, it is not necessary to use these or even to understand them. Only under certain circumstances involving arrays of numbers, usually for frequency analysis, are complex numbers useful, so they will not be discussed further in this book.

Table 1.3 *Types of Integers*

Integer Type	LabVIEW Abbreviation	Range
Signed 32-bit	I32	-2,147,483,648 to 2,147,483,647
Signed 16-bit	I16	-32,768 to 32,767
Signed 8-bit	I8	-128 to 127
Unsigned 32-bit	U32	0 to 4,294,967,295
Unsigned 16-bit	U16	0 to 65,535
Unsigned 8-bit	U8	0 to 255

Figure 1–7
This dangerous conversion will give you an incorrect number if `i32` is not in the range of -32,768 to 32,767.

As a rule, use wires having data types large enough to accommodate any size number that could possibly travel along them.

Floating-point numbers also have precision issues, but unlike integers, we often need not concern ourselves with their precision in the field of DAQ when dealing with data. Here's one big exception—when dealing with times and dates in LabVIEW's standard Time & Date format, we will need to use double precision floating-point numbers (DBL, eight bytes) rather than single precision floating-point numbers (SGL, four bytes). In this case, the SGL data type does not have enough significant figures, or digits of precision, to accurately represent time values. The SGL data type is usually appropriate for analog DAQ data. The other type of floating-point data is called extended precision floating-point (EXT, platform-dependent number of bytes). The EXT data type is generally not used in DAQ unless you're doing some sort of math that requires extreme floating-point precision, such as certain types of iterative numeric analysis. See the end of Appendix A for more floating-point details.

In LabVIEW, these different numeric data types are called the number's *representation*. You can pop up (by right-clicking) on a numeric control, indicator, or constant and change its representation with its **Representation** menu item.

The other two basic data types in LabVIEW are Boolean and string. A Boolean data type can have a value of either True or False (just one bit—not even a byte). A string data type contains an arbitrary number of bytes, which can be considered 8-bit integers. Often, strings contain human-readable text, wherein each byte corresponds to a letter, number, symbol character, or "formatting character" like a space, tab, or return character.

Two more data types should be mentioned at this point—the array and the cluster. These are both ways of grouping other LabVIEW data types (including themselves!) so that many pieces of data can conceptually flow along a wire in a single data packet. For example, you could have a group of 100 numbers in an array and visualize it moving along a wire as one data packet. Or you could have two numbers, a Boolean, and three strings grouped as a cluster, also moving on a wire as one data packet. Almost any combination of data can be grouped by using the array and cluster data types to flow along a LabVIEW wire. These will be discussed in more detail later.

Wire colors change depending on your data type. I'm afraid this black & white book won't show you many colors, unless you spill something on it, but observe the various wire colors as you build your block diagrams.

Table 1.4 provides a quick summary of what the colors usually mean on the block diagram for wires and terminals.

Table 1.4 *Data Type Colors on a Block Diagram*

Color	Data Type
Blue	Integer (Numeric)
Orange	Floating-point (Numeric)
Magenta (hot pink)	Cluster containing any non-numeric data type (Booleans, other clusters, arrays, strings, etc.)
Green	Boolean
Brown	Cluster containing only numerics

As of version 6i, LabVIEW introduced *a new data type* called the Waveform (see Figure 1–8), which I capitalize in this book to distinguish it from a generic waveform, which is a generic sequence of analog data points. LabVIEW's Waveform is like a cluster containing three parts, described in Table 1.5.

Table 1.5 *LabVIEW's Waveform*

Waveform component	Description
t0	This is the time corresponding to the initial point in the Waveform's data, Y. Its data type is DBL and is defined as a time, just like the output of LabVIEW's Get Date/Time In Seconds function.
dt	If the Waveform's data, Y, has more than one data point, these points are equally spaced in time by an interval of dt seconds.
Y	This is the Waveform's data, which is also described by the previous two parameters. Y may be an array or a single number.

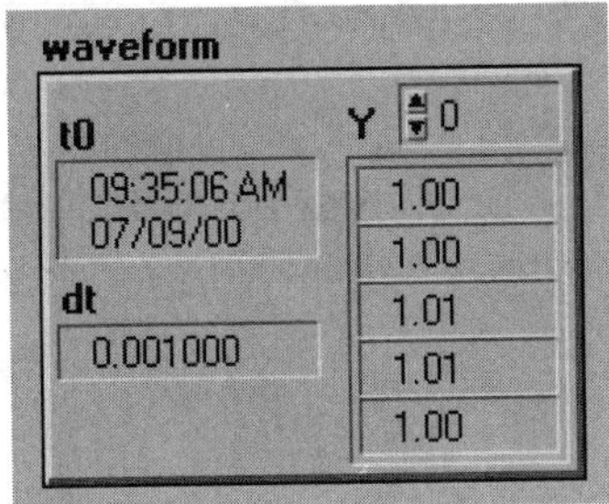

Figure 1–8
A front panel Waveform object with sample data.

Although cluster-like, the Waveform data type cannot be manipulated by normal Cluster functions—it has its own special set of functions.

For those DAQ VIs designed to handle waveforms of just one point, there's another version of the Waveform data type called the *single-point* Waveform, which is just like the normal Waveform, but certain components are hidden (see Figure 1–9).

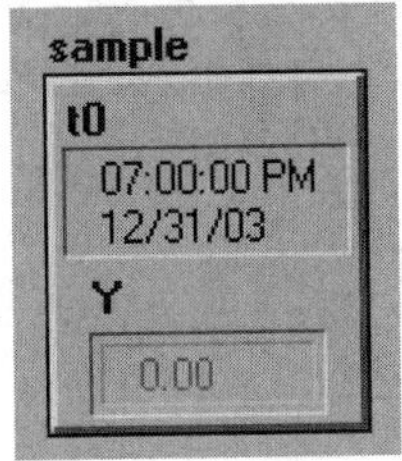

Figure 1–9
The single-point Waveform.

1.2.4 Creating Your First Functional VI

We will now build our first functional VI. Use LabVIEW to open `Sum of Squares.vi`, which you should have created in an earlier section. If you see one or two smaller windows opening up with your front panel containing tools and controls, close them. This book will not use these palettes as ever-present windows (many users find them annoying in a cluttering sort of way), but if you want to, they can be turned on and off in the **Windows** menu. Rather than using the **Tools** palette, *close it,* then select your tools by hitting the <Tab> key on your keyboard.

Position your cursor (with the mouse) anywhere over the front panel and verify that you scroll through the following tools when you hit the <Tab> key: , , , .

If you are able to continuously <Tab> through these four tools, you are in edit mode and ready to edit. If you ever find yourself tabbing through these tools , , on the block diagram, this means you are in run mode. To get out of run mode, first make sure your VI is unlocked by hitting <Ctrl-I> and adjusting the ***Security*** *item under the* ***Category*** *menu's window to "unlocked"—otherwise you will be stuck in run mode. Secondly, make sure you are in edit mode, as opposed to run mode, by checking whether an item in the* ***Operate*** *menu says* ***Change to Edit Mode*** *or* ***Change to Run Mode****. <Ctrl-M> toggles between these two modes, unless your VI is "locked."*

Suppose you try to run the VI, and the block diagram stops and looks something like Figure 1–10 with a red border surrounding the block diagram.

This means you are in run mode, and when you had this particular tool , you clicked the block diagram. Unless you're familiar with this debugging tool and you did this on purpose, get this tool again (which now may look like this) by getting to run mode (if not there already), clicking the block diagram so the red rectangle disappears, getting back to edit mode, then proceeding as normal. Yes, this can be confusing.

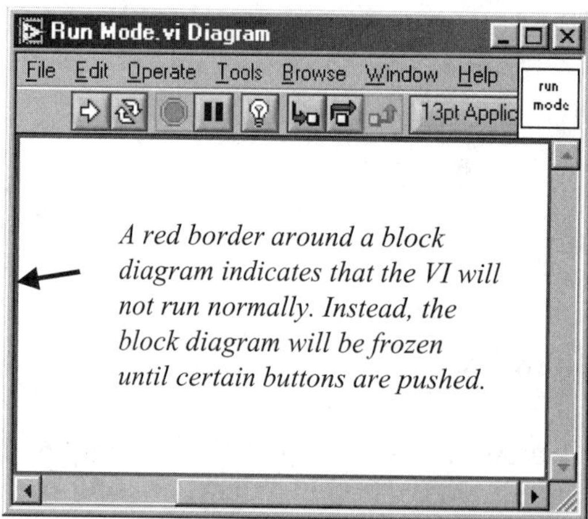
Run Mode.vi Diagram
File Edit Operate Tools Browse Window Help
13pt Applic
run mode
A red border around a block diagram indicates that the VI will not run normally. Instead, the block diagram will be frozen until certain buttons are pushed.

Figure 1–10
A red border around the diagram indicates run mode.

Be careful to click only when told to in the next paragraph. If you make a mistake, you can undo it with the **Edit»Undo...** menu item or <Ctrl-Z>. The Undo feature is very well implemented in LabVIEW and should be useful quite frequently, unless you never make misstakes, like me. See Appendix E for more information on Undo.

Refer to the `Sum of Squares.vi` front panel shown in Figure 1–2 while building this front panel. To create your first control, select any tool other than the Coloring tool (hit the <Tab> key until you see this other tool) and pop up on (right-click) the front panel, thus bringing up a temporary **Controls** palette. This is the way we will always access the **Controls** palette in this book, so if you already had the "permanent" **Controls** palette window (a little window with the word **Controls** in its title bar) in the background, close it now. If you tried to pop up with the Coloring tool , you would not have gotten the **Controls** palette. With the **Controls** palette still showing from your recent right-click, move your cursor in the **Controls** palette and over the upper left icon (**Numeric**). The **Numeric** subpalette will pop up. Then, move the cursor over the upper left icon in the new subpalette (**Digital Control**). See Figure 1–11.

*In this book, I am keeping the **Controls** palette window closed so that you must pop up on the front panel to see it. This is a personal preference, as I like uncluttered computer screens. To open it once it's closed, use the **Window»Show Controls Palette** menu item. If you want to skip ahead and get a quick description of the many items in the **Controls** palette, see Figure 1–29 and Table 1.6 (some of this won't make sense until you get there). The **Controls** palette will not open and close per VI—it stays open or closed for LabVIEW in general.*

When you click this, your next click on the front panel will create a new Digital Control (a box with a number you can edit) wherever you click. This sort of action will be hereafter succinctly described as "dropping the Digital Control from the **Controls»Numeric** palette" (when you see **Controls»** preceding the name of a palette or other object, this means to start with the **Controls** palette, found by popping up on the front panel). Click slightly towards the upper left of your front panel, thus creating your first control:

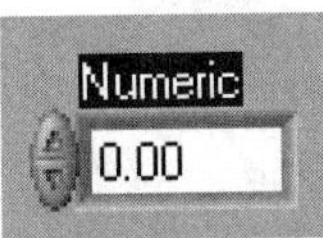

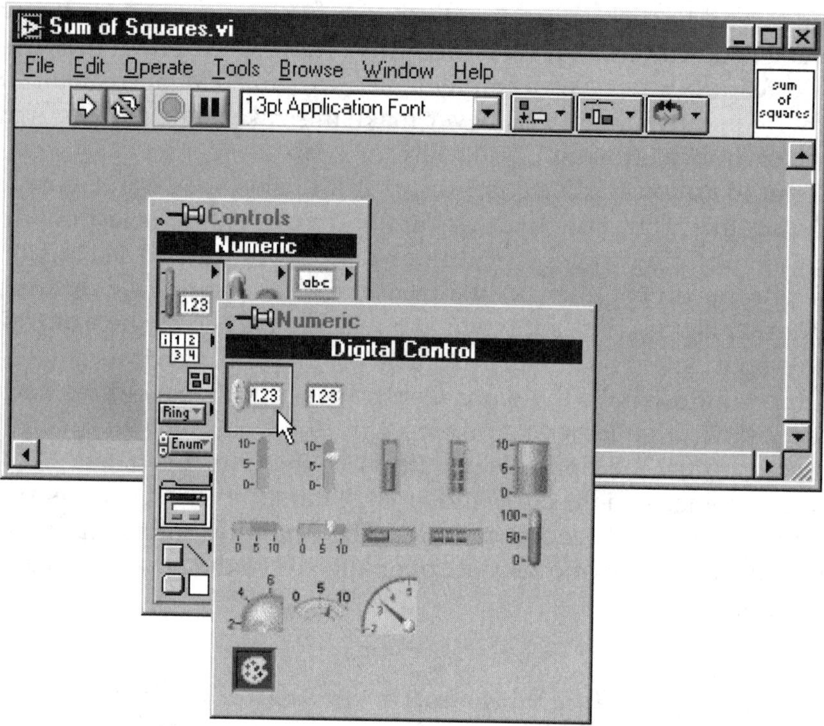

Figure 1–11
The `Sum of Squares.vi` front panel with the Numeric and Digital Control subpalettes—you will not create the icon (the white box saying "sum of squares") until later.

Before clicking again, while the label is still highlighted, type a lowercase `a`, and your new control has a label:

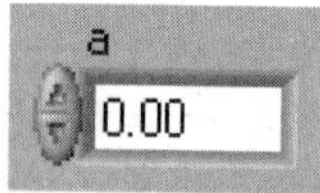

If you were to click elsewhere before you began typing the label, the label would lose its highlighting, so you would need to use the Labeling tool to edit the label.

If any control or other front panel object doesn't show up quite where you want, you can use the Positioning tool to move it around.

Remember how to use this **Controls** palette, as it's the most common way to drop new objects on the front panel. It will be discussed in detail in a later section.

Create a similar control just below this one, and label it `b`.

Finally, create another digital control to the right, and label it `sum of squares`. Pop up on it and change it to an indicator. A shortcut here would have been to select the Digital Indicator rather than the Digital Control in the **Controls»Numeric** palette.

Notice the difference between a numeric control and indicator on the front panel. Go to the block diagram, and you should see three orange rectangles, which are called terminals, each of which corresponds to a control or indicator on the front panel.

First, notice that the controls and have thicker borders than indicators like ; this is common with many node types. Pop up on any terminal (terminals are on the block diagram) and select its **Find Control** or **Find Indicator** menu item. Notice how this helps you find terminals' front panel objects! The opposite sort of mechanism works from these front panel objects with their **Find Terminal** menu items. From the block diagram, you can also double-click terminals (with the Positioning tool or Operating tool) to do the same thing.

Next, use the Positioning tool on the block diagram to arrange the terminals, with `a` and `b` on the left and `sum of squares` on the right, then add the Multiply and Add functions as shown previously in the block diagram of Figure 1–1. To do this, pop up anywhere in the block diagram to see the **Functions** palette, select the **Numeric** subpalette therein, then find the appropriate functions, placing them like you did the objects on the front panel. This is the way we will always access the **Functions** palette in this book, so if you had a "permanent" **Functions** palette window (a little window with the word **Functions** in its title bar) in the foreground, you might want to close it now.

Hit the <Tab> key until you find the Wiring tool . Wiring is where you must be the most careful while building VIs. You must also have a steady hand—limit that caffeine consumption until later in the book, when instructed. First, be aware that block diagram objects will often try to wire themselves together by default if you drag them close to one another; I recommend you disable this feature until you become more acquainted with LabVIEW. To toggle this initially cumbersome automatic wiring, hit the space bar while you are dragging an object to be wired on the block diagram.

As with the ***Controls*** *palette window, I am keeping the* ***Functions*** *palette window closed so that you must pop up on the front panel to see it. This is a personal preference, as I like uncluttered computer screens. To open it once it's closed, use the* ***Window»Show Functions Palette*** *menu item. The* ***Functions*** *palette is only seen with block diagrams, while the* ***Controls*** *palette is only seen with front panels. If you want to skip ahead and get a quick description of the many items in the* ***Functions*** *palette, see Figure 1–30 and Table 1.7.*

Earlier, I recommended that you keep the ***Functions*** *and* ***Controls*** *palettes, along with the Tools palettes, closed. There is one special case where it's especially useful to have the* ***Functions*** *palette or* ***Controls*** *palette open. When open, you may click on the little magnifying glass icon that will take you to a handy browser that allows you to type key words to search for objects. This browser is called the Functions Browser for the* ***Functions*** *palette, and the Controls Browser for the* ***Controls*** *palette.*

If you botch the wiring in this section, undo your steps <Ctrl-Z> carefully until enough wires are gone. If you have too much trouble wiring, skip ahead and read Section 1.2.8. Later, we will use more efficient wiring tricks. But for now, be aware that the math functions mentioned here (the yellow triangles) have three underlying, invisible terminals arranged like this: . Whenever you move the Wiring tool over the terminal to which it is ready to connect, that terminal will flash. Make sure you don't accidentally wire two wires to the same terminal! If so, or if you make certain other wiring mistakes, you will see the wires turn into dashed lines. This means the wires are broken because they're illegally wired for some reason, and your VI won't run—use the Undo function <Ctrl-Z> and try again.

Once you have everything wired up as shown in the block diagram of Figure 1–1, you should be able to hit the **Run** button on the front panel and watch your VI work. If the **Run** button looks broken , click it and use the resulting dialog box to track down what's wrong. If you cannot find the error, you may need to recreate the VI from scratch. Once the VI is working, enter numbers other than zero in the front panel controls a and b (must have or just prior to typing) so this VI is somewhat less boring!

Yes, I know this seems like a trivial program. And it is. But building it is nontrivial for a LabVIEW newcomer. Gradually, as you realize the power of LabVIEW, you will build much more complex and useful VIs.

1.2.5 Coloring

Let's experiment with coloring. Coloring is usually used on the front panel only, not on the block diagram. First, expect to accidentally color something wrong, so remember <Ctrl-Z> undoes your last action, including coloring mistakes. Okay, now let's go—create a new VI, then drop two new Digital Controls on its front panel. To color one of these controls, go to the front panel, select the Coloring tool , then *right-click* the object you want colored. Up pops a palette from which you can select a color, as shown in Figure 1–12.

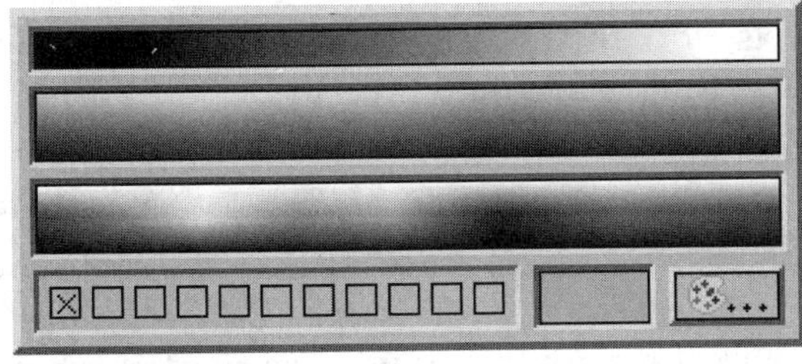

Figure 1–12
The Coloring tool palette.

I prefer to use the "More Colors" button to select a color, so I can select *exactly* the same colors for the many VIs I create. Using that button, select your favorite shade of green, if any—if not, select your least despised shade of green. On the same control, make it white again. Now try this on another control, and select something ghastly, like burnt orange.

We will now learn about the Color Copying tool . Make sure you have the Coloring tool , hold the <Ctrl> key down, and notice that your Coloring tool changes to the Color Copying tool . To use this tool, move it over an object of the color you want, and click that object with the <Ctrl> key held down. Do this to on the control that you just colored white to "suck up" the color white into the eye-dropper-looking thing, release the <Ctrl> key (so you get the Coloring tool again), then color the burnt orange section white.

To color the text itself, instead of its background, select the text area you want colored with the Labeling tool , then click the **Text Settings** menu, which looks something like this 13pt Application Font, and find the "color" section in the menu that pops up.

Undo is particularly helpful when you make a coloring mistake!

Coloring works with most front panel objects and a few block diagram objects. Like the background and foreground of text, many objects have different parts that can be colored.

1.2.6 Selecting Objects for Manipulation

In order to manipulate objects, you must first be familiar with the concepts of selection. If an object is *selected*, you are then able to perform a number of operations on it such as moving it, removing it, copying it, and so on. Multiple objects can be selected, as well.

The best way to learn about selection is to actually do it. Open your `Sum of Squares.vi` created earlier. *You must have the Positioning tool* ↖ *to select objects.* Once you do, click the a control so you see the "marching ants" (technically, *selection box*, which is the moving pattern of dotted lines around an object to show it's selected), as shown in Figure 1–13.

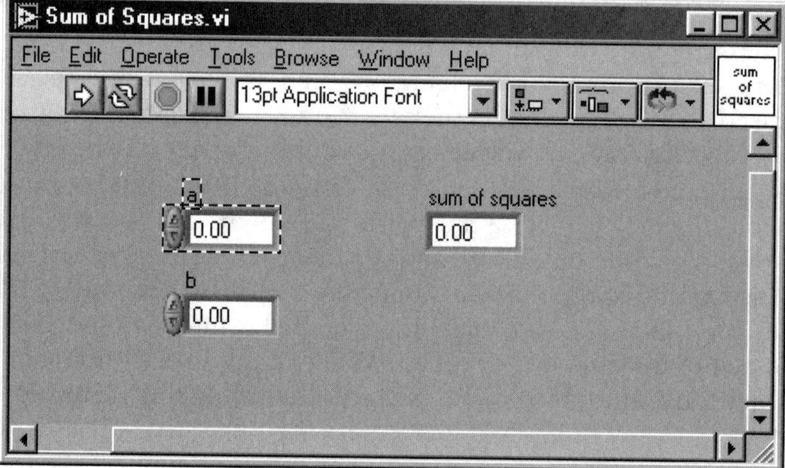

Figure 1–13
The selection box consists of moving dashes, which I call the "marching ants." This shows you what's selected.

At this point, the object is selected, and you can do things with it, like moving it and copying it, to be overviewed in the next section.

If you click an object to select it, and you hold the left mouse button down too long while moving the mouse even a tiny bit, you will accidentally drag the object. <Ctrl-Z> quickly undoes this.

To select multiple objects that are physically located near one another, you can drag a rectangle around them with the Positioning tool . See Figure 1–14.

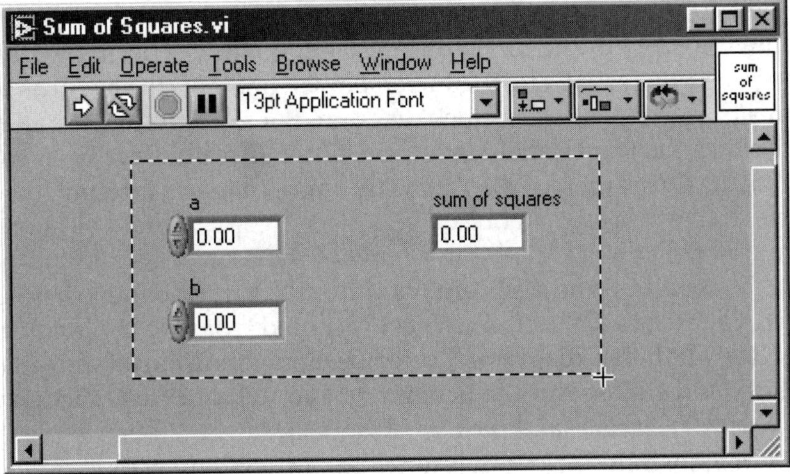

Figure 1–14
Selecting multiple objects by dragging a selection rectangle.

To select multiple objects one at a time, hold down the <Shift> key while clicking each object with the Positioning tool . This <Shift> key trick also allows you to deselect objects one at a time.

1.2.7 Manipulating Objects

This section of the book may save you the most time, especially once you become a good LabVIEW programmer, so read it carefully! The Positioning tool is used for manipulating objects, in terms of moving them around and sizing them. Several functions will be discussed in this section concerning object manipulation:

1. Cut, Copy, and Paste
2. Clear
3. Clone
4. Move
5. Resize

Everything listed above, except resizing, can be performed upon multiple objects by selecting the desired objects first, then performing the action.

The **Cut**, **Copy**, and **Paste** functions utilize LabVIEW's *clipboard*, and can be found in the **Edit** menu. LabVIEW's clipboard is a normally invisible storage space where LabVIEW can remember one object or a group of objects from a front panel and/or block diagram. For example, any or all of the three controls on the front panel of your previous VI could have been copied to the clipboard. Once there, you can view the clipboard through the **Windows** menu. When you *copy* an object, you place a copy of that object onto the clipboard. When you *cut* an object, a copy of that object goes to the clipboard as with the *copy*, but you also remove it from the front panel or block diagram. When you *paste* an object, the object is copied from the clipboard to the front panel or the block diagram. Go to the **Edit** menu now, and memorize the keystrokes for these three functions, as you will often use them (cut = <Ctrl-X>, copy = <Ctrl-C>, and paste = <Ctrl-V>).

To demonstrate these three functions in LabVIEW, first create a new VI (via the **File»New VI** menu item). Next, create any new small object on the front panel, such as a Digital Control. To cut or copy an object, you must first select it by clicking on it with the Positioning tool. Now, perform a cut (the **Edit»Cut** menu item or <Ctrl-X>), and watch the selected object disappear. Once cut, the object is residing in the clipboard. To use this object elsewhere, you can now click where you want it, and paste it (with the menu item or <Ctrl-V>). Try the same thing while copying instead of cutting (<Ctrl-C>).

When using LabVIEW, there are really two clipboards involved. One belongs to LabVIEW, and the other belongs to your operating system. LabVIEW's clipboard can remember an image or text copied from another application. This will come from your operating system's clipboard, which can only pass simple images and simple text to LabVIEW (simple in terms of format, not size). Similarly, your operating system's clipboard can only accept such simple things from LabVIEW. Unless otherwise noted, "clipboard" will hereafter refer to LabVIEW's clipboard, not your operating system's.

The **Clear** function, also found in the **Edit** menu, simply removes an object. It is similar to the **Cut** function, except clear does not involve the clipboard. The <Delete> key works similarly.

The **Cloning** function copies an object, but cannot be found in any menu. It is an action that can only be performed with the mouse *and* the keyboard! It is similar to a copy immediately followed by a paste, but the object never sees the clipboard. To clone an object, simply drag it (remember, dragging requires the Positioning tool) with the <Ctrl> key held down.

When you copy or cut front panel objects' terminals from the block diagram, the associated front panel objects are placed in the clipboard, but you cannot clone them from the block diagram. When you cut a front panel objects' terminal from the block diagram, it is not deleted. To delete that front panel object, you must do it from the front panel.

Moving an object can be done in two general ways. The first way is to *drag* the object (click it with the Positioning tool, move the mouse with the left mouse button held down, then release this button when finished moving). The second way to move it is to select it, so that the marching ants are seen, then use the <Arrow> keys (for one-pixel moves) or <Shift-Arrow> keys (for eight-pixel moves). Objects can be precisely arranged by using the **Align Objects** and **Distribute Objects** menus in the tool bar . First, select multiple objects, then select your alignment or distribution method, using those menus in the tool bar. Holding the <Shift> key down while dragging an object limits the cursor movement to horizontal or vertical, depending on the initial direction of movement.

Resizing objects is often difficult even for a pro, because you must first move your Positioning tool very precisely over certain corners of only certain objects. For example, take your `Sum of Squares.vi`, select the Positioning tool, and move the mouse over the lower right-hand corner of any front panel object until your cursor changes into two little angled black lines ( ), which look like Figure 1–15 over a Digital Control.

Figure 1–15
Cursor appearance when resizing objects.

When you have this tool, you will be able to drag the corner of a Digital Control or Digital Indicator in a horizontal direction, thus changing its size. Many other objects allow similar resizing in a vertical directional as well as horizontal. Holding the <Shift> key down while resizing limits the cursor movement to horizontal, vertical, or diagonal, depending on the initial direction of movement; however, not all objects support all of these directions.

1.2.8 Wiring in Detail: Dragon Alert

For you LabVIEW newcomers Here There Be Dragons. Prepare yourself. Wiring in LabVIEW is *very* tricky at first. Most objects on the block diagram, other than labels, have terminals to which you can attach wires. This is the basis of LabVIEW programming, so you must be skilled in wiring.

Figure 1–16 shows a few examples of common block diagram objects.

Figure 1–16
Common block diagram objects.

For some objects, or nodes (such as these), you can pop up on them and select **Show»Terminals** so that you see the terminals (where you can attach the wires, no more than one wire per terminal), shown in Figure 1–17.

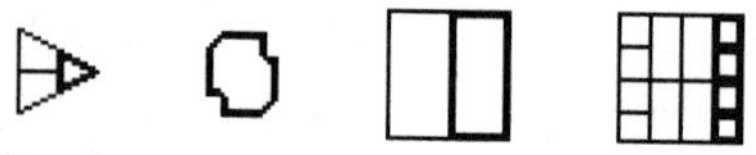

Figure 1–17
Inherent terminal patterns (hidden) from Figure 1–16.

All front panel objects with block diagram terminals have nodes with just one terminal, such as the three objects in `Sum of Squares.vi`, shown earlier. The data flowing to or from such terminals may be a single number, or it may be an array, or it may be quite a complex data structure—you will be able to determine this data type before the end of this chapter.

Wiring difficulty in LabVIEW stems from the inherent difficulty of attaching the wire to the right place; the (sneaky, invisible) terminal. Usually, you can't see the terminal explicitly, so it's easy to attach a wire to the wrong terminal. If ever your LabVIEW program doesn't work as expected, miswiring could be the reason.

For example, suppose you were wiring to the Add function in `Sum of Squares.vi`. Also suppose you attempt to wire the two input terminals on the left of the Add function, and suddenly you have broken wires and a program that won't run, as in Figure 1–18.

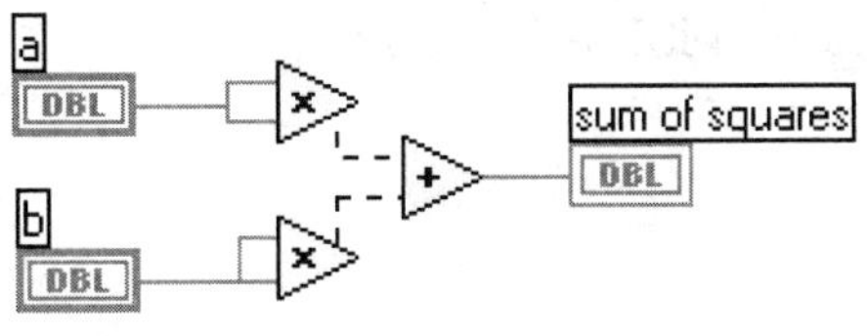

Figure 1–18
It is not clear why these wires are broken.

Notice that the **Run** button is broken; this means your VI won't run. What's going on here? Even a pro might think there should be no broken wires here, but in this case, you have two wires wired to the *same* terminal, either the top left or the bottom left terminal of the Add function. To verify this, you could triple-click either broken wire and see that they're actually the same wire, as the marching ants will indicate in Figure 1–19.

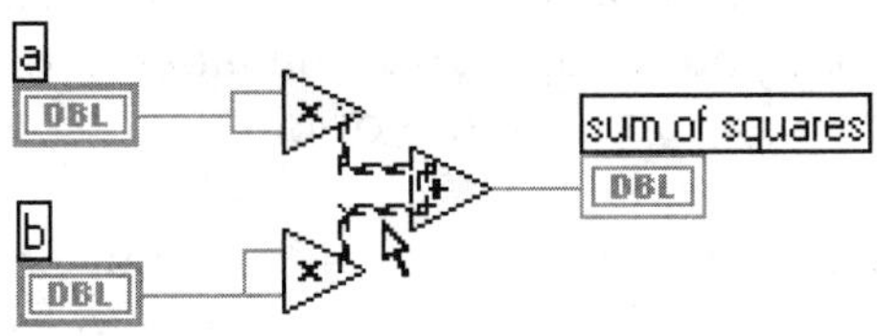

Figure 1–19
After triple-clicking the broken wires, it becomes clear why they are broken.

As a rule, once you have broken wires, hit <Ctrl-Z> to undo your last wire, then try again. If you continue to fail, hit <Ctrl-B> to delete all broken wires (some little ones might be hidden), and if that doesn't fix things, then start from scratch. Depending on the situation, once you get more wiring experience, you may find faster ways to remedy your wiring problems, such as by deleting certain wire segments and rewiring carefully.

If you click your broken Run button, and all your errors are wiring-related, a single <Ctrl-B> (Remove Broken Wires) often fixes everything.

A *wire tree* is simply all wires connected together. For example, Figure 1–20 shows two wire trees (terminals a and b belong to Digital Controls on the front panel and have a representation of I32, a 32-bit integer).

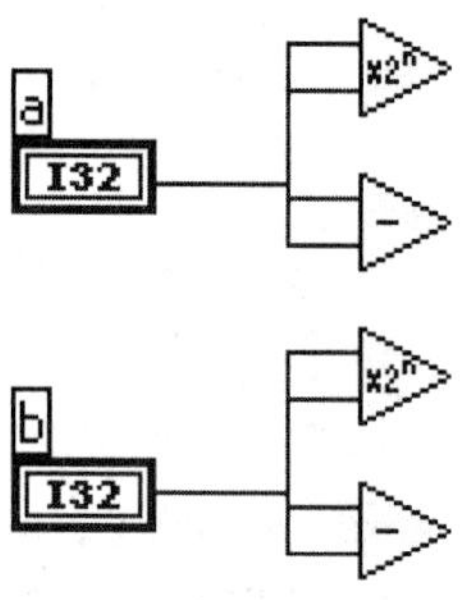

Figure 1–20
Two valid wire trees.

In order for your VI to run, a fundamental requirement is that any wire tree must have exactly one source and one or more destinations. If you were to wire the above two trees together, you would have one large tree, shown in Figure 1–21, but it would not be legal because it would have two sources (a and b).

If you now try to run the VI by clicking on the broken **Run** button, you'll see the message box shown in Figure 1–22, which is designed to help you eliminate your errors.

It may be a little tough to decipher these messages without experience, but you shlould pay attention to this line (wording subject to change with future versions of LabVIEW): "This wire connects more than one data source." Click on that line, and in the lower box, you will see the message shown in Figure 1–23.

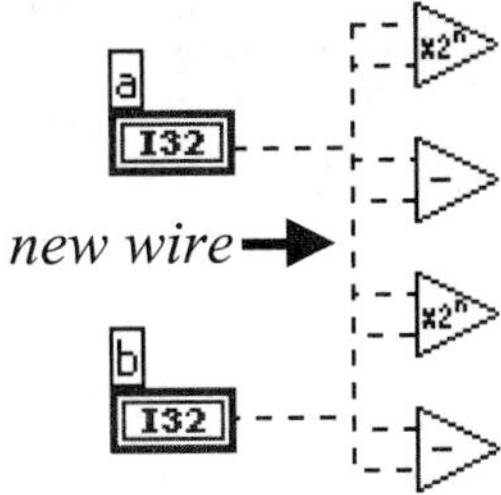

Figure 1–21
One invalid wire tree, which should be two trees. (I drew the "new wire" text and the arrow.)

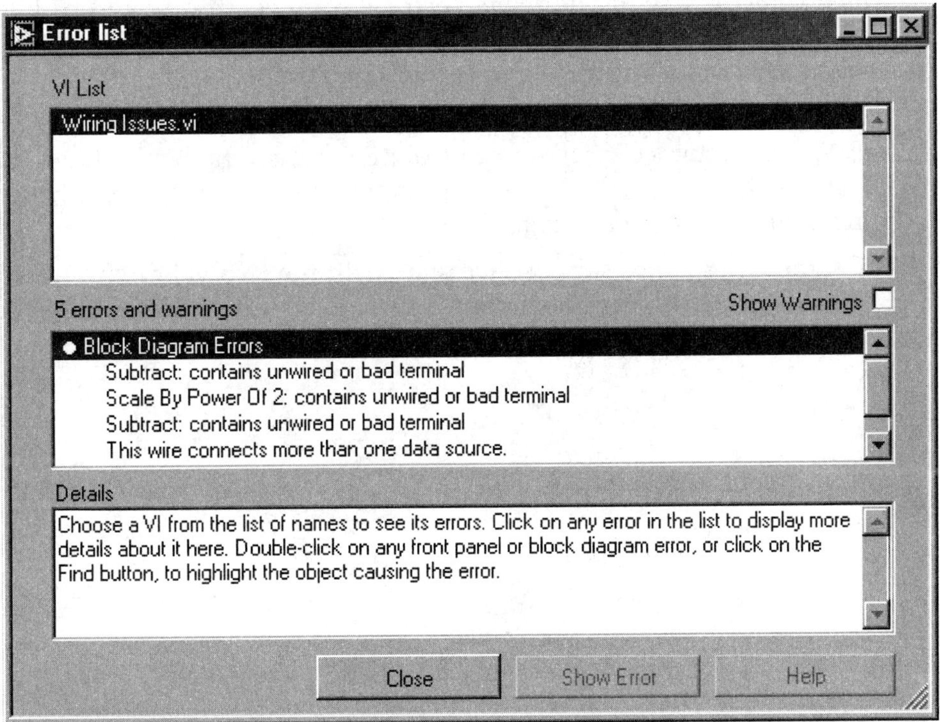

Figure 1–22
LabVIEW's Error list box, from clicking a broken Run button.

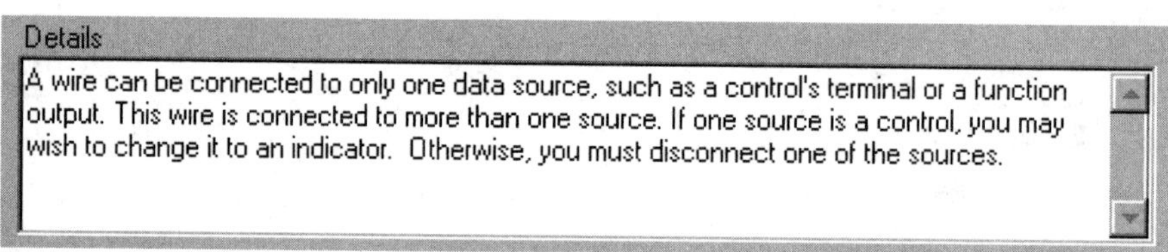

Figure 1–23
Result of clicking the "This wire connects more than one data source" line from Figure 1–22.

That means that your wire tree has more than one source, as in our wire tree in Figure 1–21. In this case, these sources are a and b, both of which are trying to set the value of the wire tree—the tree has no way of deciding which value to take. If you're an electrical engineer, this would be like connecting two different voltage sources to the same wire. Luckily, this wiring mistake won't physically fry your computer.

These error messages may be reworded from version to version of LabVIEW, but the underlying problems are the same. The rest of the messages are complaints from x2ⁿ and -, both whining that they don't have valid wires connected to their inputs.

On the other hand, suppose we start with the two valid wire trees, as in Figure 1–20, then change the upper control to an indicator, so we get broken wires (and, of course, a broken **Run** button). See Figure 1–24.

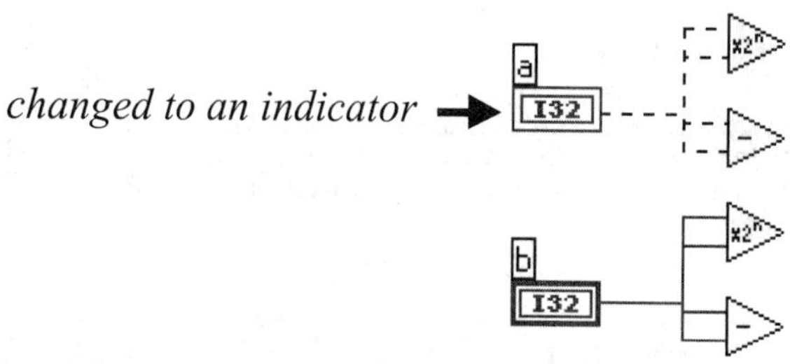

Figure 1–24
Another way to break a wire tree. (I drew the "changed to an indicator" text and the arrow.)

This time, the error box looks like Figure 1–25 (after I clicked on the appropriate line).

Read the message under **Details,** which describes the problem very well. The source of this data is the a indicator. You can fix the block diagram by changing it back to a control.

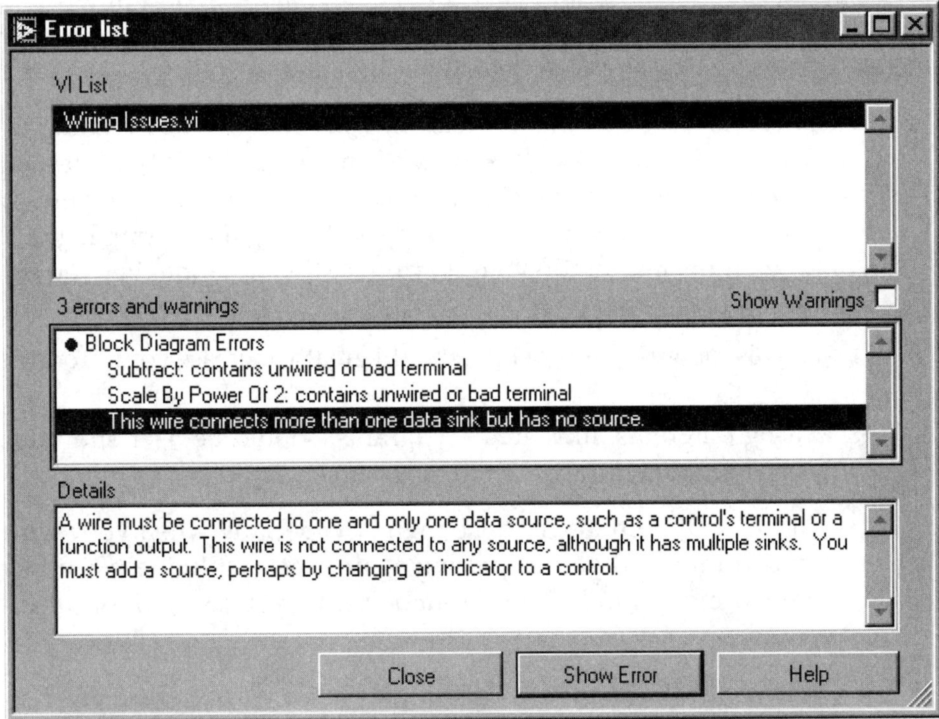

Figure 1–25
Error message resulting from Figure 1–24.

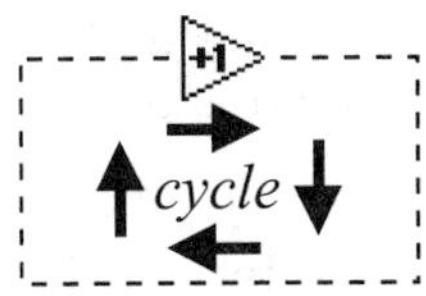

Figure 1–26
A cycle (wiring error).

The last major class of wiring error you may (rarely) run into is a *cycle*. Figure 1–26 shows the simplest example of a cycle.

If LabVIEW allowed this type of construct, which it doesn't, your program would never end, as it would get caught in an *infinite loop*. A number would come into the Increment function from the left, it would be increased by one, then it would go back into the Increment function, and so on forever.

Whenever you see the word "cycle" in your error list box relating to wiring, you have a situation like this, which is not legal. Usually, cycles happen in much more complex wiring situations and are correspondingly more difficult to spot. As soon as you run into a cycle, check your wiring logic. If you can't find your mistake, undo your way out of this situation then rewire very carefully.

Here is a list of wiring tips, which should greatly increase your productivity:

1. Wiring diagrams, like circuit diagrams, should be neat and tidy, with a minimal number of crossing wires.
2. Hit <Ctrl-B> to remove broken wires, especially when your wiring is complex. If you try wiring again, and it still doesn't work, remove more valid wires around where you're working, then rewire.
3. With the Positioning tool , single-click a wire to select one linear segment, double-click to get one branch of a wire tree, and triple-click to get the entire wire tree.
4. To determine exactly which terminal (or terminals) of a block diagram node your wire is connected to, triple-click the wire to let the marching ants lead you to its hidden connections.
5. To fine tune the positioning of your wires, single-click a wire segment, then move it with the arrow keys in a direction perpendicular to that segment's direction.
6. Let's hope NI doesn't start charging us for wire by the inch, which is all the more reason to *minimize your total amount of wire*. For example, suppose you're happily wiring along, then encounter this message: "You have run out of wire again. Please visit our Web site with your credit card ready to buy more wire."

1.2.9 Free Labels

If you left-click with the Labeling tool somewhere other than on editable text, you can place a *free label* anywhere on the front panel or block diagram. On the front panel, these are one of the few objects without block diagram terminals, like the objects in the **Controls»Decorations** palette, thus they have no block diagram data associated with them. They are simply used to clarify and describe things on the front panel or the block diagram. On the block diagram, they are useful for helping newcomers, including *you* many months later, to figure out your code; there they are often called *comments,* keeping consistent with general programming lingo. On the block diagram, I prefer to color my comments' backgrounds yellow with black text, as this prints out clearly on most color printers.

I find it convenient to not color anything else but comments on the block diagram, in the spirit of consistent programming.

As real estate is at a premium on block diagrams, keep your comments few and at a high level (don't get too detailed). I won't be making many such comments in this book, in order to save space.

1.3 LabVIEW Help

There are many components being introduced in this chapter, as I'm sure you've noticed. When you're building the block diagrams, you need not bother to pop up on each component to read the LabVIEW's help information, unless it is key to your understanding. Unless you have a serious brain capacity, you would be swamped!

Within LabVIEW, there are several convenient help utilities. Outside of LabVIEW's own help resources, you can get help through various newsgroups, Internet resources, and from NI—these external options are detailed in the preface. For this chapter, let's discuss your help options *inside* LabVIEW, then when you need further help, refer to the preface for your external options.

1.3.1 The Help Window

Let's look at the Help window contents of a simple function, the Subtract function. Open up a new VI, drop the Subtract function on your block diagram, then, while your cursor is over the function, show the Help window (<Ctrl-H> is a nice shortcut for showing and hiding the Help window). Figure 1–27 shows what it looks like.

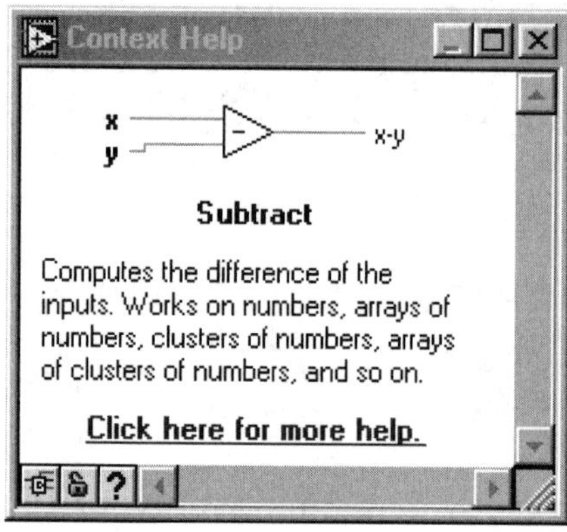

Figure 1–27
An example of LabVIEW's Help window.

In general, the Help window shows you a description of a function's inputs and outputs, the name of the funtion, and an overall description. The contents of the Help window will change as you move your cursor over certain objects on the block diagram. If these contents are sluggish when changing the Help window, wiggle your cursor a bit over the relevant object, or click the object.

See the three boxes in the lower left area of the Help window? The first box will show (or hide) normally hidden input and output wires (some functions with many inputs and outputs will hide those of lesser importance). The second box will lock the Help window onto its current contents, despite where you move your cursor on the block diagram. The third box will bring up the information in LabVIEW's *help files* (described soon) concerning the topic of the Help window, as will the <u>*Click here for more help*</u> link.

1.3.2 LabVIEW's Help

In addition to LabVIEW's Help window, LabVIEW has its own help in the format of standard Windows Help and PDF files. This is an even larger store of knowledge than the Help window. For example, if you had clicked on the ? box or the *Click here for more help* link in the Subtract function's Help window, a window would appear with a different description of the Subtract function. Clicking on this window's Index button, you would see a standard help interface, as shown in Figure 1–28.

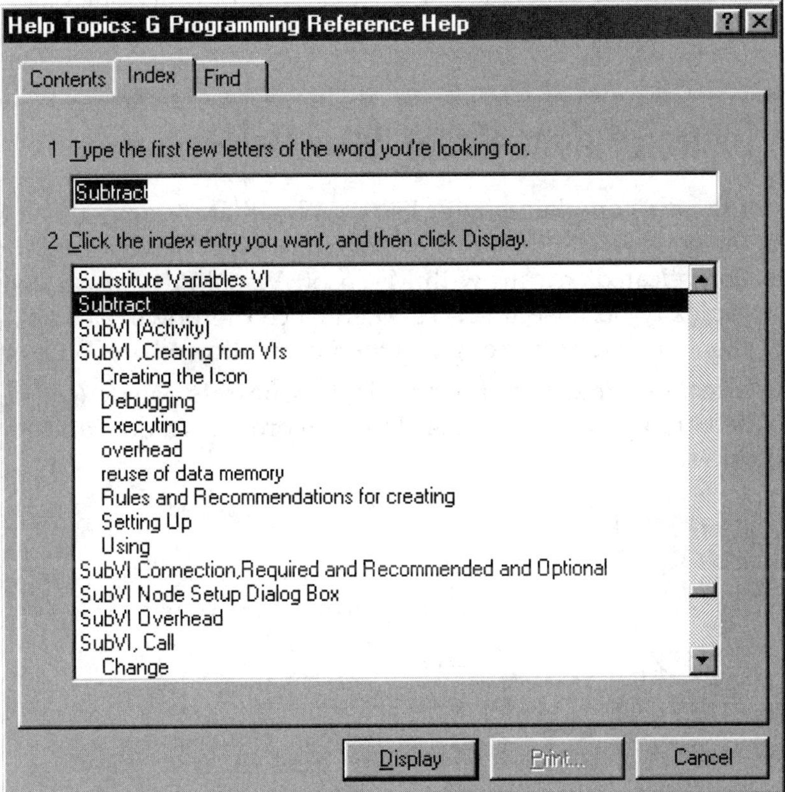

Figure 1–28
Another LabVIEW help window.

This help window can also be accessed via the **Help»Contents And Index** menu item. There are also some very helpful and well-written PDF files to be found from the **Help»View Printed Manuals...** menu item. You may want to read these to supplement this book.

1.3.3 Help Per VI

Finally, VIs can contain descriptive information in their **File»VI Properties...»Documentation** box, easily accessed with <Ctrl-I>, and in many of their objects' **Description and Tip...** boxes. If a VI is used as a *subVI* within another VI's block diagram (subVIs will be described later), these pieces of information can be available in the Help window of the subVI.

1.4 The Controls Palette (Front Panel Only)

One of my *few* complaints with LabVIEW is that it's sometimes too complicated for newcomers. But in my experience with newcomers, it ties for the "least complicated" award with Microsoft Visual Basic! As a veteran, I love the complexity because it serves to enhance the power of LabVIEW, but it takes a long time to sift through all the information when you're starting out.

A perfect example of complexity is the **Controls** palette (see Figure 1–29), described briefly in Section 1.2.4. Table 1.6 provides a quick description of its subpalettes.

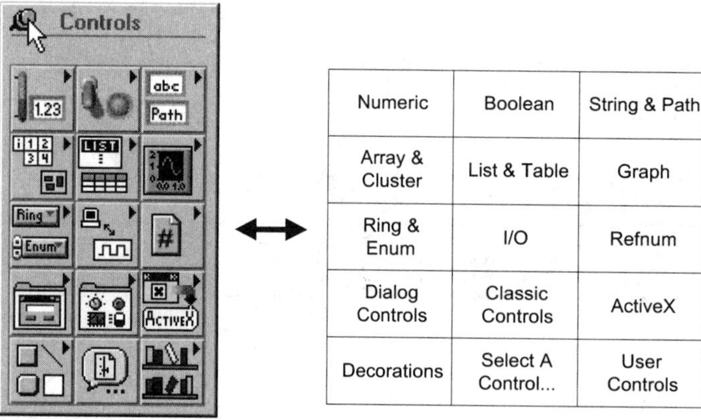

Controls
1.23
abc
Path
LIST
Ring
Enum
ACTIVEX
Numeric
Boolean
String & Path
Array & Cluster
List & Table
Graph
Ring & Enum
I/O
Refnum
Dialog Controls
Classic Controls
ActiveX
Decorations
Select A Control...
User Controls

Figure 1–29
The Controls palette.

For every item in the **Controls** palette (listed in Table 1.6) that has associated data (most of them, but not the **Controls»Decorations** objects), a corresponding terminal will be created on the block diagram once you create the front panel object. Also, when you see the word *control* in Table 1.6, an indicator will sometimes drop, depending on the particular object you select within the subpalette. You can change any control to an indicator, or vice versa, after it's been dropped.

Table 1.6 *Controls Palette Summary*

Subpalette	Description	Data Type
Numeric	A wide variety of numeric controls.	Numeric
Boolean	A wide variety of Boolean controls.	Boolean
String & Path	A string can display text or hold generic data bytes, and a path points to a filename, like `C:\Folder\My File.txt`; either type can hold an arbitrary number of bytes/characters.	String and Path
Array & Cluster	Objects that allow you to group other controls.	Array and Cluster
List & Table	Lists allow you to display multiple text items, all in one control, and tables are grids of strings (like a spreadsheet).	Numeric and 2D Array of Strings
Graph	Rectangular objects that allow you to show your data graphically (pictures).	Various
Ring & Enum	Like a list, but acts like a pop-up menu as only one item is displayed at a time.	Numeric
I/O	Many controls related to input/output, including DAQ Channel Name and Waveform.	I/O (DAQ Channel Name), Waveform, etc.
Refnum	Reference numbers for open connections of various types, such as the Byte Stream File Refnum for open files.	Different types of Refnums like the Byte Stream File Refnum

Table 1.6 Controls Palette Summary (continued)

Subpalette	Description	Data Type
Dialog Controls	A variety of controls appropriate for dialogs (pop-up windows that do not allow you to click other windows).	Various
Classic Controls	A variety of controls used in previous versions of LabVIEW; I prefer these when screen space is an issue, as they're more space-efficient!	Various
ActiveX	Controls that allow you to integrate ActiveX components (Windows only).	Various
Decorations	Front panel objects that are just for looks; they control no data, thus they have no block diagram nodes.	None
Select a Control...	Lets you select a custom control that you or somebody else may have built.	Various
User Controls	User-defined controls, which will be discussed later in this book as *custom controls;* see help files documentation on how to get them into this palette.	Whatever you want

Open a new VI, and drop controls from each of the these subpalettes just to see what they look like. With the Positioning tool, you can move them around and resize most of them. With the Coloring tool, you can color different parts of them. For a few of the controls, pop up on them and select the **Find Terminal** menu item to see what the terminals look like. Different colors of the terminals represent different data types, as described in Section 1.2.3.

Once you've seen at least one control from each of the subpalettes in Table 1.6, close your cluttered, useless VI <Ctrl-W> without saving it.

1.5 THE FUNCTIONS PALETTE (BLOCK DIAGRAM ONLY)

Figure 1–30 shows what the **Functions** palette looks like (used only on the block diagram), and Table 1.7 summarizes its basic subpalettes.

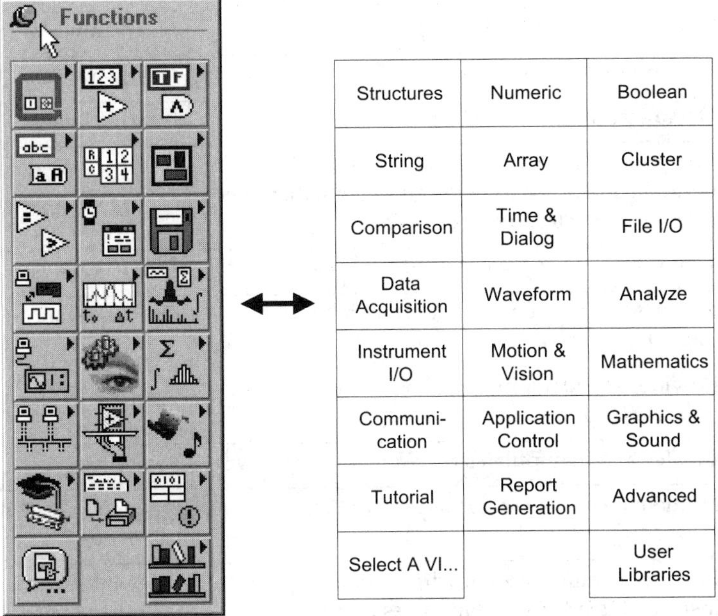

Figure 1–30
The Functions palette.

Unlike **Controls** palette items, which usually place a corresponding node on the block diagram, the reverse is not true. These **Functions** palette items never place an object on the front panel—they only place an object on the block diagram.

If you have the **Functions** palette permanently open, you can click the **Options** button in the **Functions** palette toolbar to access the Function Browser. This lets you customize the **Functions** palette.

Like the previous section, this section is only intended to give you an overview, and the important details will be explained later in the book.

Table 1.7 Functions Palette Summary

Subpalette	Description	DAQ relevance
Structures	High-level program control.	Nearly always
Numeric	Operations on numeric data types.	Always
Boolean	Operations on Boolean data types.	Nearly always
String	Operations on string data types.	Nearly always
Array	Operations on array data types.	Nearly always
Cluster	Operations on cluster data types.	Nearly always
Comparison	Compares many types of data, but usually numeric data, such as determining whether one number is greater than the another.	Nearly always
Time & Dialog	Timing and pop-up windows (dialog boxes).	Nearly always
File I/O	Reads and writes data to files (typically hard disks, floppy disks, etc.).	Often relevant
Data Acquisition	Provides functionality for all of NI's data acquis-tion (DAQ) products.	No comment
Waveform	Creates and manipulates the Waveform data type.	Sometimes
Analyze	Mathematics for Waveforms and arrays.	Frequency, uh, I mean frequently
Instrument I/O	Provides communication with a wide range of instruments having certain types of interfaces.	Relevant if you consider these instruments "DAQ"
Motion & Vision	VIs for NI's motion and vision hardware (control-ling motors and video cameras).	Not officially considered DAQ, but related
Mathematics	Analysis for arrays of numeric data, such as aver-aging, standard deviation, frequency response, etc.	Often relevant
Communication	Provides communication with other computers, other applications on your computer, etc.	Sometimes relevant
Application Control	More powerful control of your application.	Sometimes relevant
Graphics & Sound	Very fancy graphics controls, plus controls for a standard PC sound card.	Often relevant

Table 1.7 Functions Palette Summary (continued)

Subpalette	Description	DAQ relevance
Tutorial	VIs used in LabVIEW's built-in tutorial, which is the purpose of this chapter as well; try it!	Only if you need a more detailed tutorial than this
Report Generation	Helps generate printed reports from LabIEW.	Often relevant
Advanced	Advanced data manipulation.	Often relevant for advanced projects
Select A VI...	Required for subVIs.	Nearly always
User Libraries	For your own LabVIEW subVIs; see help files documentation on how to get them into this palette.	Could be relevant, if you want

Go ahead and drop some of these objects on a block diagram, but do not save anything if asked. Some of these functions (generally the non-yellow ones) are actually subVIs; they often come from LabVIEW's folders, such as `vi.lib`. If you change these subVIs and save your changes, you could easily foul up your copy of LabVIEW, so be careful—see Appendix C for details.

Notice the Compound Arithmetic function in the **Functions»Numeric** palette (when you see the name of a palette or other object starting with **Functions»**, this means to start with the **Functions** palette, found by popping up on the block diagram). If you wanted to add more than two elements, you could use this function. You can operate on as many elements as you want, by growing it with the positioning tool. You can also negate any incoming element. For example, you could configure the Compound Arithmetic function to behave as the Subtract function, so that these two icons are equivalent:  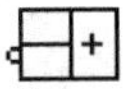 . The little circle shown on the lower input of the Compound Arithmetic function here is created by popping up on that terminal and selecting **Invert**; when adding numbers, this inversion circle takes the negative of the incoming number.

It would be tempting to describe all such useful functions here, but this is only one book and should not weigh 50 pounds.

1.6 Simple DAQ

Since this is a DAQ book, let's create an extremely simple DAQ VI. This section assumes you have a DAQ device installed with analog input, and that MAX (Measurement & Automation Explorer from National Instruments) is installed and recognizes your DAQ device. If you launch MAX and expand the Device and Interfaces item, as in Figure 1–31, you should see your DAQ device with a number next to it. Set your Device Number to 1, if it is not already. In my case, my DAQ device is an AT-MIO-16DE-10 and my device number is one. Select your device from MAX and test with the Test Panel button, shown in Figure 1–31, to see analog input.

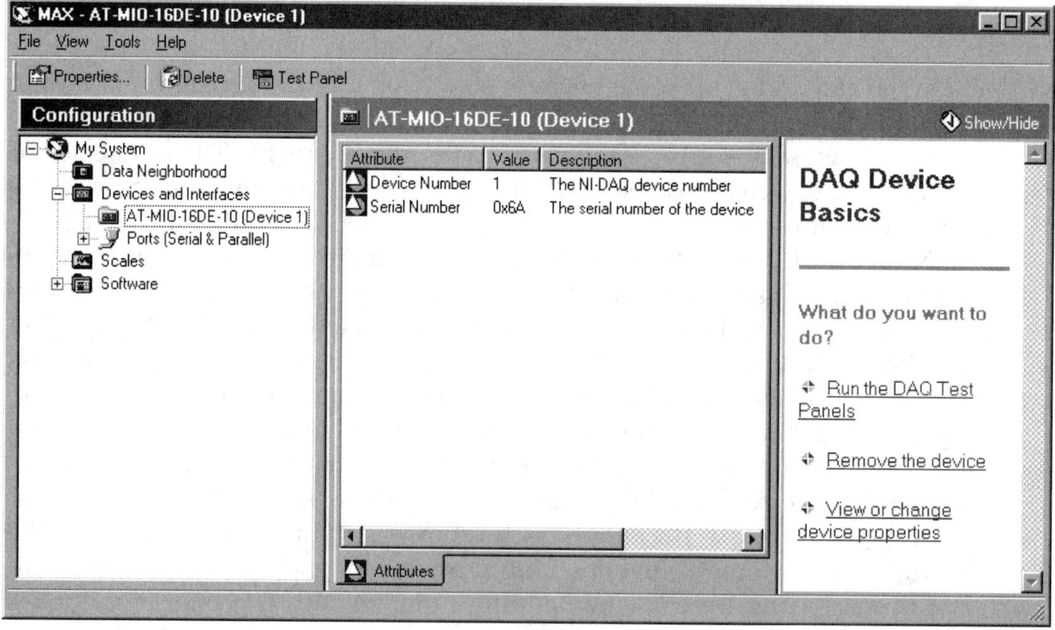

Figure 1–31
The "Devices and Interfaces" screen of MAX is shown here.

Build the VI shown in Figure 1–32, starting with the front panel, using these tips:

1. The Waveform Chart is found in the **Controls»Graph** palette. Pop up on the chart's Plot Legend (labeled `Plot 0`) and select the chart type shown, which is little squares connected with lines. The chart can accept the Waveform data type, which was covered in Section 1.2.3, but its simplest data type is a single number.
2. The Stop Button is found in the **Controls»Boolean** palette.
3. The While Loop is the big gray rectangle shown in Figure 1–32 on the block diagram, and it is found in the **Functions»Structures** palette. Loops will be covered in detail later, but they basically are rectangles in which all functions are executed repeatedly. To draw the While Loop, select it, left click the block diagram where you want one corner of the While Loop, and drag your mouse to the diagonally opposite corner.
4. Pop up on the little green square inside the While Loop and set it to **Stop If True**. This will be explained later in this chapter.
5. The box labeled "AI ONE PT" is found in the **Functions»Data Acquisition»Analog Input** palette. Be sure to choose `AI Sample Channel.vi`, *not* `AI Sample Channels.vi`.
6. The box with the metronome is called "Wait Until Next ms Multiple" function, and is in the **Functions»Time & Dialog** palette. Pop up on its left terminal and select **Create Constant**, then type in 250 before clicking anywhere.

Run the VI. If you get an error and you cannot figure it out, you'll find sources of help in the preface of this book. If you do not get an error and you see data on the screen that might look something like Figure 1–33, you have successfully performed DAQ with LabVIEW!

We must wait until Chapter 3 before covering DAQ in LabVIEW extensively, but I decided to let you know right away how it's done. Technically, we also performed DAQ with MAX when we tested the analog input channel, but LabVIEW greatly expands your ability to do just about anything you want with the data.

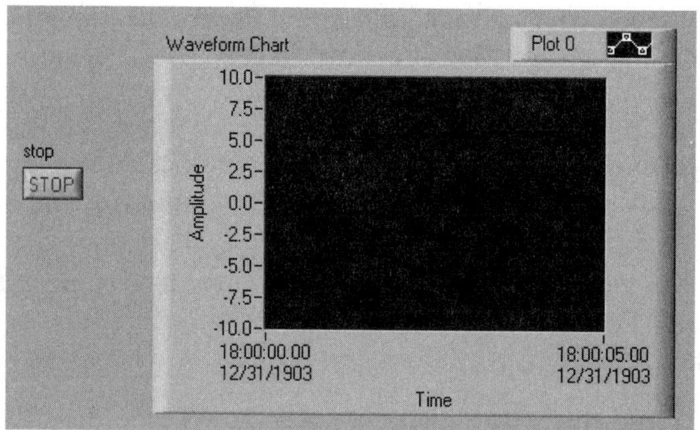

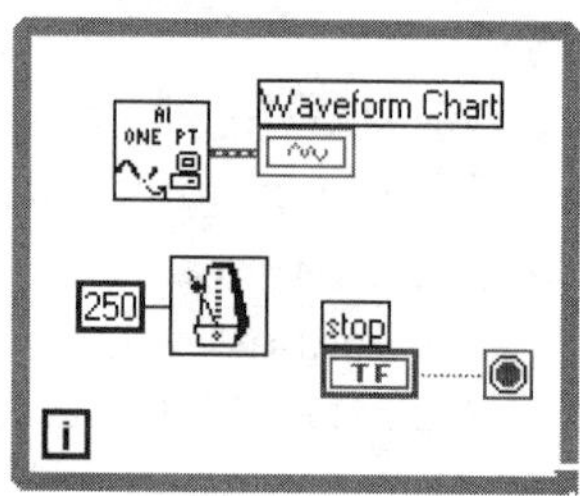

Figure 1–32
A very basic DAQ program is shown here.

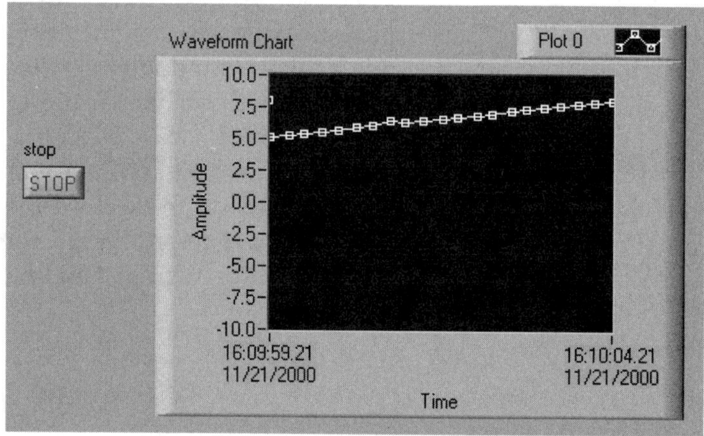

Figure 1–33
Example of a successful DAQ.

1.7 Decisions, Decisions

As of this section, we begin our dive into those aspects of LabVIEW that make it a real programming language, rather than just a fancy presentation. In every program, at some level, the program must make a decision. For those of you who have programmed before in text languages, this is sometimes called a *conditional statement* or an *if/then/else statement*. Decisions must be based upon conditions, so we will be using the conditional functions in the **Functions»Comparison** palette. We will often use the Case Structure, soon to be described.

1.7.1 A Simple Example

With that said, let's start building a VI to illustrate decisions using conditional functions. Create a new VI. Use the Digital Control from the **Controls»Numeric** palette and the Round LED indicator from the **Controls»Boolean** palette to create the front panel shown in Figure 1–34. Do not worry about the exact appearance of anything outside the main gray area throughout this book, unless instructed.

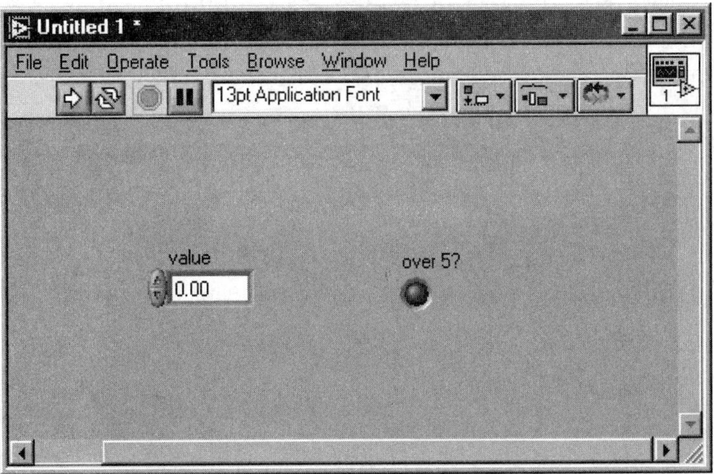

Figure 1–34
Creating a new VI's front panel.

Next, move on to the block diagram and wire it as shown in Figure 1–35 (if you have trouble, read the paragraph after Figure 1–36).

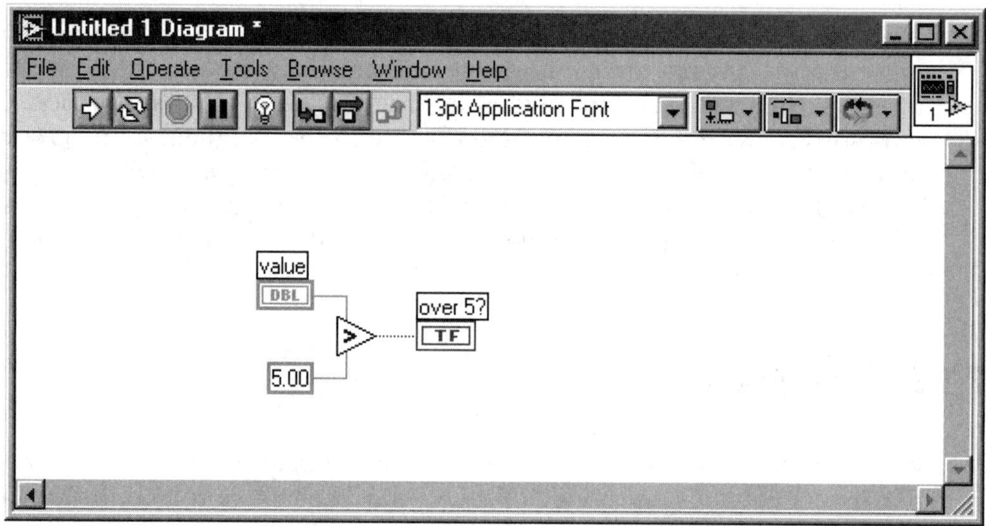

Figure 1–35
Creating a new VI's block diagram.

To save space in the rest of the book, we will be showing only the inside areas of the front panels and block diagrams of our VIs, and sometimes the icon/connector (described later) when it's important. Let's redisplay the above VI more efficiently, as in Figure 1–36.

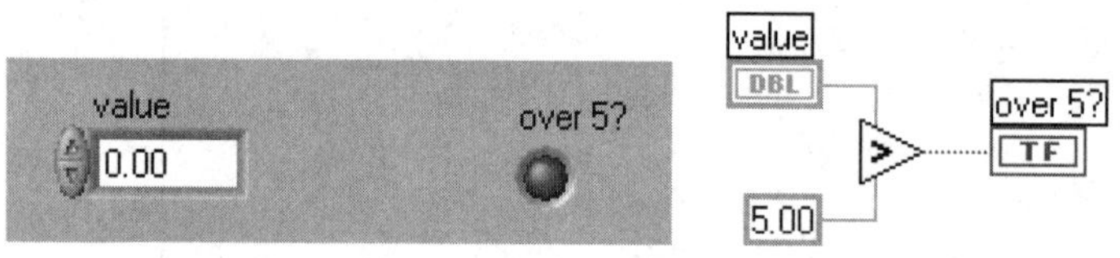

Figure 1–36
"Shorthand" for showing both a front panel and block diagram, previously seen in Figures 1–34 and 1–35.

If you've successfully built this VI, skip this paragraph. Otherwise, the `value` and `over 5?` objects should be there when you first see the block diagram. The box with the `5.00` 5.00 and the Greater Than function > can be created from the **Functions** palette. The box with the `5.00` is a Numeric Constant from the **Functions»Numeric** palette. If you immediately type in `5.00` after it's dropped, you'll have a number with the DBL data type. Otherwise, if you had just typed `5`, your number would have the I32 data type; you can change that data type to DBL by popping up on the `5.00` and selecting DBL through its **Representation** menu item. The Greater Than function comes from the **Functions»Comparison** palette. If you have difficulty wiring, review Section 1.2.8, "Wiring in Detail." To start from scratch on the block diagram, select everything (drag a rectangle around it with the Positioning tool) and delete it with the <Delete> key—only the terminals of front panel objects will remain.

With the VI successfully built (the **Run** button will be unbroken), go back to the front panel, type a 6 into the `value` box, then hit the **Run** button . You should see the `over 5?` box lighten. This light color means True, and in the case of this VI, it specifically means that "6 is greater than 5" is True. Next, type a 4 into the `value` box, hit that same **Run** button, then you should see the `over 5?` box darken. This dark color means False, or 4 is *not* greater than 5. Finally, go to the block diagram, hit the **Highlight Execution** button so that it turns yellow and has little lines emanating from it , then watch the block diagram after you hit the **Run** button so you can see in detail what's happening. Notice how this *execution highlighting* slows down your VI, so click it again to turn it off, because we'll soon run it at high-speed.

After this section, we will learn how to build a *loop*. In software terms, a loop is a means of repeatedly doing a similar thing. We could build a VI, using a loop, that monitors some data by repeatedly reading that data, and possibly acting on the data if it meets certain conditions. For now, go to the front panel again, and we will see what the **Run Continuously** button is about. In general, this is a bad button for beginners, because it can lead to great confusion—but it's nice to know about it.

Go ahead—push the button and see what happens to the buttons on the tool bar (see Figure 1–37).

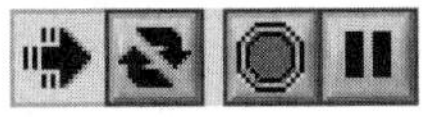

Figure 1–37
The result of pushing the Run Continuously button.

This is a sign that you are in continuous run mode. We have effectively created a loop, since the VI runs repeatedly. While this VI is continuously running, go to the front panel and change `value` to different values, some greater than 5, and some less than 5.

If you type in the numbers rather than using the little up/down arrows on the `value` control, LabVIEW doesn't use the new value until you hit the <Enter> key or click elsewhere. This little weirdness applies to most text-entry objects in LabVIEW, except the String Control, which you can configure via a pop-up menu item to have LabVIEW read the string data as you type.

As you change `value` to different numbers above and below 5, you should see the `over 5?` box change from True to False (while you are in continuous run mode).

The **Run Continuously** button, which normally looks like this, should now be darkened. Click it, so as to stop continuously running, then go to the block diagram.

A more elegant way to create a loop in LabVIEW is to use the built-in looping structures; either the While Loop or the For Loop. These will be presented in a later chapter.

Next, we will modify our VI further to illustrate more basics of conditional programming. The front panel will soon look like Figure 1–38.

Figure 1–38
Building an example front panel.

To create this, drop a String Indicator on the front panel and label it `text`. The String Indicator is found in the **Controls»String & Path** palette.

1.7.2 The Select Function

Next, modify the block diagram as in Figure 1–39.

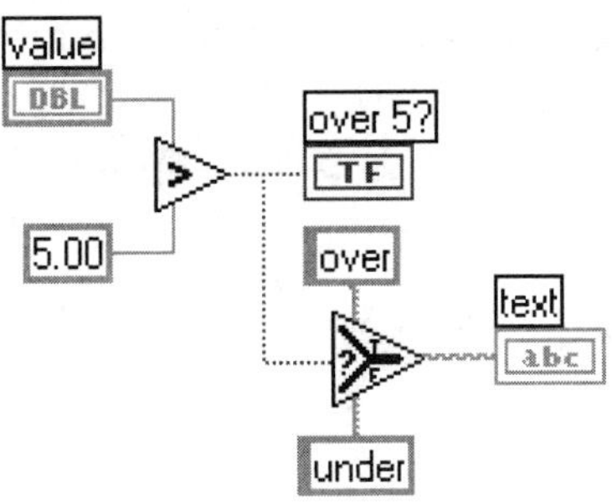

Figure 1–39
Building an example block diagram.

The new items are found in the palettes described in Table 1.8.

Table 1.8

Item	***Appearance***	***Palette***
String Constants (after you've typed "over" and "under")	over, under	**Functions»String**
Select function		**Functions»Comparison**

After wiring the VI as shown in Figure 1–39, go to the front panel and run an experiment similar to what you did before, using the **Run Continuously** button and entering values greater than, then less than 5 while the VI is running. As you should see in the `text` indicator, the Select function, presented here, selects either the data on its top wire or its bottom wire, depending on whether the data on its middle wire is True or False, respectively. To clarify this process, look at the block diagram with execution highlighting on.

This Select function is one of the two major ways to make a decision in LabVIEW. The other is the Case Structure.

1.7.3 The Case Structure

On the block diagram we've been having fun with, we'll soon replace the Select function with the Case Structure, which will do almost exactly what the Select function did. Later, we'll compare the Case Structure to the Select function.

Select the Select function with the Positioning tool (see Figure 1–40).

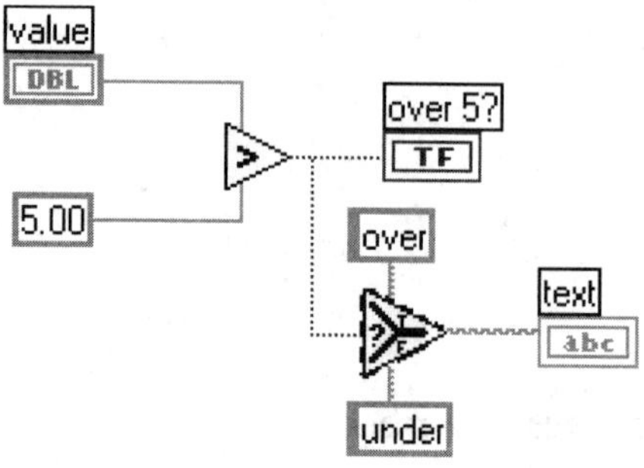

Figure 1–40
The Select function is illustrated.

Delete it with the <Delete> key, delete any bad wires remaining with <Ctrl-B>, then arrange your objects as in Figure 1–41.

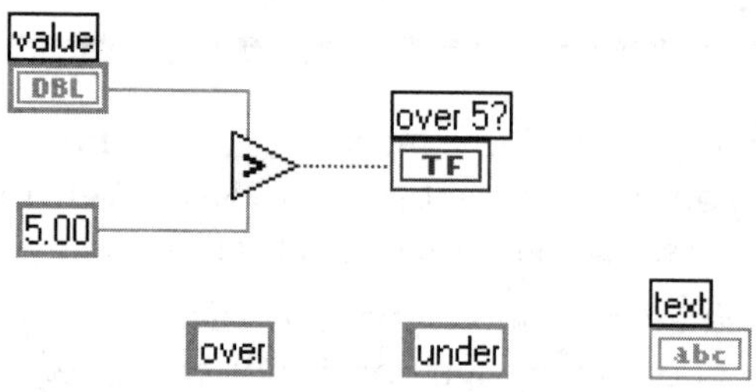

Figure 1–41
A few items from Figure 1–40 are deleted.

Select the Case Structure from the **Functions»Structures** palette (called "Case" there), and while your cursor still looks like this, , drag an outline exactly where you see the Case Structure in Figure 1–42—as soon as you release your mouse button, the Case Structure will be created there. Pop up on the Case Structure (its wall, not the interior) and select **Make This Case False,** so it looks like Figure 1–42, thus our True/False logic will be correct.

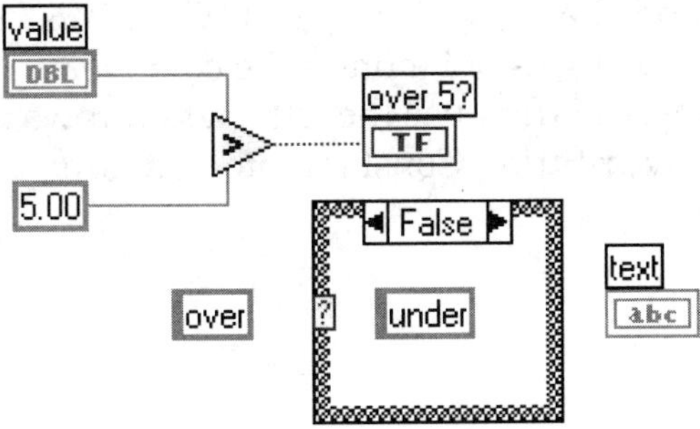

Figure 1–42
The VI from Figures 1–40 and 1–41 is being modified.

If you need to adjust the position and/or size of the Case Structure, you can use the Positioning tool to drag any of its corners.

It's very easy to accidentally hide other objects behind structures. The hidden object is still there, but you cannot see it. Simply move the structure if you think there may be something behind it.

By default, a Case Structure drops with two cases, False and True. Click one of the little left/right arrows to see the True case, notice that it's empty, then look at your False case again by using these same arrows. Wire the `under` String Constant through the Case Structure's wall to the `text` indicator, as shown below. Notice a *tunnel* is created, as in Figure 1–43.

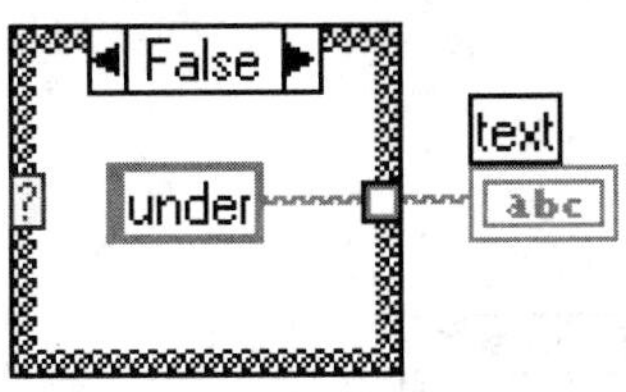

Figure 1–43
A Case Structure tunnel.

A Case Structure tunnel allows a wire to pass through the wall of the structure; we will see similar tunnels on loops soon. A Case Structure tunnel must be wired from *all* cases within the Case Structure. If it is not, it then looks like a white rectangle, and the VI will not run. When it is correctly wired from all cases, it will change to a solid, non-white color. First, wire the True case *incorrectly,* as in Figure 1–44 (first use the left/right arrows to get to your True case, then move the `over` String Constant into this case):

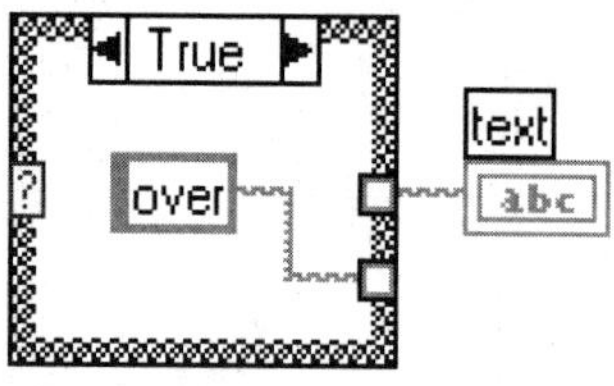

Figure 1–44
The True case wired incorrectly.

If you wanted the VI to run, you would have wired the `over` constant to the existing tunnel, not to another place on the Case Structure's wall. This is a common mistake, especially when those two white tunnels are almost on top of one another—you will think you have one tunnel that's wired from both cases that should be a solid, non-white color, when in reality you have two, so it looks like one that's white.

Let's wire it correctly now. First, the little rectangle [?] on the left wall of any Case Structure is the all-important selection terminal—the value it receives determines which case is executed. Delete the wire coming from the `over` constant, and wire your diagram so it looks like Figure 1–45 (notice I've shown both cases here, while you can only see one case at a time).

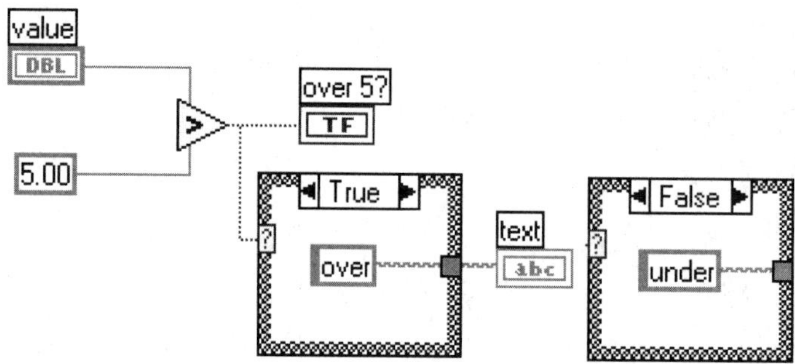

Figure 1–45
The True case wired correctly.

We have used the Case Structure to do exactly the same thing the Select function did previously. So why use the Case Structure? Its major advantage is that the objects in the cases that are not executing will not be executed. Although it won't make much of a difference in this example, some LabVIEW code may take significant time to execute. A secondary advantage is that the Case Structure saves some block diagram real estate. This could also be considered a disadvantage in that the objects in the cases that are not showing are hidden. To illustrate a case not executing, build the following block diagram, following these tips:

1. Take care *not* to hit the **Run Continuously** button  while building this one.
2. The new object inside the rectangle here is the One Button Dialog function, found in the **Functions»Time & Dialog** palette.
3. You can move from one case to the other by clicking on the little left/right arrows on the top of the Case Structure. See Figure 1–46.

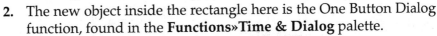

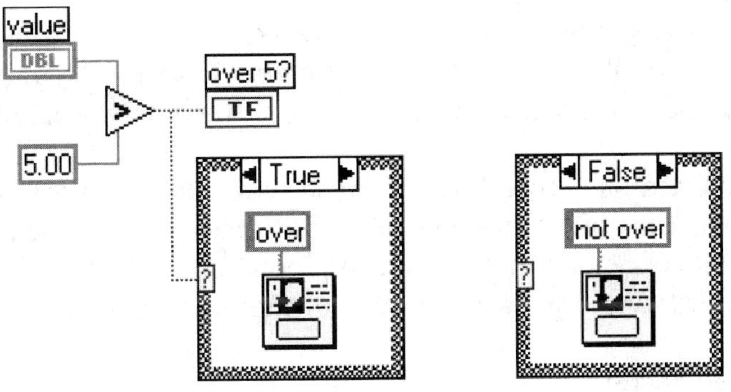

Figure 1–46
Move from one case to the other by clicking on the left/right arrows on the Case Structure.

As before, you will not see both cases of the Case Structure at the same time, as is shown above. But in this book, it is most convenient to portray all Case Structures' cases in the same figure, since you cannot very well click on the pages of this book and expect to switch cases. If this happens, you have discovered my super-secret magic edition of this book!

When you run the above VI, it should be clear that only one of the two cases executes per run, since you only see one dialog box pop up. To clarify this VI further, show both the block diagram and the front panel (there are a couple of **Windows»Tile...** menu items that do this automatically). Turn on execution highlighting , run the VI from the block diagram, then you can easily see what's happening.

Figure 1–47 shows an example of an *integer* wired to a Case Structure, as compared to the *Booleans* we've seen before. Booleans only allow two cases, but integers allow multiple cases:

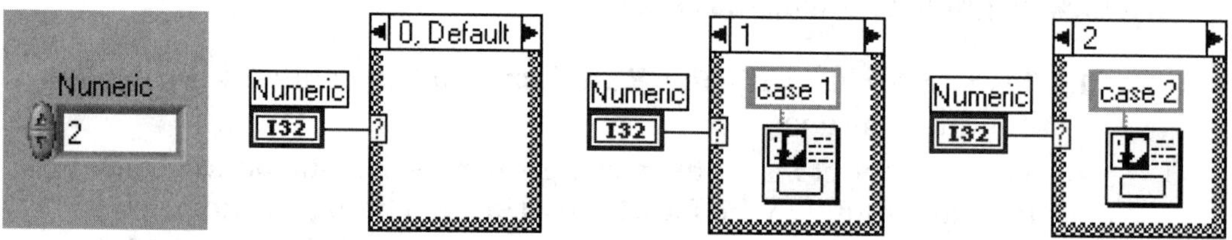

Figure 1–47
An integer wired to a Case Structure.

To create the additional case, pop up on the Case Structure and select the appropriate menu item. If you're clever, you'll create case 1 then duplicate it by popping up on the wall of the Case Structure with case 1 showing, selecting the Duplicate Case item, then editing the text of the newly created String Constant.

Note that every Case Structure (except those with Boolean inputs) has a default case that executes when none of the other cases' conditions are met. Case Structures with Boolean inputs should generally have exactly two cases. Here is a summary of the basic features of the Case Structure:

1. Case Structures allow a choice of executing one of two or more portions of LabVIEW code, based upon incoming data.
2. Depending on the data that reaches the selection terminal ?, only one of the cases will be executed.
3. You must be careful to click on the wall of the Case Structure, not inside it, if you want to pop up on it.
4. The box in the top center with the two arrows lets you move from one case to another; pop up on the wall of the Case Structure for many more options.
5. You may pass a Boolean, integer numeric, string, or Enum (detailed soon) to a Case Structure's selection terminal ?.

6. Every Case Structure (except those with Boolean inputs) has a default case, which will execute if none of the other cases are a match to the input.
7. You may have two or more cases in a Case Structure. Although one case is possible, it is pointless. Case Structures with Boolean inputs use only two of their cases.

The Enum Constant object is found in the **Functions»Numeric** palette. It is a *ring*, and as with all LabVIEW text rings, you can type any number of distinct text values, and use the pop-up menu or <Shift-Enter> to add new values. The Enum is especially useful with the Case Structure, because it displays its text values at the top of each case—but its data type is really numeric, not text.

To summarize, decisions in LabVIEW are fundamentally handled in two different ways: (1) by the Select function, for very simple two-way decisions, and (2) by the Case Structure, for nearly all other decisions. Very seldom has any useful LabVIEW program been written without the Case Structure, as decision-making is fundamental to all but the most simple of programs.

1.8 The Sequence Structure

In LabVIEW, there are only two ways to control the order of events: dataflow (wiring nodes together) or the Sequence Structure. If you were to build the block diagram shown in Figure 1–48, you would have no guarantee that `message 1` would appear before `message 2`.

Figure 1–48
This block diagram allows no control over the order of events.

The fact that one block diagram node is placed to the left of the other has nothing to do with the order of execution. Since these two nodes are not wired to one another, LabVIEW may arbitrarily choose to pop up either box before the other. The Sequence Structure, found in the **Functions»Structures** palette, has *frames* that act like the frames of a film; the first frame (number 0) executes, then the second frame (number 1), then the third frame (number 2), and so on. The Sequence Structure may have one or more frames, but not zero frames.

Create a new VI. Drop a Sequence Structure on the block diagram (like you dropped the Case Structure earlier), then create the `message 1` String Constant (in the **Functions»String** palette) wired to a One Button Dialog (in the **Functions»Time & Dialog** palette) as shown in Figure 1–49 within the Sequence Structure. Next, pop up on the Sequence Structure (like any structure, you must hit its edge, not its innards), so you can add a second frame with identical contents.

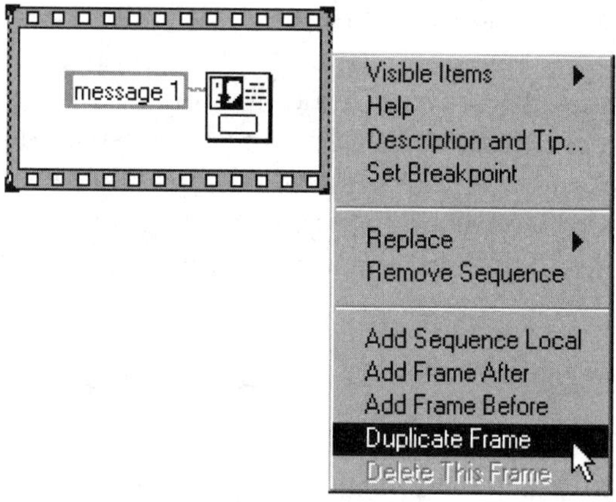

Figure 1–49
Use the pop-up on the Sequence Structure to add a second frame with identical contents.

Move to the second frame if needed by clicking the little right arrow with the Operating tool, then change the string in the second frame of your sequence to `message 2`, so your two frames look like Figure 1–50.

Figure 1–50
Two frames of a Sequence Structure are shown here at once, although you'll only see one at a time on a block diagram.

Sequence Structure frames are always ordered 0, 1, 2, and so on, unlike Case Structure cases, in which you may skip numbers.

You are now guaranteed to have `message 1` pop up before `message 2`, because the frames of a Sequence Structure always happen in order. Why did we wait until now to introduce the Sequence Structure? We recently introduced the One Button Dialog function, which is not usually wired to other parts of the block diagram, so it is inherently more difficult to control exactly when it executes in relation to the rest of the block diagram. Most of the other functions in LabVIEW are usually wired to one another, and wiring *also* controls the order in which nodes occur, thanks to dataflow. Since it's usually unnatural to wire these One Button Dialog functions to other parts of any block diagram, a Sequence Structure is often appropriate with these functions.

One common use for the Sequence Structure is timing sections of the block diagram. First, let me introduce the Get Date/Time In Seconds function, whose Help window contents are shown in Figure 1–51.

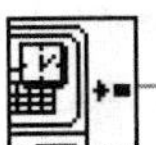

seconds since 1Jan1904

Get Date/Time In Seconds

Returns the number of seconds that have expired since 12:00 am, Friday, January 1, 1904 Universal Time.

Figure 1–51
The Get Date/Time In Seconds function Help window.

This function gives you a special number used often in LabVIEW to represent the current time and date. This was described in Section 1.2.3 as LabVIEW's standard Time & Date format. Why not always represent the time and date with a string? Computers manipulate numbers much more efficiently than strings. When it comes time to display this time and date to the user, *that's* when the number is converted to a string.

Build the VI shown in Figure 1–52, using these tips:

1. Create the front panel Digital Indicator `elapsed time` at the very end, by popping up on the Subtract function's output and selecting **Create»Indicator**.
2. As with the Case Structure, we are showing multiple frames of a Sequence Structure. There is really only one Sequence Structure in the block diagram.
3. That little box on the bottom wall of the Sequence Structure is a Sequence Local, produced by popping up on the bottom wall and selecting **Add Sequence Local**.
4. The two new functions shown are found in the **Functions»Time & Dialog** palette.

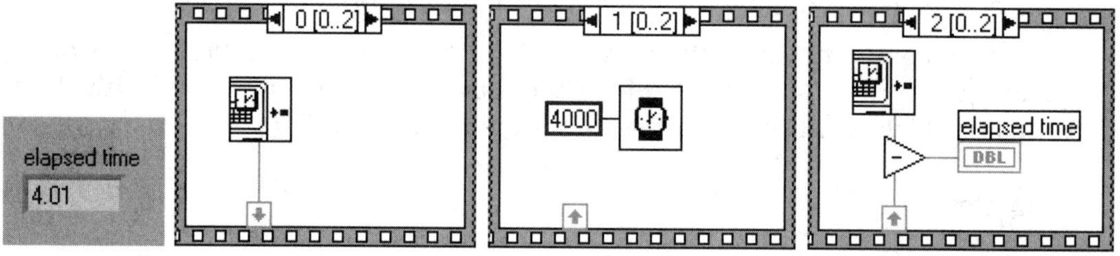

Figure 1–52
The Sequence Structure is used for timing purposes.

Run this one with execution highlighting, keeping in mind this will lengthen your `elapsed time` a bit. Even without this, your timing will likely be quite inaccurate unless you're using a real-time operating system, in which case it will be less inaccurate.

1.9 Arrays, Loops, Graphs, and Charts

The last section focused on how to make decisions; we now shift gears and begin to talk about manipulating and displaying arrays of data.

1.9.1 Arrays: A Quick Summary

In general, even outside of LabVIEW, an array is a group of objects of the same type. In LabVIEW, an array is a group of objects having an identical data type. For example, you could have an array of numbers

[8, 6, 7, 5, 3, 0, 9]

or an array of strings

["remora", "tick", "leech", "Democrat", "telemarketer", "parasite", "lawyer"]

but not an array of *both* number *and* string data types. A *cluster* can handle different data types, but we'll just have to wait for the cluster.

It is helpful for most people (me, for example) to think of the array objects as being in a row, as in Figure 1–53, rather than scattered about at random as in Figure 1–54.

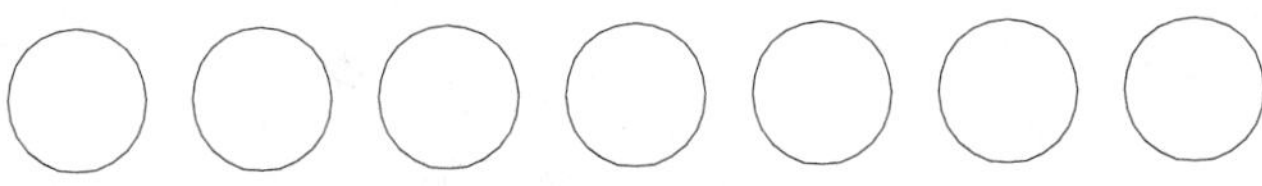

Figure 1–53
An array of generic objects, circles in this case, is shown.

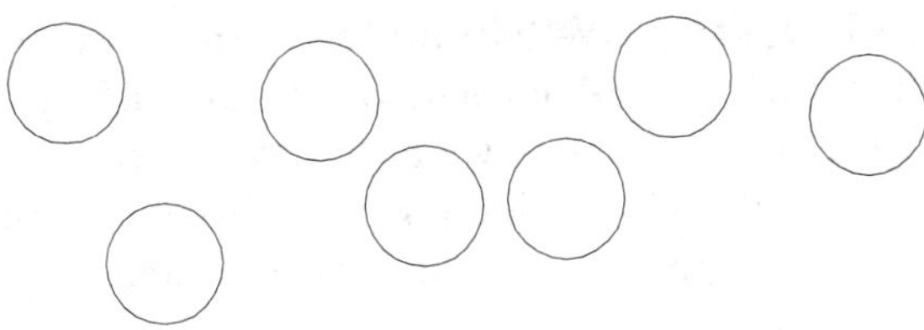

Figure 1–54
The same array from Figure 1–53 is shown, but scattered about—this can be confusing.

This is especially true when thinking about 2D arrays, which could be portrayed as in Figure 1–55.

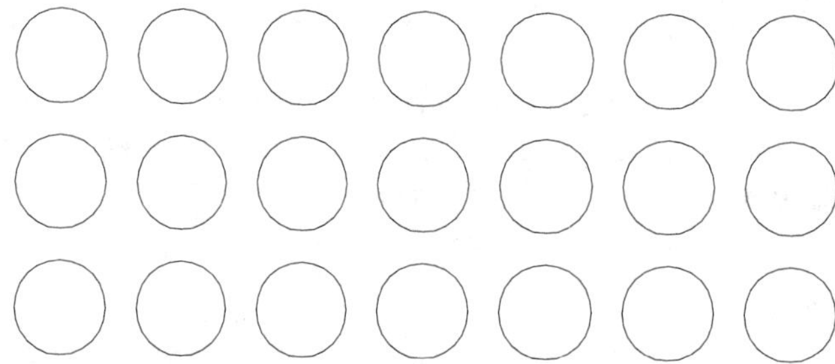

Figure 1–55
A two-dimensional array of generic objects, circles in this case, is shown.

To efficiently handle the arrays within LabVIEW (or any programming language), each element of any array is assigned an index. Taking for example our 1D (meaning one-dimensional; 2D will mean two-dimensional) array of numbers above—the indexes (also correctly called indices) would be as follows:

The array: 8, 6, 7, 5, 3, 0, 9
Its indexes: 0, 1, 2, 3, 4, 5, 6

Whenever you say "element N," then "N" is implicitly the index of that element, *not the element itself.* So in this array, element 0 is 8, element 1 is 6, element 2 is 7, and so on. This array has a size, or length, of 7.

Now, let's start using arrays in the context of LabVIEW. They can be used on the front panel or the block diagram. We'll build the numeric array shown above, and soon illustrate the tricky parts of these front panel arrays. Create a new VI, closing any others you might have without saving them. Drop an Array onto the front panel from the **Controls»Array & Cluster** palette. You should see the image shown in Figure 1–56 on your front panel.

You should also see this on your block diagram: Array [ ] . This is an empty array, so we need to make it an array of something before we can use it. We could make an array of almost any LabVIEW data type, like strings or Booleans, but we'll stick to numbers for a while, since they make the best demonstration. To make this an array of numbers, simply drop a Digital Control into the large gray square portion of the front panel array. As soon as you drop the Digital Control into that large gray square area, the array shrinks itself around that control to look like Figure 1–57.

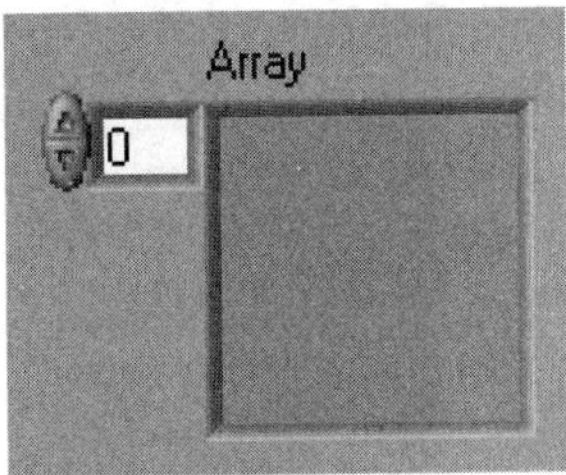

Figure 1–56
An array dropped onto the front panel from the Array & Cluster palette.

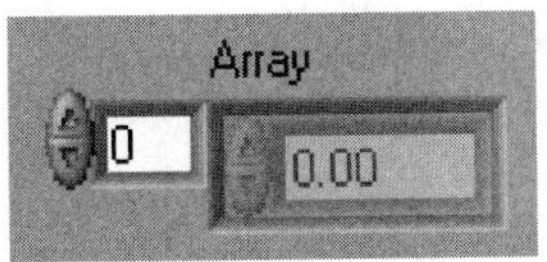

Figure 1–57
An array of numerics is shown on the front panel.

Arrays are sensitive about where they're clicked. If you pop up on an element inside the array, the pop-up menu refers to that element, not to the array. You might think you were popping up on the array, and wonder why array-relevant menu items were not present in the pop-up menu.

I've used a drawing program to blacken the rectangular area of the array in Figure 1–57 that is not relevant to the array where you pop up or click—this black area is relevant to the array's *element* (see Figure 1–58).

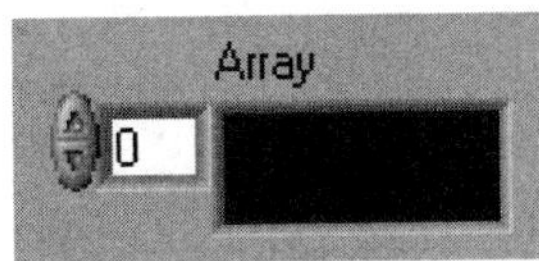

Figure 1–58
You'll probably never actually see this in "real" LabVIEW, but the "element" portion of a front panel array is drawn in black.

These front panel arrays can be confusing, so follow these instructions carefully.

1. With the Positioning tool, wiggle your cursor over the lower right-hand corner of the *element inside the array,* not the array, until the cursor looks like Figure 1–59.

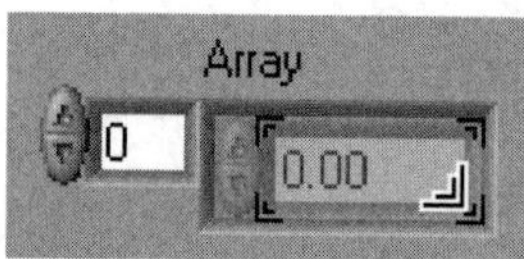

Figure 1–59
The cursor is placed over a corner of an *element* of the array, not the array itself.

2. While the cursor still looks like that, *drag* that corner of the array's element to the left a bit so your array looks like Figure 1–60.

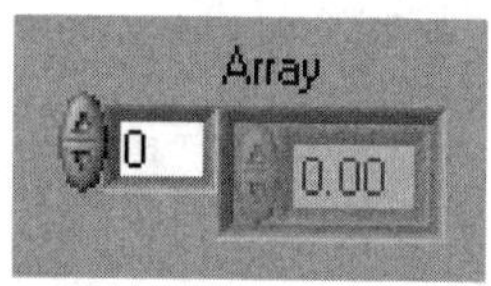

Figure 1–60
The cursor shown in Figure 1–59 was dragged to the left a bit to shrink the size of the array's element.

3. With the Positioning tool, wiggle your cursor over the lower right hand corner of the *array,* not the element inside the array, until the cursor looks like Figure 1–61.

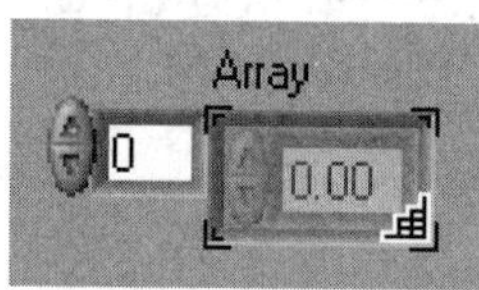

Figure 1–61
In contrast to Figure 1–59, the cursor is placed over a corner of the array, not its element. Very slight mouse movement will differentiate these two cursor appearances!

4. With the cursor looking like it does just above, drag the array corner to the right so you have eight elements showing, as in Figure 1–62.

Figure 1–62
If the cursor shown in Figure 1–61 were dragged to the right, more elements would appear.

5. Switch to the Operating tool or the Labeling tool, and enter the values as shown in Figure 1–63.

Figure 1–63
The empty array from Figure 1–62 is now filled in.

6. Make these values default, as you'll use them throughout this chapter, by using the array's **Data Operations** pop-up menu item.

You have just created your first array! Now let's use it.

To get a quick idea of the things you can do with arrays, clone this array just below itself. You must take care to click on the array, not the element inside, as illustrated earlier in Figure 1–58. After you have the clone, `Array 2`, change it to an indicator (by popping up on it *or* one of its elements), so that you have the image shown in Figure 1–64.

Figure 1–64
A "control" array is contrasted to an "indictor" array.

Suppose you were to try to grow the array to show eight elements with the element selected instead of the array. In other words, if your cursor looked like this ⌟ instead of this when you did the drag to show eight elements, you might wind up with something like this:

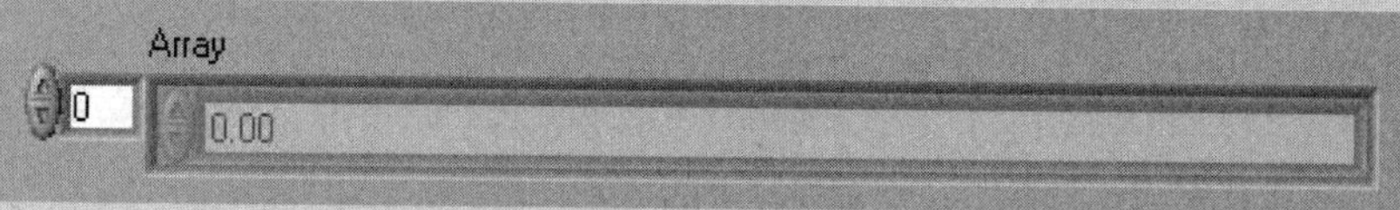

When this happens, which it will eventually if you continue to use LabVIEW, just use the undo feature, then go back and do it right.

Suppose you accidentally typed in eight elements instead of seven, like this:

If you noticed this mistake right away, you could undo it, but if you didn't notice it until much later, you could use the array's ***Data Operations»Empty Array*** *pop-up menu item to empty the array, then type the numbers in again. If the array were much larger, you could use the array's* ***Data Operations»Cut Data*** *pop-up menu (read the LabVIEW manual or Section 1.9.4 to find out how it works).*

Now, on the block diagram, arrange and wire your two array terminals as shown in Figure 1–65 (hint: you can wire *first*, then pop up on the middle of the wire, then select the Increment function from the **Insert»Numeric** palette).

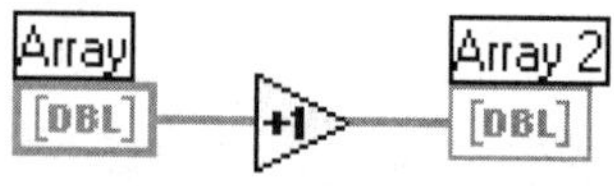

Figure 1–65
This will increment each element of the array.

Looking at `Array 2` on the front panel, run the VI and notice all the elements of `Array` are incremented by one and placed into `Array 2`, as shown in Figure 1–66.

Figure 1–66
Each element of the upper array in Figure 1–64 are incremented by one.

Once you get this working, save the VI as `Increment 1D Array.vi`.

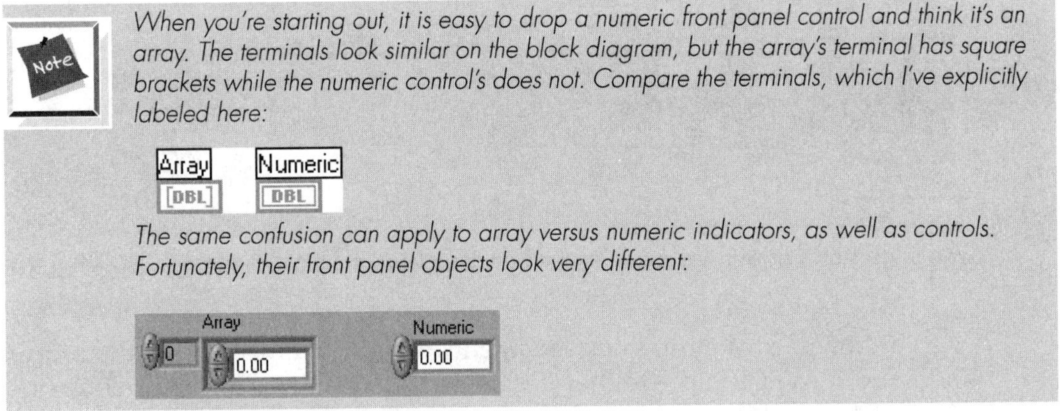

When you're starting out, it is easy to drop a numeric front panel control and think it's an array. The terminals look similar on the block diagram, but the array's terminal has square brackets while the numeric control's does not. Compare the terminals, which I've explicitly labeled here:

The same confusion can apply to array versus numeric indicators, as well as controls. Fortunately, their front panel objects look very different:

Notice how these array wires (as in Figure 1–65) are **two** pixels thick rather than one—this is indicative of a 1D array. Also, notice the square brackets around the DBL data type; this is also indicative of an array. If you had an empty array, it would look like this: Array [ ], which is useless until you drop some sort of control or indicator into it on the front panel.

Arrays can also be created as constants on the block diagram; the Array Constant is in the **Functions»Array** palette. You drop objects into this Array Constant on the block diagram much as you did on the front panel, so it might look like Figure 1–67.

You may want to use 2D arrays. Popping up on an array control on the front panel, or an array constant on the block diagram, you will see an **Add Dimension** menu item. Figure 1–68 shows examples of 2×3 2D arrays (2 rows by 3 columns of data); one is on the front panel, the other is a constant on the block diagram.

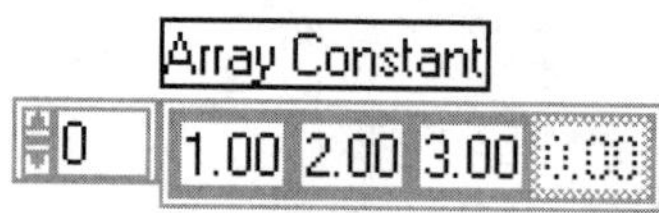

Figure 1–67
An array on the block diagram is shown, in contrast to the previously-shown front panel arrays.

Figure 1–68
Two-dimensional arrays are shown on the front panel and the block diagram.

I've shown extra empty array elements on either dimension in the arrays in Figure 1–68 just to prove they are 2 × 3. Without these elements showing, we might be looking at subsets of much larger arrays.

There are many array functions at your disposal in the **Functions»Array** palette. Take some time to experiment with them as they're described in this section; many of them should come in quite handy for even your most basic programs. If you have shelled out an insane amount of money for your version of LabVIEW, rather than just a large amount, you may have "advanced analysis" functions, which are prepackaged statistical, filtering, and mathematical functions for your arrays. If available, these advanced analysis functions would be found in the **Functions»Mathematics** palette.

Here are some of the more commonly used array functions (whose Help panels I've copied directly from LabVIEW) with which you should become familiar. There are quite a few of them, but they are very fundamental to any LabVIEW program, so you should certainly experiment with them as you read about them. To do this easily, create a new VI per experiment, drop the array function being described, then pop up on the various terminals, selecting **Create Control** for the inputs and **Create Indicator** for the outputs. This technique will automatically create a front panel object with an appropriate data type.

Remember that as you're working through this book, you may close all VIs after each section if you want, or your screen will become quite cluttered.

First, the Array Size function, shown in Figure 1–69, tells you how many elements there are in an array.

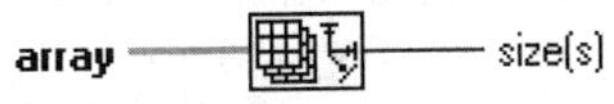

Array Size

Returns the number of elements in each dimension of input. Input can be an n-dimensional array of any type.

Figure 1–69
The Array Size function.

As an example, if you called this function on the array [8, 6, 7, 5, 3, 0, 9], it would return 7, because there are seven elements in this array.

The Index Array function, shown in Figure 1–70, can pick out any element of an array. This is taken care of automatically by passing a wire through a For Loop, provided you want to access the elements in order. Here's LabVIEW's description of the Index Array function:

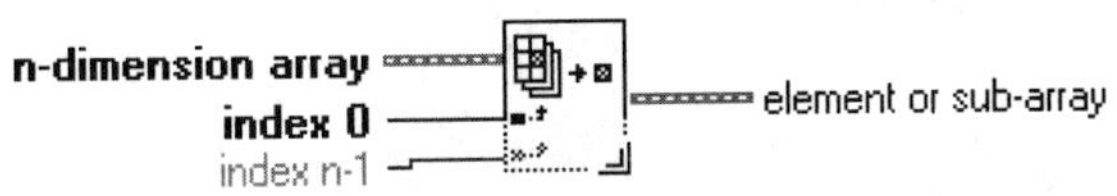

Index Array

Returns an element or subarray of array at index. Leaving an index unwired will slice out a cross-section of the array along that dimension.
The function will automatically resize to show one index per input array dimension, but you may resize by hand to show more outputs to access multiple elements in one operation. If all indices for the second or subsequent outputs are unwired, function will auto-increment the indices for the previous output in a sensible manner.

Figure 1–70
The Index Array function.

When you drop this function from the **Functions»Array** palette, it defaults to a 1D array data type. It can be grown to handle additional dimensions with the Positioning tool, as seen in Figure 1–70. As an example, if you called this function on the array [8, 6, 7, 5, 3, 0, 9], and you passed in a 0 for the index; it would return 8. If you passed in a 1 for the index, it would return 6, a 2 for the index would return 7, and so on.

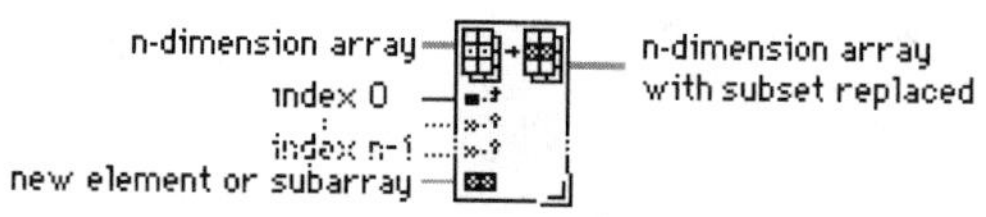

Replace Array Subset

Returns a portion of array starting at index and containing length elements. Array can be an n-dimensional array of any type. Function will automatically resize to contain as many index/length inputs as dimensions in array.

Figure 1–71
The Array Subset function.

The Array Subset function, shown in Figure 1–71, lets you pick out a subset of an array (a portion of the input array, if you want).

For example, if you had the array [8, 6, 7, 5, 3, 0, 9] and you wanted to pick out the subarray [7, 5, 3], you would need to specify an index of 2 (because 7 is the element having an index of 2) and a length of 3.

The Replace Array Subset function, shown in Figure 1–72, is fairly self-explanatory, wherein you can choose any subset *or element* of an array and replace it with something else.

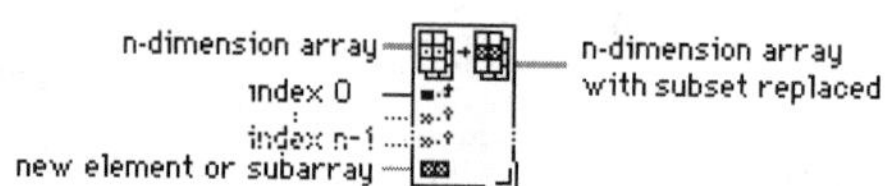

Replace Array Subset

Replaces the element or subarray in array at index. If an index is unwired, it describes a slice of the array along that dimension. For example, to select a whole column in a 2D array to be replaced, wire the column index but leave the row index unwired. The base type of the new element or subarray must be of the same type as the input array. Function never resizes original input array; subarrays which are too large to fit will be truncated.

The function automatically resizes to show one index input for each dimension in the input array, but may also be resized by hand to show multiple element/indices sets. This allows multiple portions of an array to be replaced in one operation.

Figure 1–72
The Replace Array Subset function.

The Build Array function, shown in Figure 1–73, is often useful for building arrays. Be warned that it is slow for building very large arrays element-by-element.

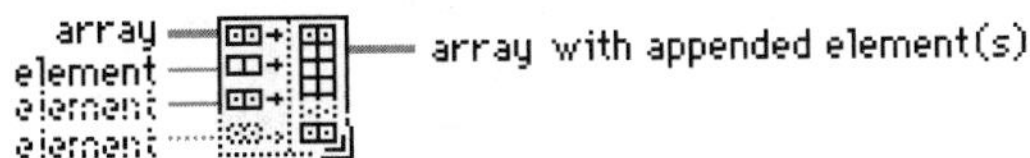

Build Array

Build an n dimensional array out of n and n-1 dimensional inputs (elements or subarrays). Inputs are concatenated in order. If all inputs are the same (n dims), you can popup and select `Concatenate Inputs' to concatenate them into a longer n-dim array, or deselect it to build an n+1 dim array.

Figure 1–73
The Build Array function.

Be sure to review all the array functions—they are really quite powerful. Arrays are inherent to most DAQ, as will be seen in Chapter 3, so the more you know about these functions, the more power you have.

The Sort Array function is especially powerful because, given an N element array, it sorts in order N·log(N) time, not N^2. This is a huge timesaver for large arrays, should they need sorting.

1.9.2 Loops: A Quick Summary

Any time you want to perform the same action, or nearly the same action, repeatedly, loops are the way to get this done. Loops often work on array elements one at a time.

Open your `Increment 1D Array.vi` from the last section, if it's not already open. On the block diagram, spread out the three objects about an inch apart horizontally, then select the For Loop from the **Functions»Structures** palette so that your cursor looks like this: . With that cursor, drag a For Loop around the Increment function, but not around the front panel terminals, as in Figure 1–74.

First, run the program with [8, 6, 7, 5, 3, 0, 9] in your front panel `Array` control, so you can see that it works, then I'll explain what's going on.

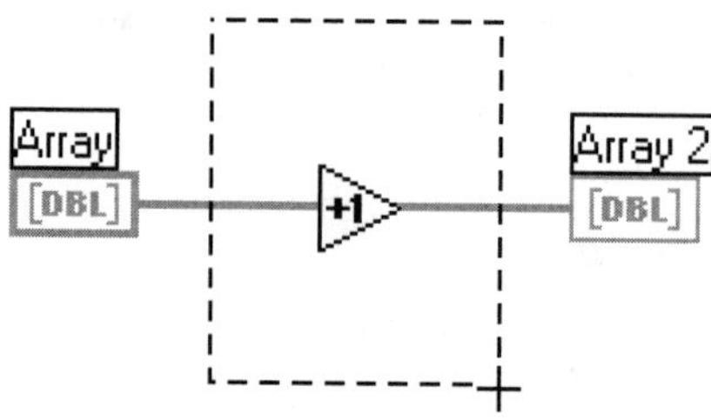

Figure 1–74
A For Loop is being drawn around the Increment function.

Then, release the mouse button so you have the image shown in Figure 1–75.

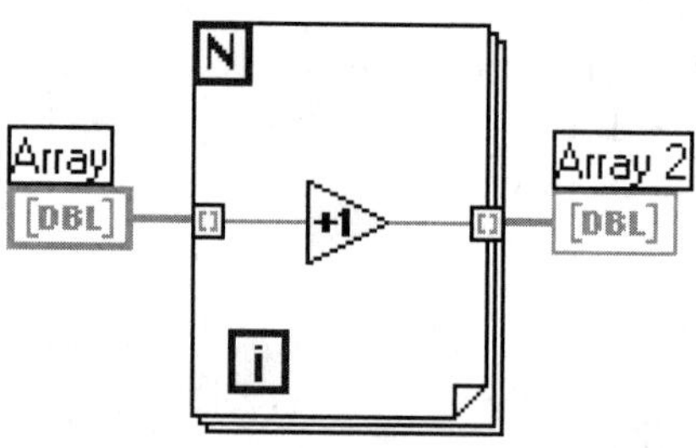

Figure 1–75
Once the mouse button is released from Figure 1–74, a For Loop is drawn.

Before you added the For Loop, the Increment function took care of *indexing* the array, meaning it *stepped through each array element* to perform the math. The For Loop explicitly indexes the arrays with its tunnels, which are the little boxes [ ] where the array passes through the wall of the For Loop. Run the VI from the block diagram with execution highlighting on, and you'll see indexing in action. You wouldn't be able to see this without the For Loop, as the indexing would happen in one step inside the Increment function.

The For Loop causes everything inside to be executed N times. In this case, N is determined by the size of our input array, which is seven. That means that everything inside the loop happens seven times. For each iteration, 0 through 6, that particular element of the input and output arrays are operated upon. The first iteration of the loop takes the 8 from element 0 of the input array, and puts the resulting 9 into element 0 of the output array. The second iteration takes the 6 from element 1 of the input array, and puts the resulting 7 into element 1 of the output array. This continues for all seven iterations, until the output array has been completely built. Close your `Increment 1D Array.vi` without saving it.

I will now describe the parts of the For Loop. The blue box with the N is the For Loop's *count terminal* [N], and the blue square with the **i** [i] is the *iteration terminal,* which starts at 0 and increments per loop iteration. If you don't have any indexing array tunnels, as shown in Figure 1–75, you can use the count terminal to determine how many iterations the For Loop will make by wiring an integer to it. If you do have indexing array tunnels coming into the For Loop, the count terminal will tell you how many times the loop will iterate. Let's build a VI to demonstrate these two For Loop terminals.

Use the **File»Save As...** menu item to save this VI as `Count Terminal Demo.vi`. Delete the front panel array control (controls can only be deleted from the front panel) and the increment function, then hit <Ctrl-B> to delete all broken wires, so you're left with the image in Figure 1–76.

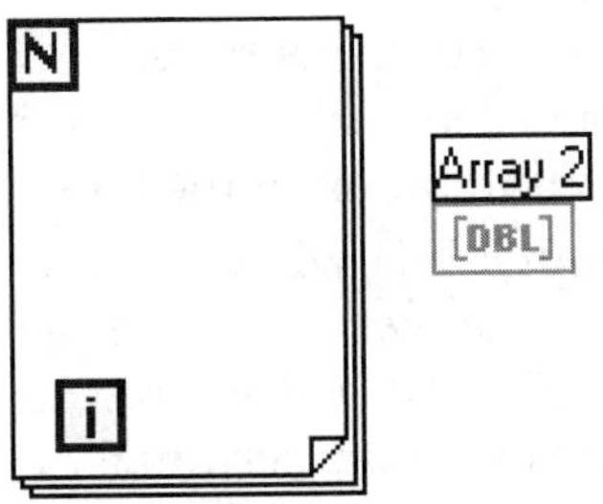

Figure 1–76
A block diagram is being built with only a For Loop and the terminal from a front panel array.

We will now build a new VI, as shown in Figure 1–77. Pop up on the count terminal [N] and select **Create Control**—label it `N`. Add the Multiply function in the middle right of the For Loop, then pop up on the upper left terminal of the Multiply function and select **Create Constant**, giving it a value of 3. Save this VI again (as `Count Terminal Demo.vi`), then run it from the front panel with `N` set to 0, 1, 2, then 3. You should see now that the number of iterations of the loop is determined by `N`, and the size of the output array as built through an indexing tunnel is determined by the number of iterations of a For Loop.

The iteration terminal starts at 0 on the first iteration, becomes 1 on the next, 2 on the next, and so on. Run this VI with execution highlighting while watching the block diagram if you're not clear on how it works. As it is a 32-bit integer, the iteration terminal will roll over to –2,147,483,648 right after

2,147,483,647, then count right up to zero again and keep repeating this cycle. Keep this in mind if you think the loop will ever get this far!

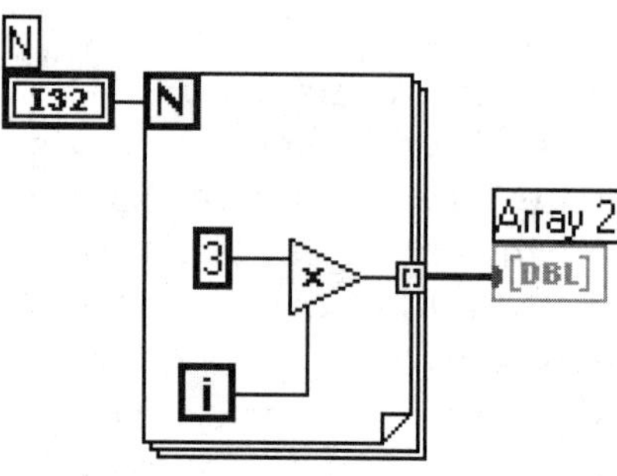

Figure 1–77
The block diagram from Figure 1–76 continues to be built.

Notice how the wire changes from one to two-pixels-thick as it passes through the right wall of the loop. This is consistent with the idea that each element of the array is being operated upon per iteration of the loop.

A For Loop must have either its count terminal **N** wired, as in Figure 1–77, or an indexing array coming into a tunnel, as shown in Figure 1–75. If the For Loop has both of these wired, the number of iterations will be the smaller of these two numbers: the number received by the count terminal or the size of the array coming into the indexing tunnel. With multiple incoming indexing array tunnels, the For Loop uses the smallest-sized incoming array to determine its number of iterations.

The other type of LabVIEW loop is the While Loop. Open your `Increment 1D Array.vi` from the last section, whose block diagram should look just like Figure 1–65, then build the block diagram shown in Figure 1–78, noting that you will need to pop up on the array tunnels and change them to indexing tunnels.

By default, passing a wire through a While Loop's tunnel does not index an array, while a For Loop's does.

Notice the broken **Run** button (because that new object inside [conditional terminal icon] is unwired). To fix it, add the objects shown in Figure 1–79 to the block diagram (notice the resizing and moving around of objects, so things don't look too crowded).

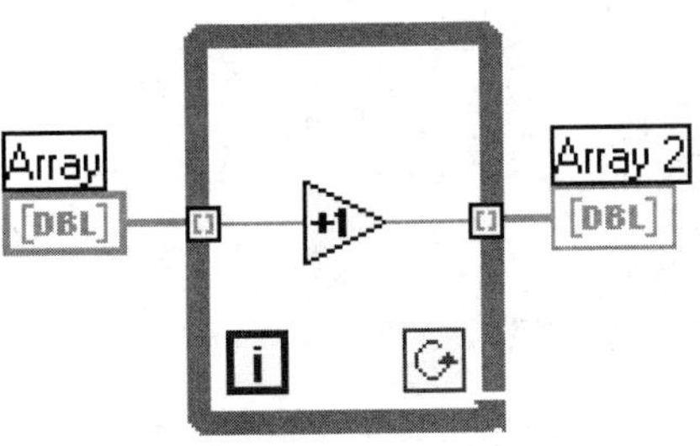

Figure 1–78
A While Loop can have indexing tunnels like a For Loop.

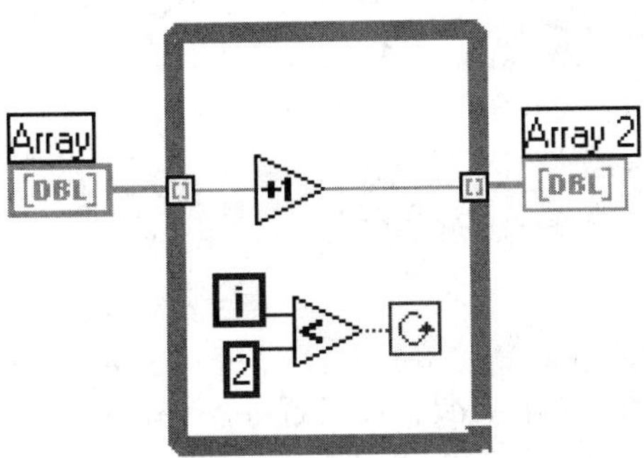

Figure 1–79
The While Loop block diagram from Figure 1–78 is now changed to a working program.

Run the program, and see that now your output only has three elements! This is because the While Loop iterates only until its *conditional terminal* is False. On iterations 0 and 1, the condition "i < 2" is True. On the third iteration, when i is 2, the value hitting the conditional terminal is False, so the loop stops there. If you have any trouble understanding this, run the loop with execution highlighting on.

Shift registers are usually the preferred way of passing data from one iteration of a loop to the next. They are created by popping up on a vertical wall of a loop, then selecting the **Add Shift Register** menu item. They work with both For Loops and While Loops. Figure 1–80 is an example of how you could sum the elements of an array using a shift register.

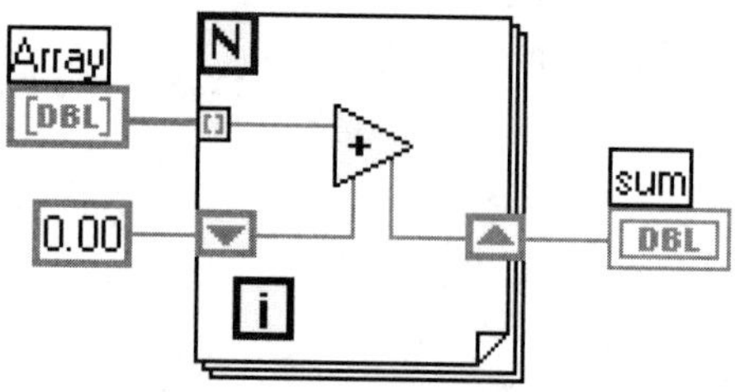

Figure 1–80
An example of how to sum the elements of an array using a shift register.

Although the shift register above appears to have a box on both sides of the loop, think of it as just one unit, because it can only hold one number at a time.

You can grow the left side of a shift register downward so it holds more numbers, but we won't cover that in this book, as it's not often used in DAQ.

Here's how that shift register works: Whatever number shows up in the right box at the end of one iteration of the loop will come out of the left box on the next iteration. The initial value of the shift register is determined by whatever is wired to the left box; in Figure 1–80, it starts out at 0.00.

Build the VI shown in Figure 1–80, and run it with execution highlighting on, and the array as shown in Figure 1–81. In case it is not obvious by now, sum is a Digital Indicator. Before running, your front panel might look something like Figure 1–81.

Figure 1–81
The block diagram from Figure 1–80 could have this front panel.

Unless you're an advanced user, you probably won't want to leave the left shift register unwired. If you do, the shift register actually remembers its number from one run of the overall VI to the next! This should seem weird, since so far, VIs have not been able to remember anything between runs. This memory feature can lead to great confusion for beginners, because the VI may behave differently every time you run it, since the uninitialized shift register may contain different values at the beginning of each run; so unless you know what you're doing, initialize all shift registers. Other common ways to remember numbers from one VI run to the next are by using local or global variables, covered later, or in files (on disk), covered even later.

The above was just a concocted example to clearly demonstrate the shift register. To sum the elements of an array, you would normally use the Add Array Elements function, Σ, oddly found in the **Functions»Numeric** palette, not in **Functions»Array**.

A 2D array can be manipulated or created by putting one loop inside the other with indexing tunnels. Figure 1–82 is the block diagram and front panel of a simple example of creating such an array.

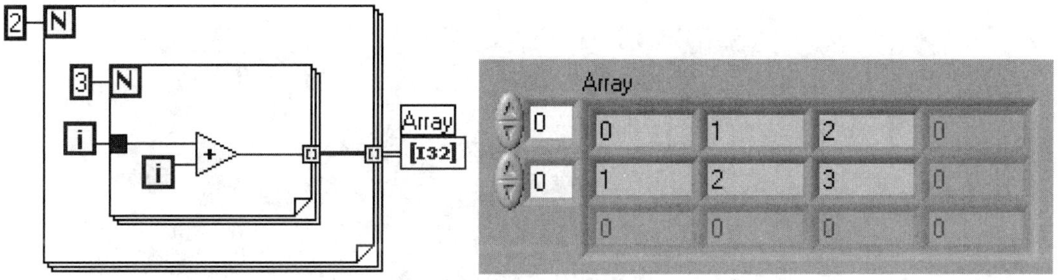

Figure 1–82
This VI illustrates a two-dimensional array.

Note the difference between an indexing tunnel [ ] and a non-indexing tunnel ■ . Arrays are sometimes smart enough to index when you want them to and not index when you don't, whenever you pass a wire through their walls. But if you wanted to force indexing one way or the other, you could pop up on any loop's tunnel and do so, as seen with the While Loop previously.

I generated the 2D array on the front panel by popping up on the far right tunnel [ ] and selecting **Create»Indicator**.

Build such a VI, then run it with execution highlighting turned on. On the first iteration of the outer For Loop, the wire entering the left indexing tunnel ■ from the left is a simple number, not an array. After the inner For Loop executes three times, the 1D array [0, 1, 2] appears on the wire between the two tunnels, and fills up the first row of the 2D output `Array`. The second (and final) iteration of the outer For Loop places the 1D array [1, 2, 3] on the wire between the two tunnels, which fills up the second row of the 2D output `Array`.

A good way to build a really large array in LabVIEW is to first create a full-sized array of zeros or NaNs, then replace each element with the data. This circumvents the inherent slowness of building a large array element-by-element with the Build Array function. This VI, although it could perform its duty without the shift register or Replace Array Subset function, demonstrates how to replace each element of an array with the numbers 0, 1, 2, and so on. Build the VI shown in Figure 1–83, using the Replace Array Subset function on the block diagram and popping up on the *wall* of the For Loop to create the shift register, and save it as `Simple Array Shift Register.vi`.

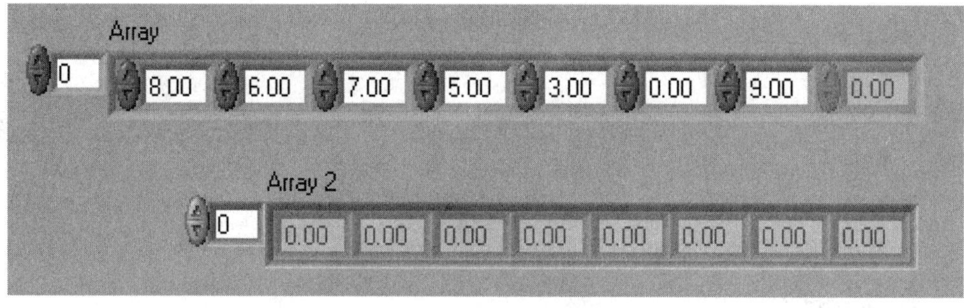

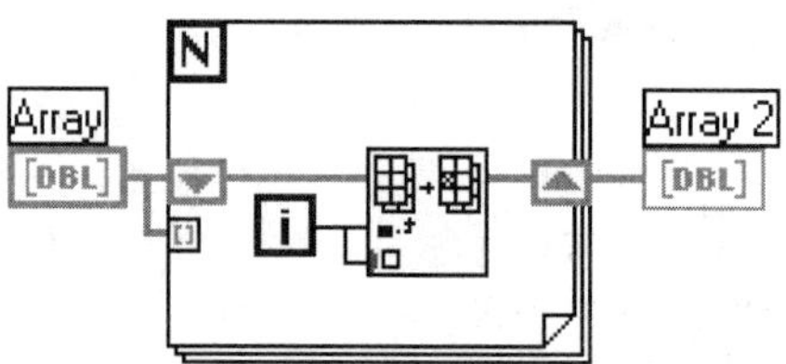

Figure 1–83
This VI illustrates the Replace Array Subset function and a shift register.

Shift registers are especially helpful when you pass an entire array into them.

If you're an experienced programmer who's new to LabVIEW, you might think this is a bad thing, as you're visualizing many numbers being moved around every iteration of the loop; but such is not necessarily the case. If LabVIEW can perform its array-in-a-shift-register duties without moving the array, it will, as in the VI shown in Figure 1–83.

The contents of the shift register after each iteration of the For Loop is illustrated here:

[0, 6, 7, 5, 3, 0, 9]
[0, 1, 7, 5, 3, 0, 9]
[0, 1, 2, 5, 3, 0, 9]
[0, 1, 2, 3, 3, 0, 9]
[0, 1, 2, 3, 4, 0, 9]
[0, 1, 2, 3, 4, 5, 9]
[0, 1, 2, 3, 4, 5, 6]

It would be impossible to use execution highlighting to see these values, since execution highlighting will not show arrays.

Suppose we had not used the shift register, but wired the array directly through the For Loop wall, then disabled indexing on the two tunnels. As soon as we disable indexing, the For Loop doesn't know how many times to iterate, so we could add another tunnel whose only purpose is to give the For Loop a count. Our block diagram would look like Figure 1–84.

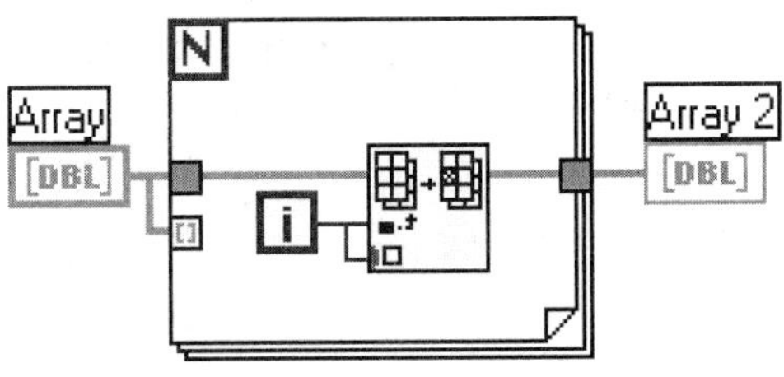

Figure 1–84
A nonsense VI, illustrating why the Shift Register of Figure 1–83 is useful.

If we ran this nonsense VI, it would not do the same thing as our last VI because the same input array, [8, 6, 7, 5, 3, 0, 9], would appear on the upper left tunnel on each iteration of the loop. For this reason, only the last element of the incoming array would be replaced in "Array 2." Remember that the shift register is the best way to pass any information from one iteration of a loop to the next, and that includes entire arrays.

Exercise 1.1

Brace yourself: This exercise covers quite a bit of material.

Suppose you have an array of seven 32-bit integers, [8, 6, 7, 5, 3, 0, 9], on a front panel. Shown in Figure 1–85 is a front panel in which the elements of such an array control are reversed and placed into an array indicator; I've changed the array elements' representation to I32, a 32-bit integer, so you don't see any decimal points.

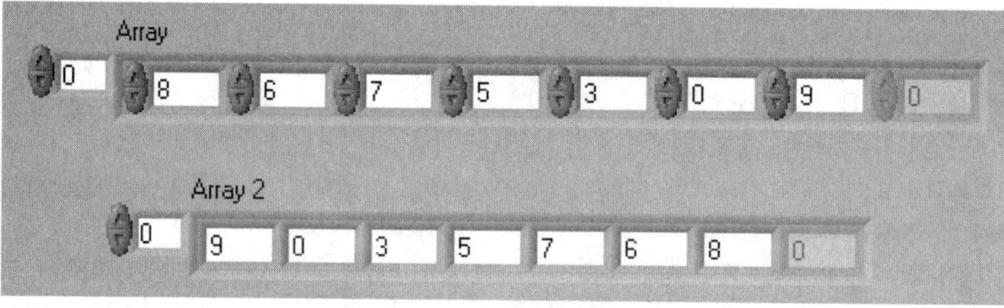

Figure 1–85
Two front panel arrays with I32 elements.

The simplest way to reverse the order of elements of this array would be to use the Reverse Array function, so that you get the above result from the very simple block diagram shown in Figure 1–86.

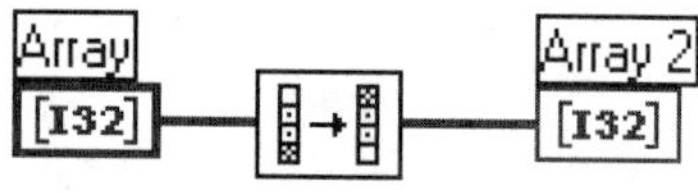

Figure 1–86
The Reverse Array function is illustrated.

But suppose LabVIEW did not have the Reverse Array function. Could you build a LabVIEW VI that would reverse the array, as shown in Figure 1–85, without using the Reverse Array function? Try to do this in LabVIEW before reading the following hints.

1. Use a For Loop that iterates once per array element.
2. Although you could use a shift register, use tunnels on the For Loop—this will simplify your VI.
3. Use arithmetic that can perform addition and subtraction.
4. You'll want to index the array, either with an indexing tunnel or the Array Index function.
5. You will need to determine the Array's size, either with the Array Size function, or the count terminal in conjunction with an indexing tunnel.

Two Solutions

There are many ways to solve this problem! Figure 1–87 shows the easiest solution to understand, in my opinion (or least difficult, if you're brand new to LabVIEW).

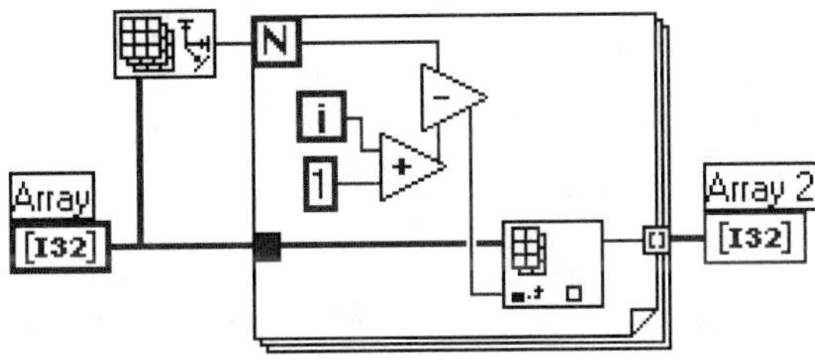

Figure 1–87
One solution to this exercise is shown.

The Array Size function is used to tell the loop to iterate seven times here, since our input array has seven elements. For each array iteration i, where i ranges from 0 to 6, the Index Array function picks the element number (7–i–1), which is the element needed to build a reversed array in the tunnel on the right. The Add Function and Subtract function are responsible for computing (7–i–1). Figure 1–88 shows another way to solve the problem, optimized for size, not clarity.

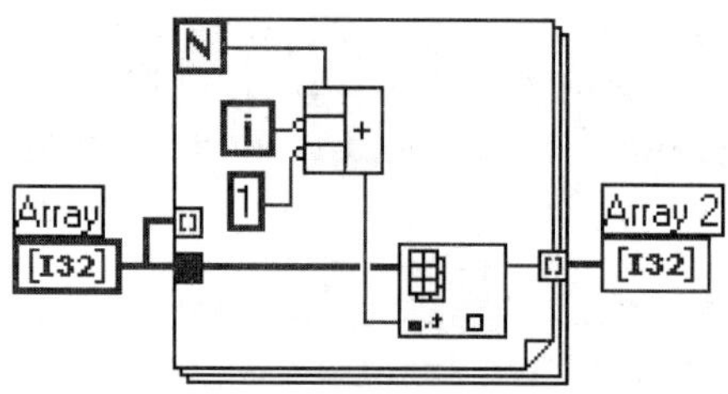

Figure 1–88
A second solution to this exercise is shown.

Both of the solutions above may be described with the picture in Figure 1–89, where the input array is on top and the output array underneath.

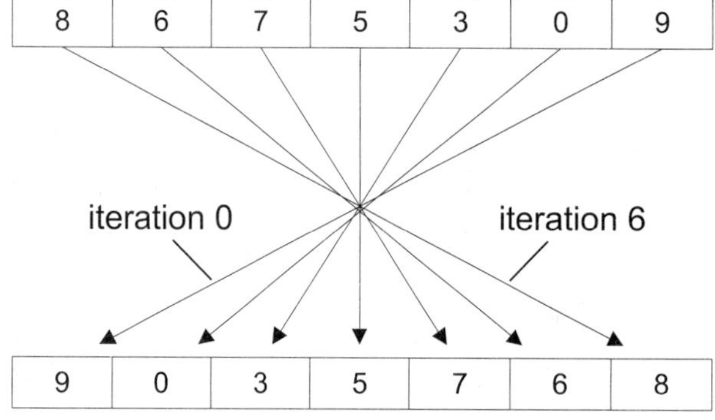

Figure 1–89
A graphical illustration of the underlying data manipulation of this exercise is shown.

While I was building this VI, I found it helpful to wire a 1 (a Numeric Constant) to the For Loop's count terminal, then run the VI with execution highlighting to make sure my math was right. I did this again with a 2 instead of a 1. Obviously, with just one tiny math error anywhere, this VI probably won't work.

1.9.3 Graphs and Charts: A Quick Summary

Now that we have, I hope, a general understanding of arrays and loops, it's time to discuss LabVIEW's graphs and charts. Create a new VI with an array of Digital Controls as shown in Figure 1–90 (you can drag an array from one of last section's front panels), and a Waveform Graph (found in the **Controls»Graph** palette; ignore the plot on the graph for a moment).

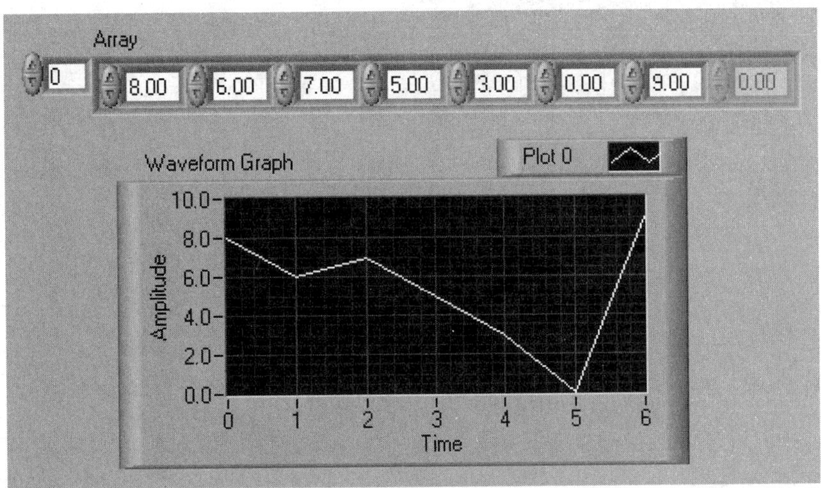

Figure 1–90
A Waveform Graph (with an arbitrary plot) is shown.

On the block diagram, wire the array directly to the Waveform Graph, as shown in Figure 1–91.

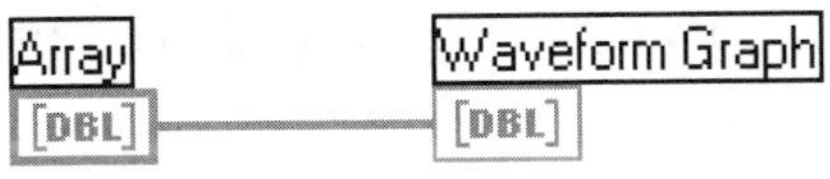

Figure 1–91
A block diagram used to show the graph in Figure 1–90 is shown.

Run the VI, and you should see something like the plot in Figure 1–90 on your front panel. I've shrunk my graph a bit, so my numbers on the axes might be different than yours. With graphs, you have little control over the exact spacing of these numbers.

The entire array is sent from the numeric array to the graph. You can change some of the numbers in your array and run it again to verify that it works.

This item, Plot 0, is the graph's *Plot Legend*. Pop up on it, and pick the point style shown in Figure 1–92.

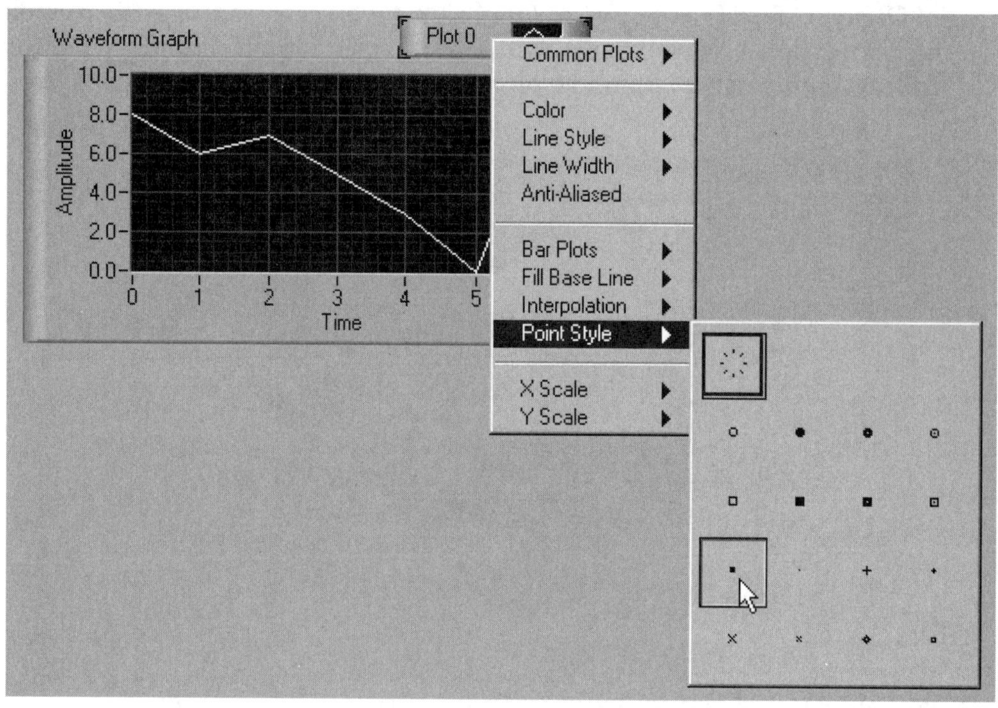

Figure 1–92
My favorite point style is this one—it is as small as possible while still being easily visible.

As Figure 1–93 shows, you can now easily see each individual point.

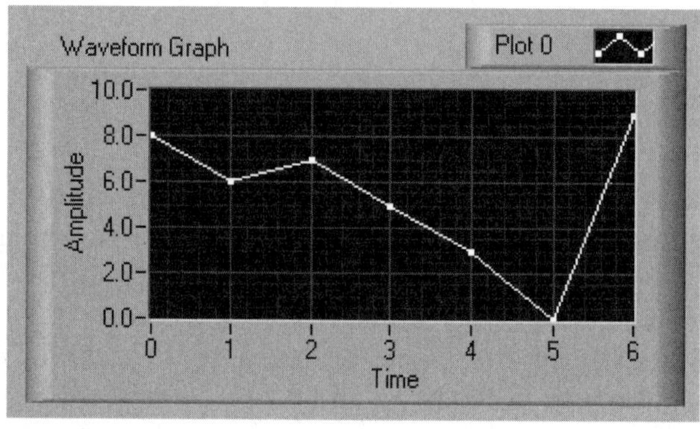

Figure 1–93
The point style chosen in Figure 1–92 is perfect for most applications.

A chart is very similar to a graph, except it is fundamentally designed to take one point of data at a time rather than an entire array. There are many exceptions to this rule, but that's the basic idea.

Go to your block diagram and add the items shown in Figure 1–94 by dragging a new For Loop across the wire and inserting the Wait Until Next ms Multiple function from the **Functions»Time & Dialog** palette.

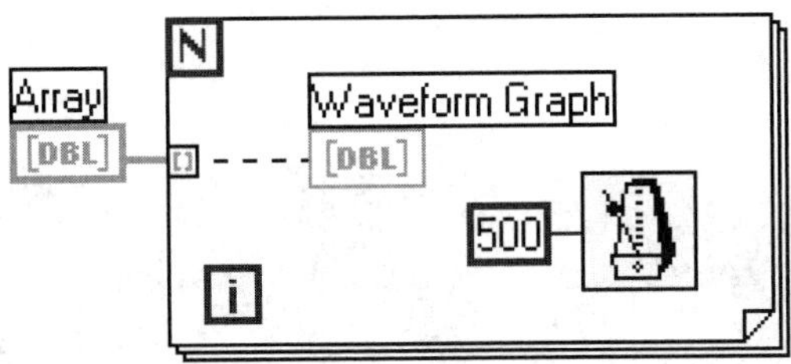

Figure 1–94
Preparing to build a simple DAQ application.

The Wait Until Next ms Multiple function is designed to work inside loops to make them run at the given rate. Realize that unless you're using a real-time LabVIEW setup, you cannot count on this timing to be accurate, since most operating systems are not truly real-time.

On the front panel, replace the Waveform Graph with a Waveform Chart by popping up on it and using the **Replace** menu item, and finding the Waveform Chart in the **Controls»Graph** palette. By doing this, you do not need to rewire your block diagram, and the broken wire you see above will fix itself. This is because a Graph does not accept simple numbers, whereas a chart does. The graph, as mentioned above, is geared towards arrays, not numbers.

Run the new VI, and notice how the points are sent to the chart one at a time. If you did not have the timing function in the block diagram, the For Loop would run full speed, and unless you're running on a 1 MHz computer or slower, all of the points would appear to show up on the chart at the same time.

1.9.4 Arrays and Loops: Details and Features

Front Panel Issues

The box with the 0 in it, towards the upper left in Figure 1–95, is the array's *index*. It specifies the index of the first array element shown (always the closest element to the index)—in this case, it means that the 8.00 has index 0. Both arrays shown in Figure 1–95 contain *exactly the same data*, but if you didn't notice that the index was 3 in the second array, you might think that its data was simply 5, 3, 0, 9.

Figure 1–95
The index of an array can be very confusing when not zero.

You may cut, copy, or paste sections of the array by popping up on it and using the **Data Operations** menu, the same menu you would use to empty the array. You can select the entire array, or a subset of the array. This menu can be tricky to understand at first. In order to select a subset of an array, you must define your selection by using the array's index. First, change the index and select the start of your selection, then change the index and select the end of your selection. When you paste, the paste begins at the element shown by the index.

Block Diagram Issues

Speed is often an issue when you're using arrays, because arrays can contain an arbitrarily large amount of data. As mentioned previously, a very common mistake is to build a big array element-by-element, using the Build

Array function. Judging from its name, I can see why a beginner would choose to use the Build Array function to build an array. Here's what can happen. Suppose you're brand new to LabVIEW and you want to create an array of 1,000,000 elements (make the 1,000,000 proportionally larger for the faster computer you probably have, long after I'm typing this). Which of the methods shown in Figure 1–96 do you think is faster?

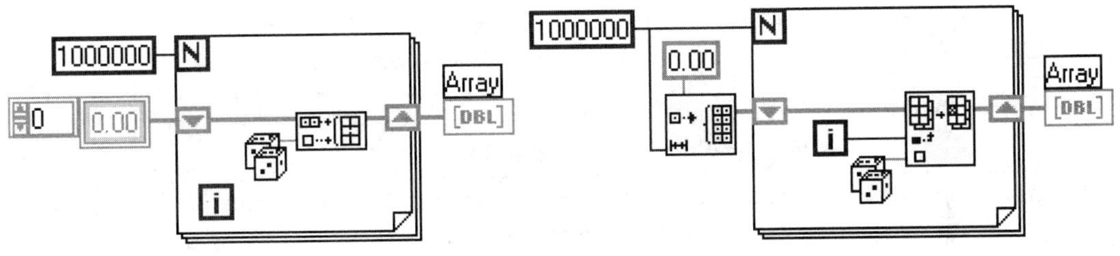

Figure 1–96
Two ways to create an array of 1,000,000 elements

On my Pentium II, 400 MHz computer, the method on the left takes about 11 seconds, compared to 2 seconds for the method on the right! This is because the array is repeatedly being grown, and this is slower due to the underlying memory management mechanism. This problem is not specific to LabVIEW, but to all programming languages. The Build Array function is often fine for smaller arrays of a few hundred elements or less, but be aware of this speed issue with large arrays.

Complicated arrays can also slow things down, particularly when you have array elements which themselves contain arrays (or strings). In general, the more complex your data type, the more likely it is to decrease your execution speed.

Graphs and Charts: Details and Features

Much of what is said about graphs in this section will also apply to charts. It's cumbersome to always say "graphs or charts," so this section will often just say "graphs" when referring to either a graph or a chart. A chart is really just a special type of graph.

There are quite a few other issues that we will briefly touch upon concerning graphs. First, realize they are *polymorphic*, meaning you can wire different data types to them. LabVIEW's help utilities document the specific permissible data types per graph. A sneaky way to see at least one of your graph's data types is to go to the block diagram, pop up on the graph's terminal (even if unwired), then select **Create»Constant**.

Table 1.9 describes the basic types of graphs.

Table 1.9 *Basic Types of Graphs*

Graph Type	Description
Waveform Chart	Fundamentally displays one point, or group of points, at a time; has internal memory called *history*.
Waveform Graph	Displays an entire array or arrays of points with equally spaced x-values; has no internal history.
XY Graph	Displays an entire array or arrays of points with arbitrarily spaced x-values; has no internal history.
Intensity Chart	Takes a 1D array of data points in which each data point corresponds to a rectangular area of the chart; has history.
Intensity Graph	Takes a 2D array of data points in which each data point corresponds to a rectangular area of the graph; has no history.
Digital Waveform Graph	Useful for displaying digital data, which is a type of DAQ signal to be discussed in a later chapter.
(many others)	Will vary widely with future versions of LabVIEW, I'm sure.

Suppose you were acquiring data in some experiment, one point every second, and displaying it on a chart. Since we have not yet discussed how data is acquired, we'll simply use LabVIEW's random number generator to give us simulated data, where each data point will be in the range 0.0 to 1.0. Go ahead and build this VI in LabVIEW, as it's being described, using these tips:

1. Begin by dropping a Waveform Chart, and shrinking it as shown in Figure 1–97.
2. That's a label on the front panel below the chart, which you can get by simply selecting the Labeling tool and typing `no autoscaling`, which will be described soon enough.

3. The random number generator is found in the **Functions »Numeric** palette.
4. The square wired to the `stop` button's terminal is the While Loop's conditional terminal, [icon], shown in Section 1.9.2, but we have inverted its logic by popping up on it and changing it to **Stop if True**. In other words, the loop stops if this terminal receives a Boolean value of True, instead of False as it did before.

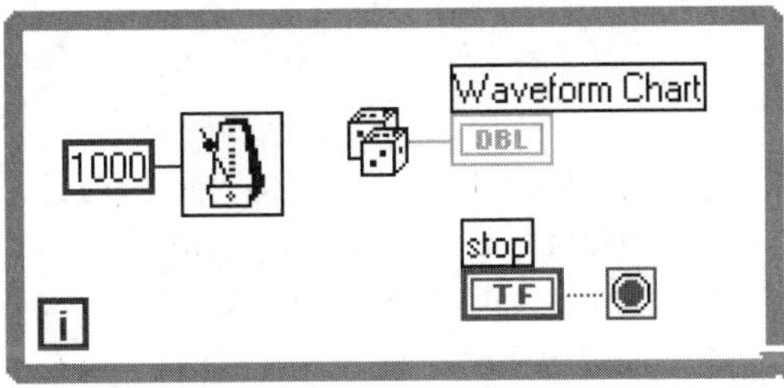

Figure 1–97
This is the format of the simplest LabVIEW DAQ application, but here we simulate voltage with a random number.

If this were done in a DAQ context, you might want to pass the Waveform data type instead of a numeric data type, as in Figure 1–32.

Pop up on the `stop` button on the front panel and notice that its **Mechanical Action** is set to one of the *latching* types. Latching is only useful in loops. Latching means that once the button is read from the block diagram, it returns to its default value on the next iteration of the loop. This is necessary to pop the button back up after you press it. The logic shown above keeps the value on the conditional terminal False for each loop iteration except the one on which the `stop` button is pushed.

The `1000` wired up to the Wait Until Next ms Multiple function causes the loop to wait one second (1,000 milliseconds) between iterations.

Now run the VI for about 10 seconds, so you see something like Figure 1–98 on your front panel.

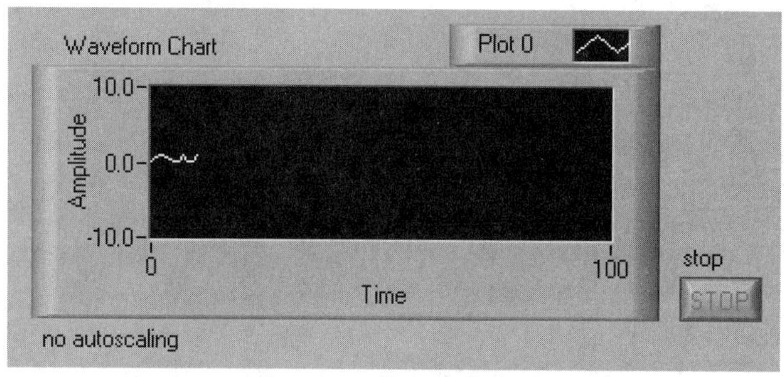

Figure 1–98
11 data points with no autoscaling.

This is a chart of about 10 points, but it's a bit tough to see this data as it's compressed into a tiny portion of the chart. First, let's expand the data on the y-axis. There are two ways to do this: You can choose the Operating tool (or the Labeling tool), and type in numbers like 2.0 and –2.0 on the top and bottom of the y-axis, as shown in Figure 1–99, and the chart will rescale a bit.

When you rescale a graph axis manually like this, it will reduce your confusion if you *only* type into the end points; leave the middle points alone. To see the x-axis a bit more clearly, type in a 12 (using the proper tool) where you see the 100, so you get the image shown in Figure 1–100.

Currently, it's not easy to see exactly where the data points are. Let's fix that—pop up on the chart's *legend*, Plot 0, and select my favorite point style, as shown in Figure 1–92, so the plot looks like Figure 1–101.

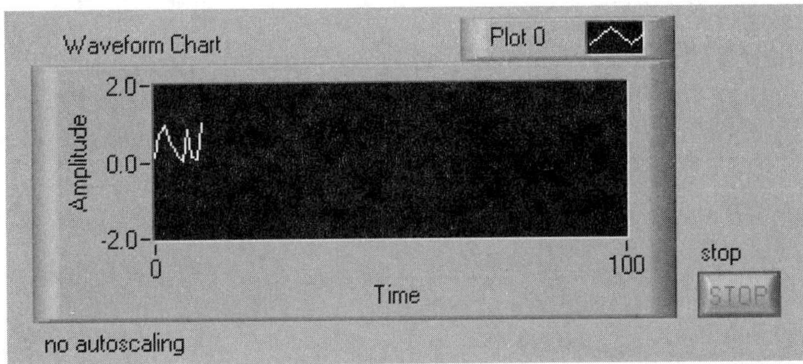

Figure 1–99
11 data points from the previous figure where we manually set the y-axis.

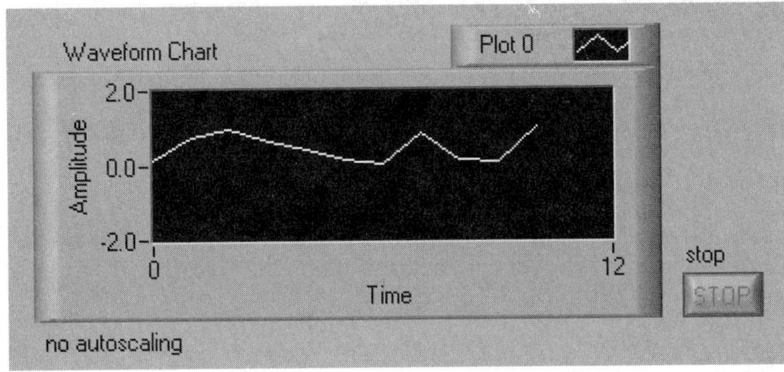

Figure 1–100
11 data points from the previous figure where we manually set the x-axis.

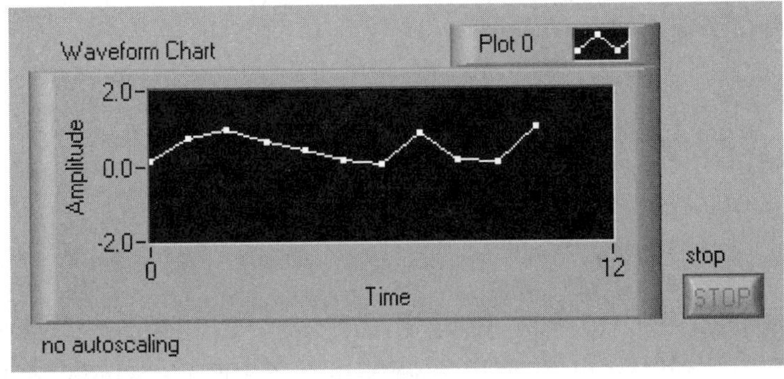

Figure 1–101
11 data points from Figure 1–100 where we set the point style.

Another popular feature of graphs is autoscaling. When autoscaling, LabVIEW automatically scales a graph's axis so you can most efficiently show the data. One way to set autoscaling is to pop up on the graph itself (not the legend, label, or many other parts you'll soon see), and select autoscaling. Turn autoscaling on for the y-axis of your chart (pop up and look in the **Y Scale** menu), then change the label below the chart, so you see something like Figure 1–102.

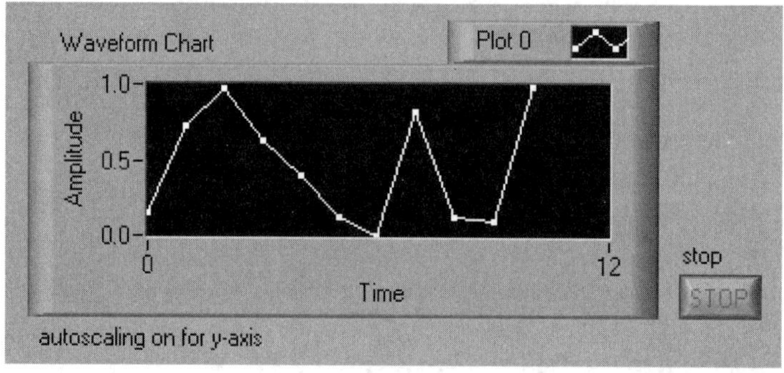

Figure 1–102
11 data points from the previous figure with autoscaling on the y-axis.

Delete your free label concerning autoscaling, because we're about to show autoscaling in a different manner. Pop up on this chart, select the **Visible Items»Scale Legend** option, and you'll see the little box shown in Figure 1–103, called the *scale legend*:

Figure 1–103
The graph's scale legend.

For each axis, you can set autoscaling by using the appropriate buttons per axis. See how the little light next to the picture of the y-axis is on (the **Autoscale Now** button), and the little padlock just to its left is closed? That means that autoscaling is turned on for the y-axis. Autoscaling is not currently on for the x-axis, as its little icons indicate. You can turn it on quite easily from this scale legend by clicking the x-axis' padlock . If you had clicked the x-axis' **Autoscale Now** button, which looks like this , the autoscaling would have occurred at the time you clicked the button, but it would not remain on as if you had pushed the padlock button.

You can control other axis properties, such as numeric precision or grids, by popping up on the X.XX and Y.YY buttons.

Next, select the chart's **Visible Items»Graph Palette** item to show the palette in Figure 1–104.

Figure 1–104
The graph's palette.

First, unlock both axes with the padlock buttons from the scale legend, so they appear like this: . Next, click the little magnifying glass icon, and experiment with zooming. If you lose sight of your data, you can always get it back by clicking the two **Autoscale Now** buttons , .

Clear your chart's data by popping up on it and selecting **Data Operations»Clear Chart.**

The next major topic will involve showing multiple plots at once. First we'll look at charts, then graphs. We now use the *cluster* data type before its "official" section of this chapter. Using the VI we've been building, drop the Bundle function from the **Functions»Cluster** palette, and build the block diagram shown in Figure 1–105.

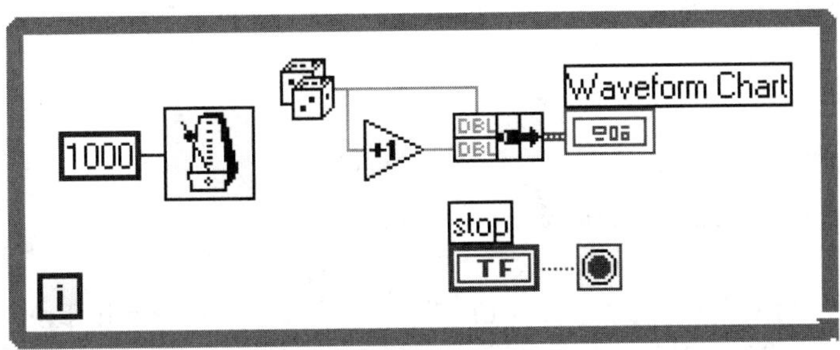

Figure 1–105
The Bundle function can be used with a Waveform Chart to create multiple plots on one chart.

The Bundle function creates a cluster data type, which the Waveform Chart can accept. Turn autoscaling on for both axes by closing both little padlocks and run your VI about 10 seconds; you should see something like Figure 1–106 (your waveforms will vary).

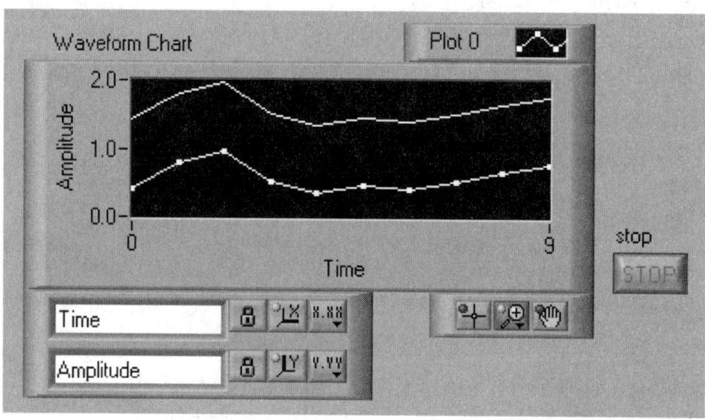

Figure 1–106
Multiple plots are shown on one chart, as created by the block diagram of Figure 1–105.

Finally, drag the Plot Legend over to the right of the chart, and using the Positioning tool , grab the legend's lower right-hand corner, and drag it down a bit, as in Figure 1–107.

Figure 1–107
A chart's legend must be grown to customize multiple plots.

Pop up on your new `Plot 1` on this Plot Legend, give it a hollow square point style and a green color (gray in this book), then your chart should look something like Figure 1–108.

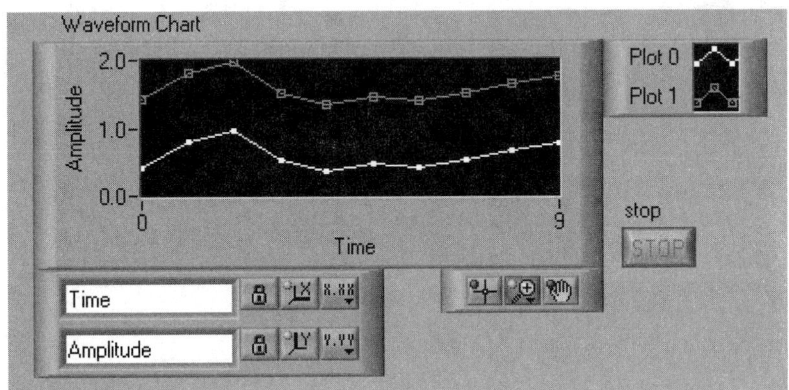

Figure 1–108
Plots are customized on this chart.

When you're building VIs and you have just one plot, you'll probably want to keep the legend hidden. But when you have more than one plot, as in Figure 1–108, you can type your own descriptive plot names where you see `Plot 0` and `Plot 1`. This chapter is long enough as is, so we won't be doing this here.

Save this VI now as `Two Plot Chart.vi`, as we'll use it in the debugging section later on.

Time for graphs! Save this VI now as `Two Plot` **`Graph`**`.vi`, ensuring that we don't overwrite our `Two Plot` **`Chart`**`.vi`. On the front panel, delete the `stop` button.

On the block diagram, get the Positioning tool, drag a big rectangle around everything, and delete it (only the front panel chart, `Waveform Chart`, should remain).

Next, build the block diagram shown in Figure 1–109.

On the front panel, make sure the chart's autoscaling is on. Run the VI a few times, and you see how the chart, being polymorphic, can also accept arrays of data. How? The chart displays them as if they were sent sequentially, one at a time. This might be useful information if you ever need to speed up your charts' charting.

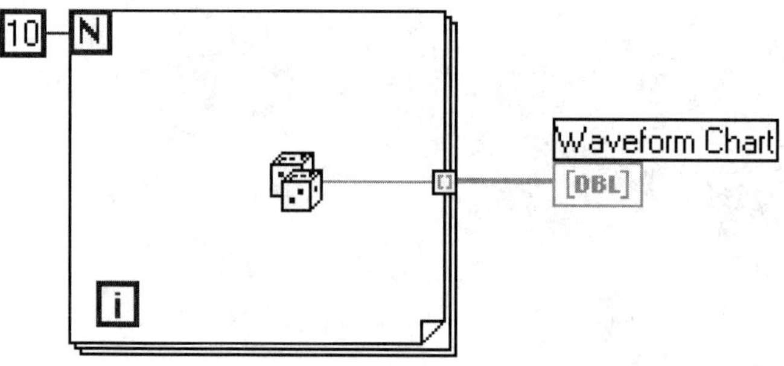

Figure 1–109
A 1D array of random numbers is created.

Graphs are quite similar to charts, except they take all of the data at once rather than one point (or points, as in our last example with the Bundle function), at a time. Graphs cannot take a simple numeric data type—they need an array (or something more complex). Go to your front panel, pop up on the chart, and *replace it* with the Waveform Graph. Notice how the **Run** button is *not* broken! This means that both the graph and the chart can accept a simple array of numbers. However, they handle them differently, as you can see by running the VI a few times. Save the VI now with its new name, so it earns its name with the word "Graph" in it.

Notice how the graph, unlike the chart, has no history, or memory, of past data points. But the graph can accept more complex forms of data than the chart.

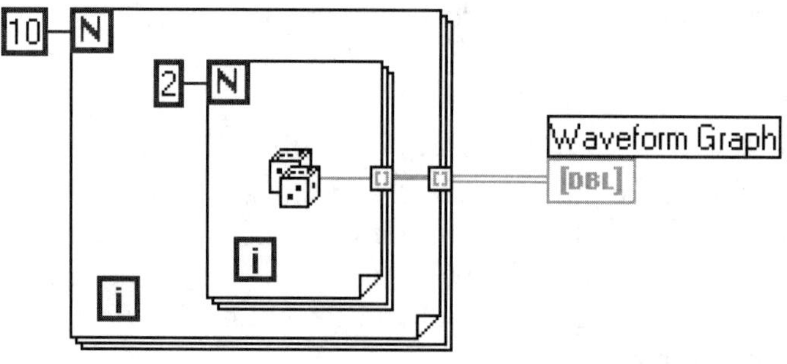

Figure 1–110
A 2D array of random numbers is created.

Let's try to produce two plots of 10 data points. Go again to your block diagram, and make it look like Figure 1–110 by dragging a For Loop around the Random Number function, then wiring a new numeric constant with a value of 2.

Run the VI, then see the sort of nonsense on the front panel shown in Figure 1–111.

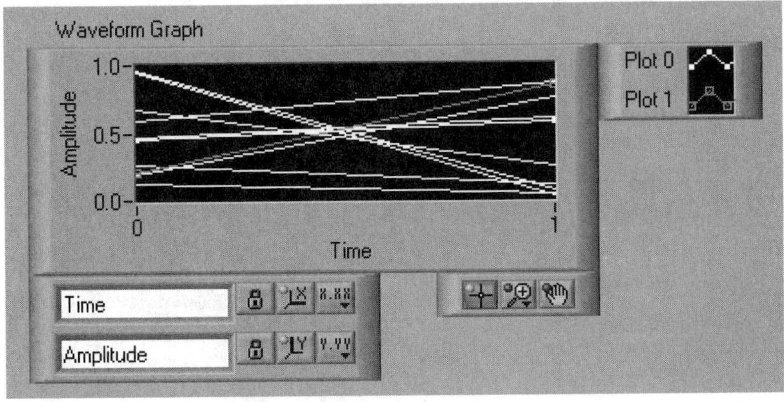

Figure 1–111
A nonsense graph, 10 plots of two points.

Pop up on the graph, select the **Transpose Array** menu item, then see something more logical looking, like Figure 1–112 (your exact waveforms will almost certainly vary, since these are random numbers).

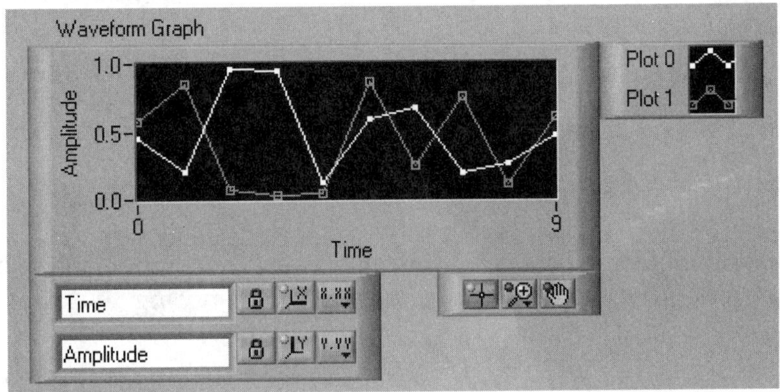

Figure 1–112
Better graph, two plots of 10 points.

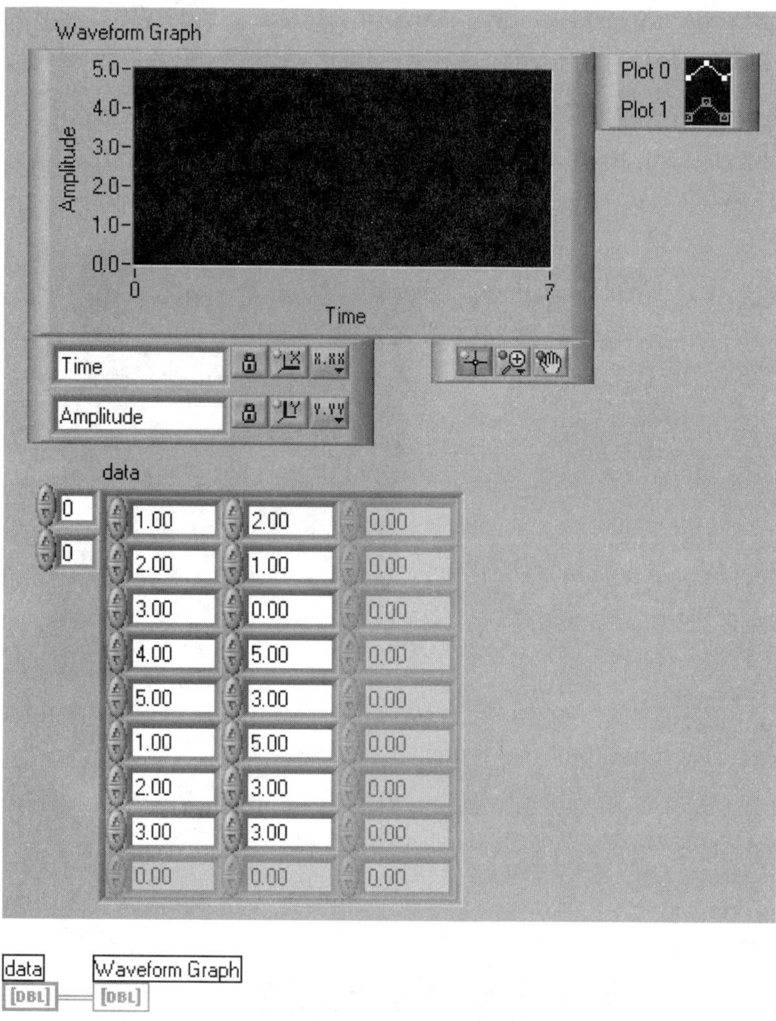

Figure 1–113
A 2D array is wired directly to a Waveform Graph.

What we are plotting here is a 2D array of numbers. In the nonsense graph above, we had 10 plots of two numbers. By transposing the array, we're showing two plots of 10 numbers.

To make sure you understand just how a 2D array of numbers gets to a graph, get the Positioning tool, then drag a rectangle around everything on the block diagram so that it's all selected. Hit your <Delete> key to erase everything, so that the only object left on the block diagram is the Waveform Graph's terminal. Now go to the front panel, grow your graph vertically somewhat as shown in Figure 1–113, then create a 2D array of Digital Controls labeled `data`, so that you have the VI in Figure 1–113.

After you run this VI, the data on your graph should appear as in Figure 1–114, which you should be able to easily correlate to the numbers in `data`. Change one number at a time slightly (keep them roughly in the +/- 10 range) to see how the graph represents its data.

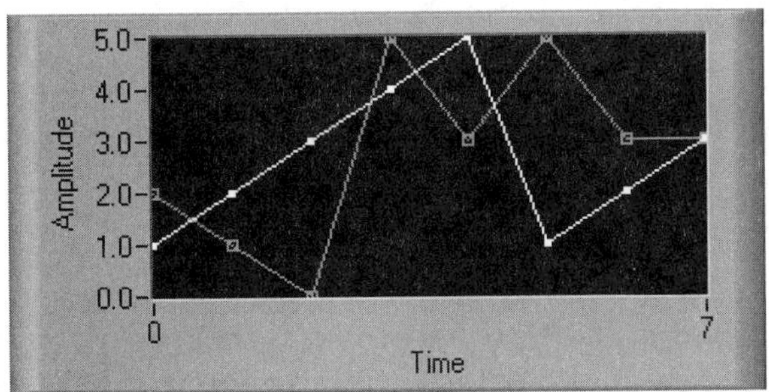

Figure 1–114
The numbers shown in the previous figure are drawn by the Waveform Graph as two plots.

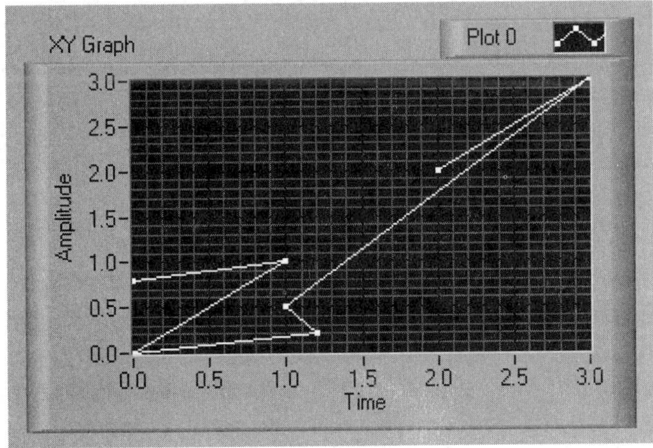

Figure 1–115
An XY Graph allows arbitrary x-coordinates, unlike a Waveform Graph or Waveform Chart.

The final sort of *very* useful graph to be described here is the XY Graph, but we need to cover **clusters** before we get into that. As a little preview, we can display any string of (x, y) coordinates, where unlike the Waveform Chart and Waveform Graph, the x-coordinates do not need to be evenly spaced.

An XY Graph example is shown in Figure 1–115.

1.10 SubVIs

Every programming language on the planet supports calls to subroutines, right? So does LabVIEW, which is the topic of this section.

1.10.1 Overview

Just poking around on the **Functions** palette, you can see that LabVIEW has tons of built-in functions. But what if LabVIEW doesn't have the function you want? If you can build it in LabVIEW as a VI, then you can make your *own* function from it, a *subVI,* which can appear on the block diagram of another VI. The two VIs are referred to as the *caller* (the parent VI) and the *subVI* (called by the parent). For example, the exercise in which we reversed the elements of an array could have been encapsulated into a single icon and used just like the Reverse Array function in any number of other VIs. You will generally not see the front panel of a subVI when it is used, just as you do not see block diagrams nor their functions when they're used. There are two major reasons you should use subVIs: (1) you are performing the same tasks at more than one place in your program, and (2) your block diagram has become too large. Let's review both cases.

The first reason to use subVIs is to perform the same set of functions at different places in your LabVIEW program. Having been a programmer for over 20 years, I see that the most common and time-wasting programming mistake is having redundant code. This is a subtle and seemingly harmless mistake, but it is by far the most damaging. If you are a text-based programmer, and you have ever found yourself copying more than a few lines of text and using them as they are, you are guilty of this mistake. In LabVIEW, if you find yourself using the same few functions in the same way in more than one place in your program, you are also guilty of this mistake! Why is this so damaging? Other than the obvious reason of having too much code, if you later come back to modify one section of code, you might forget to modify the other, when the two pieces of code should be acting identically. In the real world, this happens all the time. Code sits around for years (or even days), then the original programmer, or some other person, will fix one section of code without realizing that the other also needs fixing.

The other reason to use subVIs is that your block diagram is too large.

1.10.2 An Example to Build

Our objective here is to build a function, like the Add function, which calculates the hypotenuse of a right triangle. See Figure 1–116.

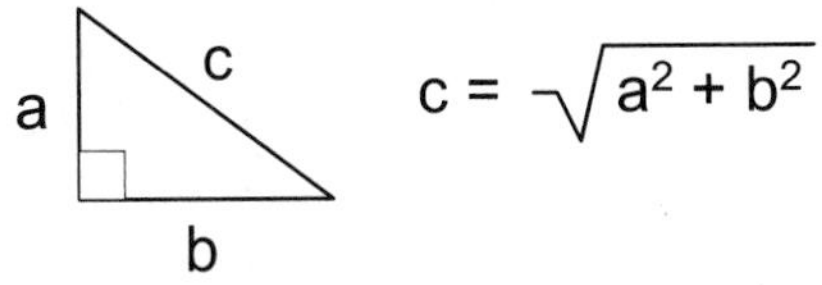

a c b

$$c = \sqrt{a^2 + b^2}$$

Figure 1–116
The Pythagorean Theorem.

The three parts of any VI are the front panel, the block diagram, and the icon/connector. The icon/connector has been largely ignored until now, but we must discuss it here, as it is the only interface between a subVI and its calling VI. The icon/connector consists of an icon (a little picture you can draw) and a connector pane (a pattern of terminals to which you can connect wires). Examples are shown in Figure 1–17. With the connector pane, we get to choose from a variety of terminal patterns to effectively define our own function. Given that our function has two inputs and one output, we will use this pattern:

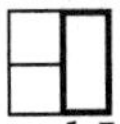

To build a subVI, here are the steps you should follow:

1. Build a VI that you will use as a subVI, using front panel controls for data input and indicators for data output.
2. Pick a pattern for the connector pane, and assign only the necessary front panel controls and/or indicators to the connector pane's terminals. With the connector pane showing, pop up on all

input terminals and select **This Connection Is»Required**. *Always do this for subVI input terminals,* unless they really must be optional, as it will save debugging time later when you forget to wire an input terminal.

3. Create an icon for your VI.
4. Save the VI.

Let's make our hypotenuse subVI following these four basic steps:

1. Build a VI that you will use as a subVI, using front panel controls for data input and indicators for data output. Create a VI with the front panel and block diagram shown in Figure 1–117. Make sure it works by setting `a` to `3.00` and `b` to `4.00`, then run it—`c`, of course, should be `5.00`.

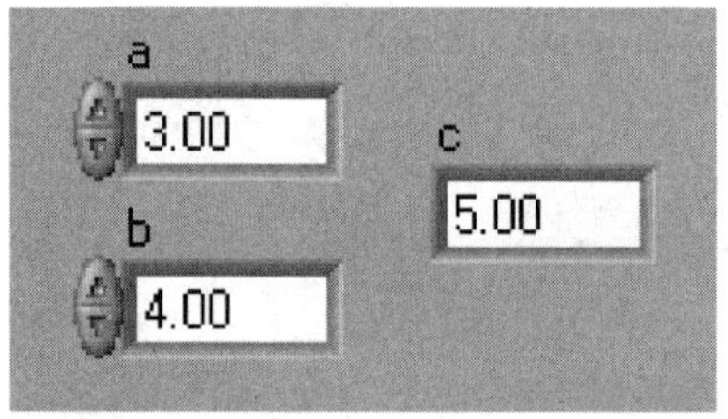

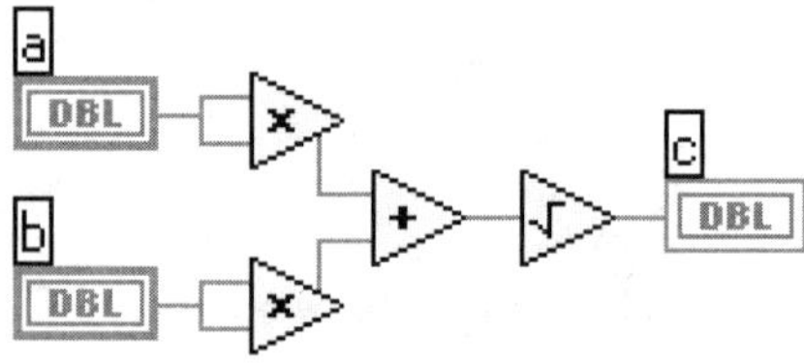

Figure 1–117
A simple VI for hypotenuse calculation.

2. Pick a pattern for the connector pane and assign only the necessary front panel controls and/or indicators to the connector pane's terminals. The icon/connector (the little square in the upper right-hand corner of your front panel or block diagram) will be showing either the icon (by default) or the connector pane. We want it to show the connector pane, so pop up on the icon/connector and select the **Show Connector** menu item. If you do not find this item, you will see the **Show Icon** menu item instead, which means the connector pane is already showing.

   Pop up on the connector pane again, and select this pattern: . To associate the front panel objects `a`, `b`, and `c` with these three terminals, you must use the Wiring tool , then click on sections of the connector pane immediately before (or immediately after) you click on the corresponding front panel object.

Notice that when you hit the <Tab> key to cycle through the tools on the front panel, the Wiring tool is not one of them. You could explicitly show the Tools palette and select the Wiring tool, but a quicker way to get the Wiring tool is to select the Operating tool , then click on any terminal in your connector pane. For this to work, the icon/connector must be showing the connector pane, not the icon.

For our example, Figure 1–118 shows a rather speedy order in which you could click on the various areas to quickly assign your connector pane terminal (the numbered black squares represent six separate mouse clicks with the Wiring tool).

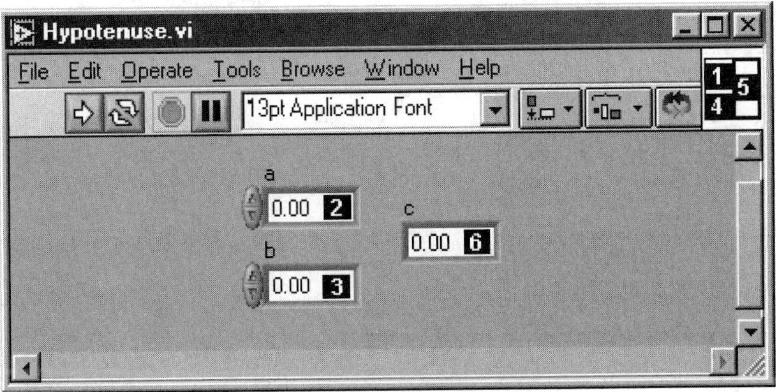

Figure 1–118
An efficient clicking sequence (white-on-black numbers drawn in by me).

3. Create an icon for your VI. When your subVI appears in the block diagram of another VI, you may control its appearance by creating an icon for this subVI. Pop up on the connector pane and select the **Show Icon** menu item if it's there. Now double-click the icon, and up pops the icon editor. You have the option here of making a picture or text. Years of experience tells me that you're better off with a text-only description, no matter how clever you think your picture is, so let's make some text. First, using the

selection tool [selection tool icon], select everything just inside the outer rectangle of the large editing window, as in Figure 1–119.

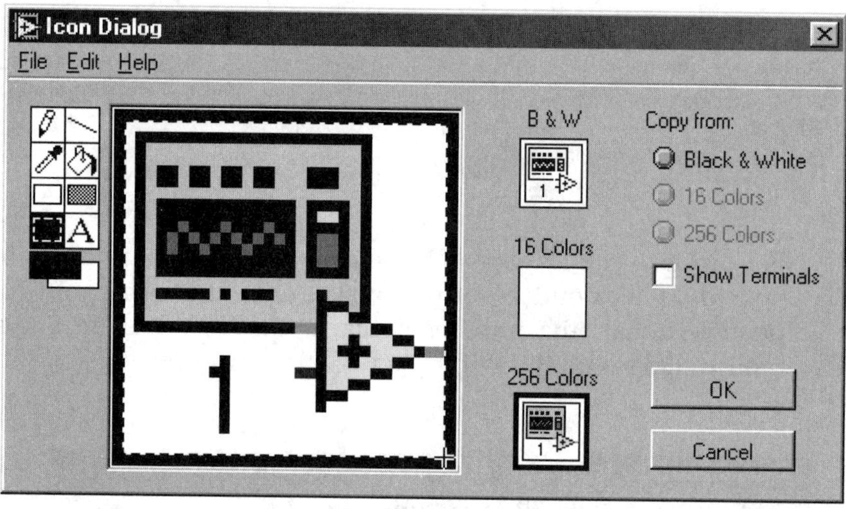

Figure 1–119
The inside of a LabVIEW VI's default icon is selected in preparation to erase it.

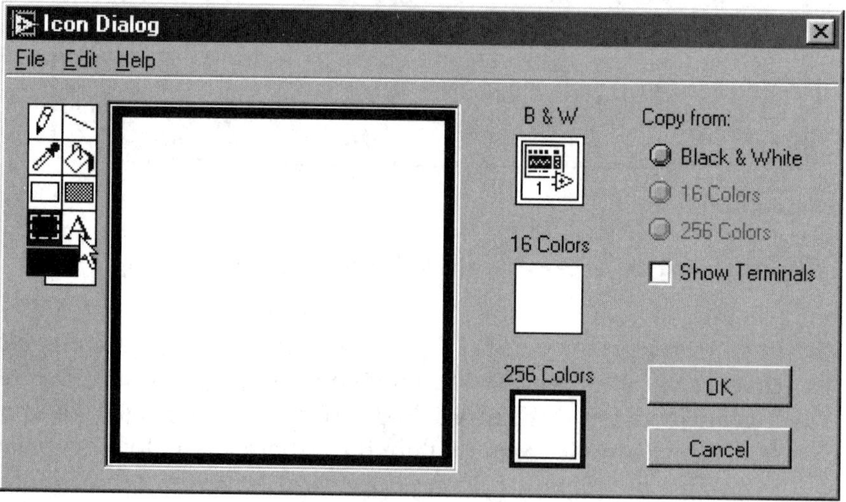

Figure 1–120
As hoped for, the inside of an icon is erased.

Next, hit the <Delete> key, so you have just the outer rectangle remaining, as in Figure 1–120.

Now you must type some text in that square. It's important to use the right font here, so double-click the text tool [A] and select the **Small Fonts** font, size 10. If you manage to exit LabVIEW without crashing, this will be your default font until it's changed again here, using this version of LabVIEW on this computer. Your objective is now to produce the icon shown in Figure 1–121 (or something that looks like it).

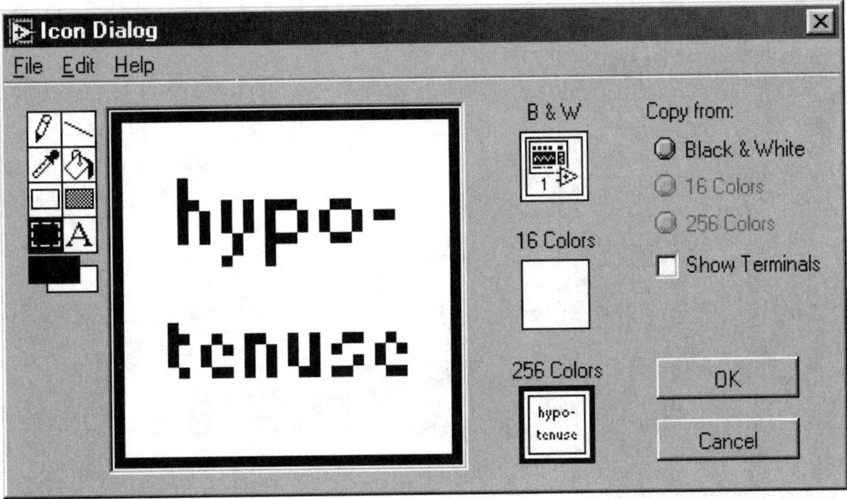

Figure 1–121
Small Fonts, size 10, is an excellent choice for icons' text.

Using the text tool, click where the lower left corner of the h is (do h's have corners?), and start typing. If you make a mistake typing or with your initial cursor position, you can use the <Backspace> key to erase your typing. You can use the selection tool [ ] to select and drag your text around in this main editing screen if your positioning is a little off. Once the icon looks good, make sure nothing is selected on your icon (no marching ants), then copy its smaller image from one of the boxes to the right by clicking on said box and hitting <Ctrl-C>. If anything had been selected in the large icon, it would have been copied—but we want the whole image. Next, paste the image into the other two boxes by clicking on them, one at a time, then hitting <Ctrl-V>. Or, you could use the items under "Copy from:".

4. Save the VI. Save this VI as `Hypotenuse.vi`.

The subVI has now been created. Next, we'll use this VI in another VI, which we'll call the *calling* VI. Create a new VI. Select the **Functions»Select A VI...** button, shown in Figure 1–122.

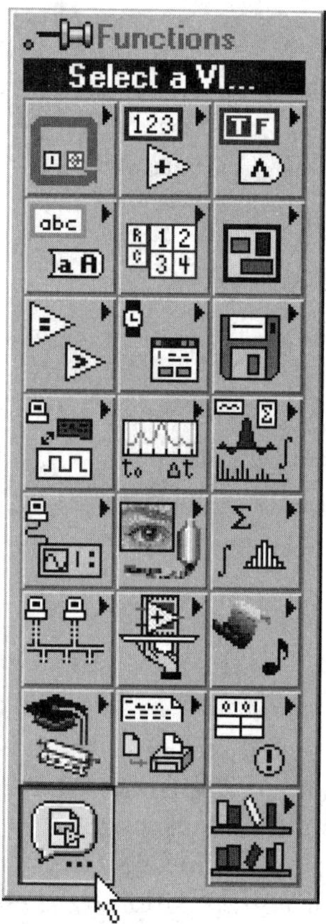

Figure 1–122
This lower left icon is one way to drop a subVI into a block diagram. It will pop up a file dialog box to let you select that subVI.

Using the file dialog box that will pop up, find your `Hypotenuse.vi`, select it, and drop it into the block diagram. On your calling VI, create front panel controls and indicators just like those on `Hypotenuse.vi`, and wire them up to all three terminals of your subVI. Running your new VI, you can see that it actually performs the function of `Hypotenuse.vi` by calling it as a subVI; this is our first subVI in action. Save your new VI as `Hypotenuse Caller.vi`, as we'll be using it to demonstrate debugging techniques later. It should look something like Figure 1–123.

Figure 1–123
Your `Hypotenuse Caller.vi` should look something like this.

In this simple example, we don't really need this subVI. But if you wrote a program that needed this hypotenuse calculation in more than one place, or if a block diagram had become crowded, the subVI would have been appropriate.

1.10.3 SubVI Details

If you have a subVI with an output that does not receive any data because it's in a Case Structure case that does not get executed, the subVI passes the default value of that output to the calling VI. For example, suppose the False case were executed in the subVI's block diagram shown in Figure 1–124. The *default value* of the `output` indicator would be passed to the calling VI, which may or may not be desirable.

If you are using many subVIs in a LabVIEW program, and you want to get a graphical overview of what VIs are calling what subVIs, use the **Browse»Show VI Hierarchy** menu item then hit <Ctrl-A>. At the very minimum, the names of a subVI's inputs and outputs, as wired on its connector pane, will appear in its Help window (toggled with <Ctrl-H>). You can also specify additional information for its Help window, as described in Section 1.3.3.

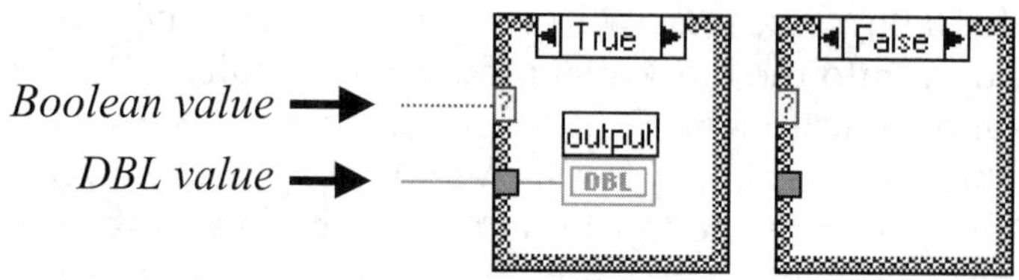

Figure 1–124
What happens if the `output` indicator is not executed, yet it's the terminal of a subVI?

1.10.4 Custom Dialog Boxes

Suppose you wanted a pop-up dialog box that's a little more complex than the ones LabVIEW gives you (the One Button Dialog and Two Button Dialog functions). For example, suppose you want a window to pop up and ask the user for a password. We will build such a dialog box as a VI, then make it a subVI. This is one you could likely use in your real applications.

Create a VI with the front panel and block diagram shown in Figure 1– 125 (make sure your **OK** button has a latching mechanical action; this button comes ready-to-go from the **Controls»Dialog Controls** palette).

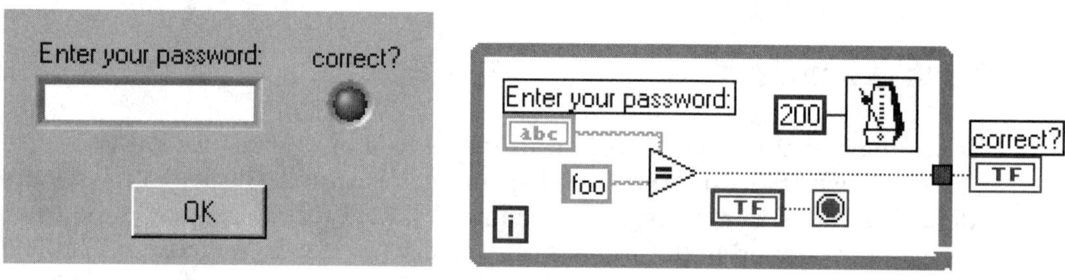

Figure 1–125
A simple VI to be used as a subVI dialog box for entering a password.

Don't forget to change the conditional terminal to **Stop If True**. Try the VI by running it a few times, typing in a password, then hitting the **OK** button. Notice that the Boolean indicator `correct?` is True when you've entered `foo` as the password; otherwise, `correct?` is False. Once you're happy with how it works, you can pop up on the String Control and select the **Password Display** menu item so that prying eyes cannot see what you're typing, unless they have a keyboard-monitoring device (hardware or software) or they're watching your fingers very carefully.

Make sure the ***OK*** *button is associated with the <Return> key via its* ***Advanced»Key Navigation...*** *pop-up menu item. The <Return> key on your keyboard may have the word "Enter" on it. If you got the* ***OK*** *button from the* ***Controls»Dialog Controls*** *palette, it should already be set that way. This allows you to hit the <Return> key after you've typed the password, which is more convenient than clicking it.*

To use this as a subVI, you must connect the `correct?` indicator to a terminal in the connector pane. Just one terminal is all you need in this connector pane. Make its icon say "password." Save this VI as `Password.vi`.

Now, how do we make this VI pop up when it's used as a subVI? Use the **File»VI Properties...** menu item to open the **VI Properties** window. Set the **Category** menu ring to **Window Appearance**, then click the **Dialog** radio button, as shown in Figure 1–126.

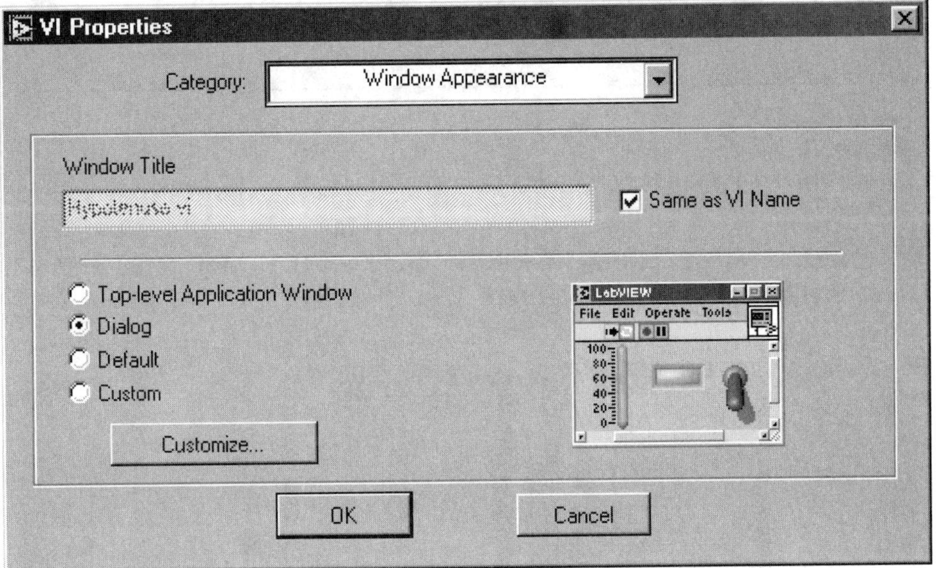

Figure 1–126
This **Dialog** button makes it easy to create a subVI dialog.

This will cause the VI to pop up like a dialog box when called as a subVI, then close itself later. If you want more control over your dialog box, you can click the **Custom** radio button, then use the **Customize...** button. I prefer these settings, shown in Figure 1–127, which are similar to the setting resulting from the **Dialog** radio button, because they don't allow the user to close the dialog box without using the **OK** button.

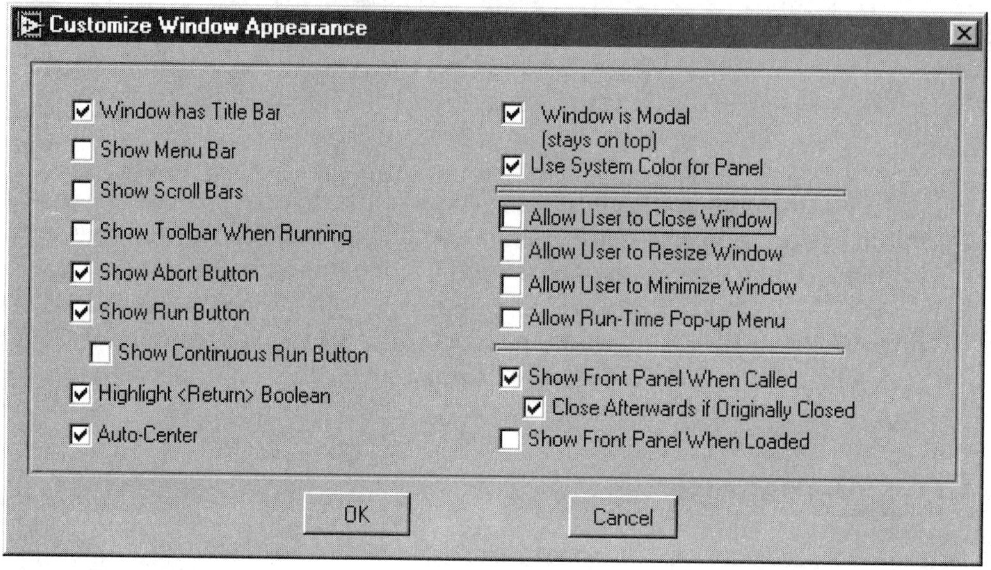

Customize Window Appearance

- [x] Window has Title Bar
- [ ] Show Menu Bar
- [ ] Show Scroll Bars
- [ ] Show Toolbar When Running
- [x] Show Abort Button
- [x] Show Run Button
  - [ ] Show Continuous Run Button
- [x] Highlight <Return> Boolean
- [x] Auto-Center
- [x] Window is Modal (stays on top)
- [x] Use System Color for Panel
- [ ] Allow User to Close Window
- [ ] Allow User to Resize Window
- [ ] Allow User to Minimize Window
- [ ] Allow Run-Time Pop-up Menu
- [x] Show Front Panel When Called
  - [x] Close Afterwards if Originally Closed
- [ ] Show Front Panel When Loaded

OK Cancel

Figure 1–127
It is better if the user is *not* allowed to close subVI dialog boxes. Otherwise, it is possible to "hang" the entire program.

Next, create a calling VI with nothing on the front panel and with the block diagram shown in Figure 1–128.

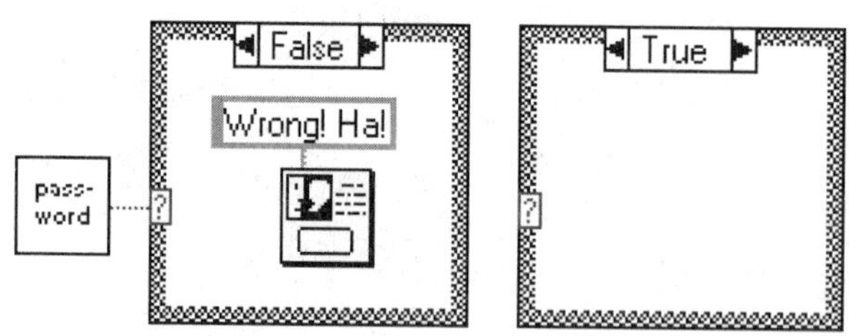

Figure 1–128
A very simple VI tests our password subVI.

Run this VI, and you'll see your newly created dialog box in action. Notice that you must select your password entry string when the password dialog box pops up, and the last thing you typed is still there! This is highly unacceptable. We will be improving `Password.vi` later in this chapter when we cover the necessary topics, thus making it fit for actual use. Save the calling VI as `Password Caller.vi`.

1.10.5 The Create SubVI Menu Item

This menu item is so incredibly useful, it deserves its own section. Its usefulness ranks right up there with <Ctrl-Z> and <Ctrl-F>, in my opinion.

Create a VI with the front panel and block diagram in Figure 1–129, using these tips:

1. Do *not* run this VI until instructed.
2. This **OK** button comes from the **Control»Dialog Controls** palette.
3. Be sure to select the Two Button Dialog function, not the One Button Dialog function.
4. The Format Into String function is in the **Functions»String** palette.

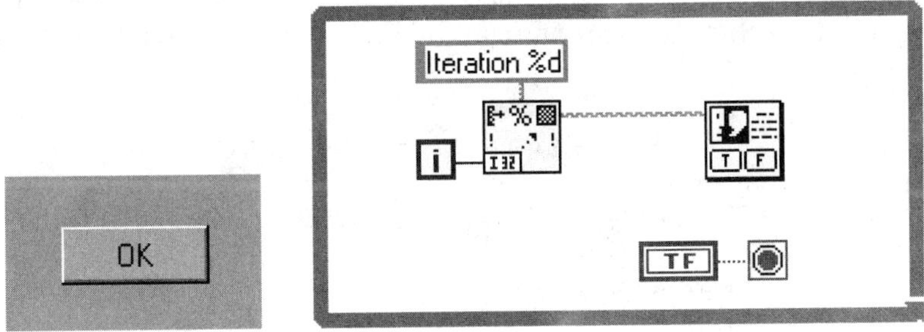

Figure 1–129
We begin to build a safe message box, although this version is anything but safe.

Save this VI as `Safe Message Box.vi`.

If you were to run this, which I do not recommend, you might not be able to easily stop it, because it pops up the same dialog box over and over so quickly that you cannot press the **OK** button. This could force you to <Ctrl-Alt-Delete> your way out of LabVIEW, or if you're not terribly bright, you might reboot your computer! We can use a safety mechanism here. Since the box that pops up has two buttons, we'll call one **Stop VI** so we can push this button to stop LabVIEW's execution, if desperate. Modify your VI as in Figure 1–130 (the little stop sign stops the VI, and can be found on the **Functions»Application Control** palette).

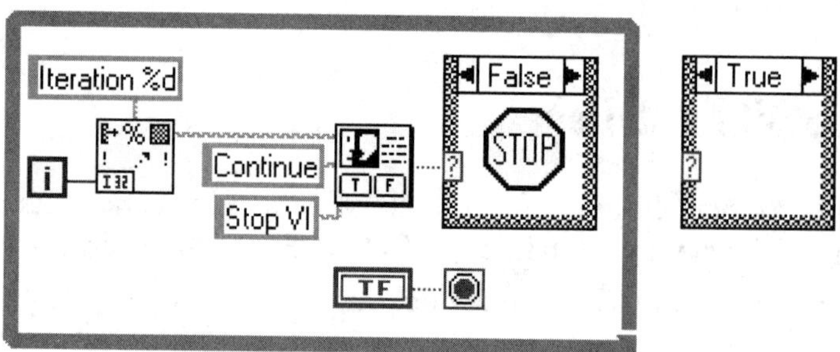

Figure 1–130
This version is much safer than the one shown in Figure 1–129.

You can create a subVI from this and use it in place of all of your One Button Dialog function calls just to make sure you never fall into the trap of the never-ending pop-up box described a bit earlier. The easiest way to change this piece of code into a subVI is to select the generically useful code by dragging a selection rectangle around it as in Figure 1–131 to select the images shown in Figure 1–132.

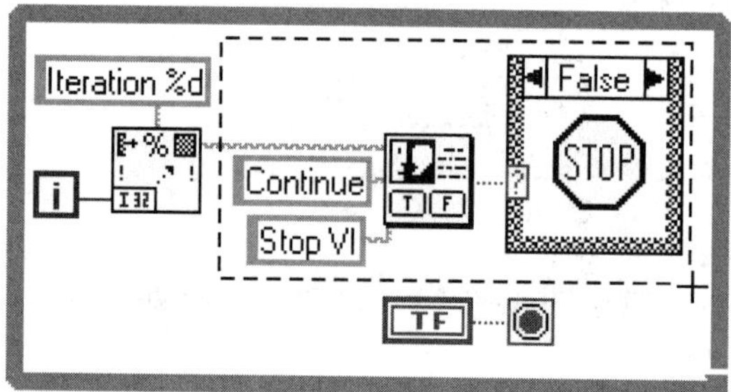

Figure 1–131
We really need only the selected items to create a safe dialog box for generic use.

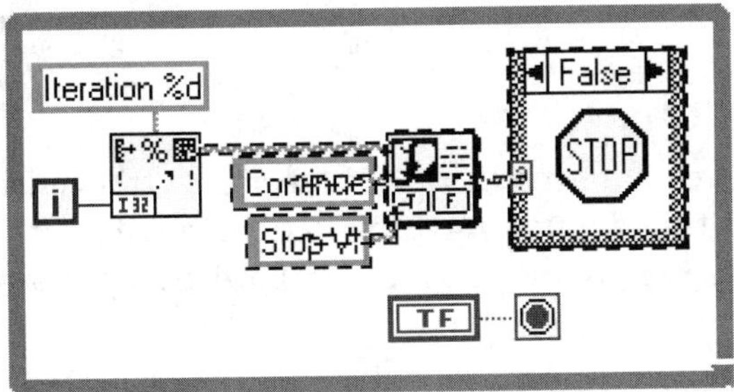

Figure 1–132
Make sure you only have these items highlighted when building this.

With the selection shown above, select the **Edit»Create SubVI** menu item. Whenever you do this, you should immediately open your newly created subVI, give it a nice icon, and save it. For this subVI, put the text `safe dialog` on its icon, and save it as `Safe Dialog.vi`. Your new block diagram should look something like Figure 1–133.

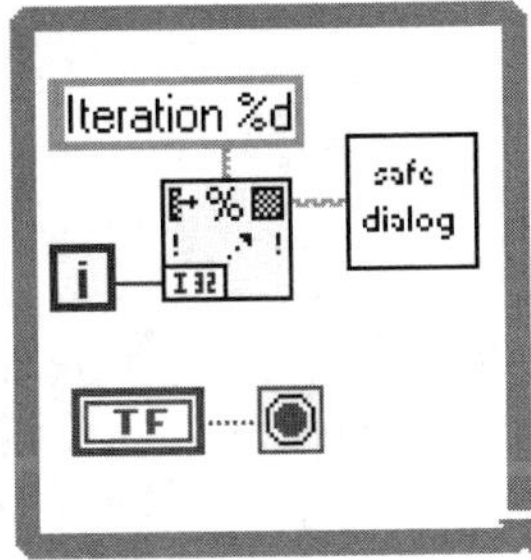

Figure 1–133
Finally, our safe dialog box appears on the block diagram of a calling VI.

Note how we didn't need to fool around with any wiring or the connector pane when creating this subVI; the **Create SubVI** menu item did that dirty work for us! This is even more convenient when multiple wiring connections are involved. On the other hand, this automatic wiring often doesn't give you exactly the connector pane you might want (too many terminals from unused front panel items are often created for you), so you may want to set up the connector pane manually as you wish.

1.11 Debugging Techniques

If you never make any mistakes, you may skip this section. Otherwise, read on. LabVIEW has many built-in debugging tools, and I have quite a few tricks up my sleeve for you as well. First, we'll cover LabVIEW's debugging tools.

1.11.1 LabVIEW's Debugging Tools

One of LabVIEW's most commonly used debugging tools is the probe. Close all VIs you may have open, then open your `Two Plot Chart.vi` from Section 1.9.4, pop up on the wire connected to the Random Number function, then select the **Probe** menu item, as in Figure 1–134.

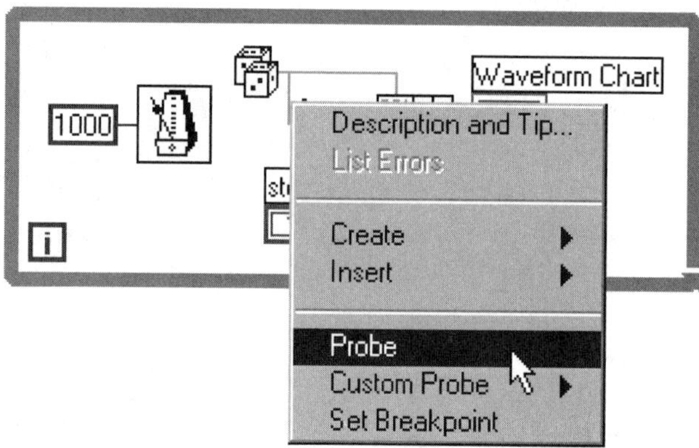

Figure 1–134
If you pop up on a wire, you can create a *probe*.

Up pops a temporary little probe window, which looks something like Figure 1–135, even when the VI is running.

This probe will display the current value of the wire being probed as the VI is running. You could have many probes on different wires.

Beyond the probe, this whole string of buttons on your block diagram's tool bar (see Figure 1–136) is devoted to debugging.

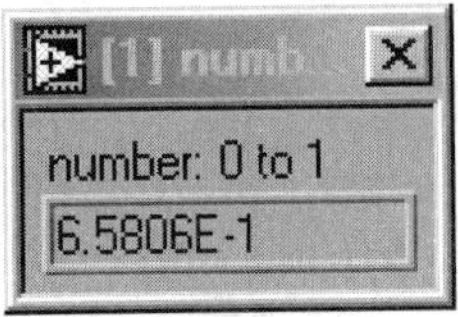

Figure 1–135
Here's a simple numeric probe.

Figure 1–136
Various other debugging buttons exist on the block diagram's tool bar.

Let's start with the **Highlight Execution** button, which is responsible for turning on execution highlighting. We've already used this several times in this chapter. Execution highlighting is most useful with numeric or string data types, as it can show the values moving along the wires in the block diagram when the VI is running, when it is activated, like this (see Figure 1–137).

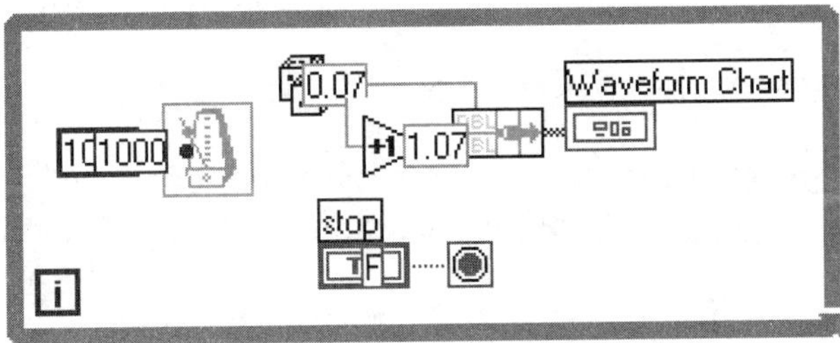

Figure 1–137
Execution highlighting can be very useful with simple data types when you can see the values.

What about those other buttons? The one on the far left is the **Abort Execution** button. When you have a VI running that you cannot stop by conventional means, like a Stop Button, this button might stop your VI.

I find the **Pause** button to be of limited usefulness. If you have this button pressed, the VI will pause when it is run. This is more useful if your VI is a subVI.

This set of buttons is a bit more useful: . They allow you to step into subVIs, step over subVIs, and step out of subVIs, respectively. These buttons treat structures, such as the Sequence Structure, the For Loop, the While Loop, and the Case Structure, like subVIs. To see these buttons at work, you can open your `Hypotenuse Caller.vi` and press the leftmost button a few times, and you'll see yourself stepping into the `Hypotenuse.vi` subVI. If you had been pressing the middle button instead , you would not have stepped into `Hypotenuse.vi`.

1.11.2 Other Debugging Techniques

Sometimes, you may need something a little more powerful than the probe. Suppose, for example, you want to look at eight elements of a 200-element array. If you place a probe on the array's wire, you will only see one element at a time, as shown in Figure 1–138.

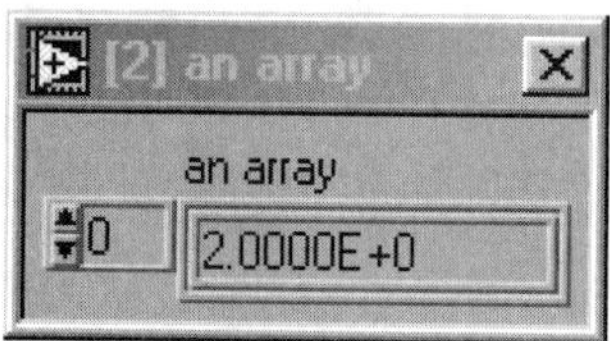

Figure 1–138
A probe only allows you to see one element of an array.

In this case, you might want to place an array of numeric indicators on the front panel, and wire these to your array in question, as in Figure 1–139.

Figure 1–139
If you want to "probe" multiple elements of an array, try a front panel array indicator like this.

You can later delete this front panel indicator when you're sure your VI works correctly.

It is sometimes helpful to pop up a dialog box with debugging information only under special conditions. When you can, use the `Safe Dia-`

`log.vi` you created earlier inside a Case Structure so that the dialog pops up *only* under your special conditions, and pass in a helpful string. You can make such a string even more helpful by using the Format Into String function, as in the pseudo-VI in Figure 1–140.

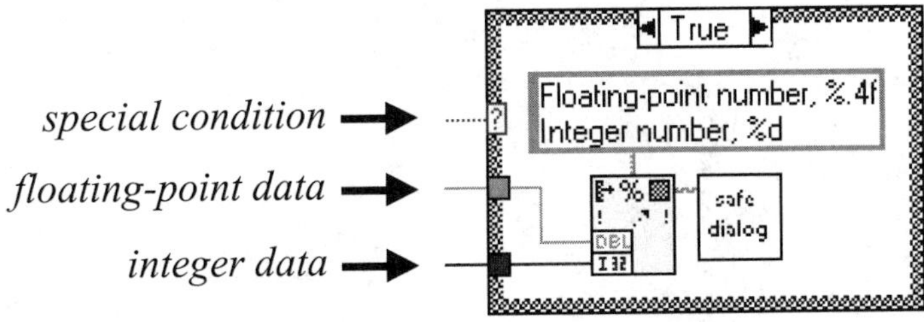

Figure 1–140
This pseudo-VI only pops up a box when a special condition is true.

Final debugging tip—you will usually have a main loop in your program. Divide this loop's iteration terminal by 100 and display the remainder on the front panel, so you can immediately tell when your VI has unexpectedly stopped (crashed). See Figure 1–141.

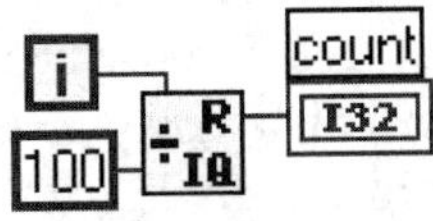

Figure 1–141
Displays a different two-digit number per loop iteration so you know the loop is looping and your VI has not crashed.

Tuck the `count` Numeric Indicator in an obscure corner of your front panel. The remainder business ensures your count will always be two digits or less, so the `count` indicator on the front panel may be physically small.

1.12 Clusters

There are two basic means of grouping objects in LabVIEW: arrays and clusters. An array is a variably sized group of objects having the same data type, whereas a cluster is a fixed-size group of objects, which may have different data types. For example, an array of numbers may change in size from four numbers to eight numbers while the VI is running, but a cluster of four numbers must always have four numbers while the VI is running. Unlike the array, a cluster could contain any assortment of data types, as in Figure 1–142.

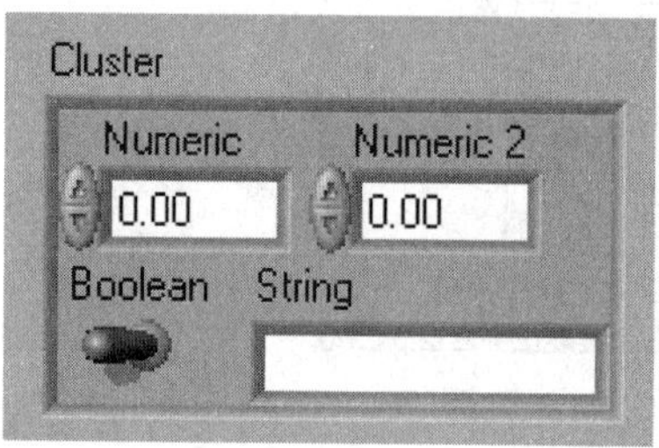

Figure 1–142
A cluster can contain arbitrary data types.

1.12.1 Using Clusters

In LabVIEW, to *bundle* means to group objects into a cluster. We will build the cluster shown above, then use it to show clusters' features. Create a new VI, then drop a Cluster onto the front panel from the **Controls»Array & Cluster** palette. Size it so that it's a bit larger than the one shown in Figure 1–142, then create and drop the objects inside, as shown. Now pop up on the cluster (be sure to get its edge, not the inside), and select the **AutoSizing»Size to Fit** menu item. Save this VI as `Cluster Example.vi`.

Cluster objects are always ordered. To see this order, pop up on the cluster (its edge) and select its **Reorder Controls In Cluster...** menu item. The screen will then look something like the screen in Figure 1–143, perhaps with a different order.

Click on the little boxes with the numbers in them [0|0] and change the order to something other than what you have. When you're finished, click the **OK** button [OK] in the tool bar if you want to keep the new order; otherwise, click the **X** button [X] to revert to the original cluster order. You will also see this on the tool bar: [Click to set to 0]. You can start out with a different number than zero for your first click, but this box will increment after each click until all cluster objects have been counted. Finally, order them as shown in Figure 1–143.

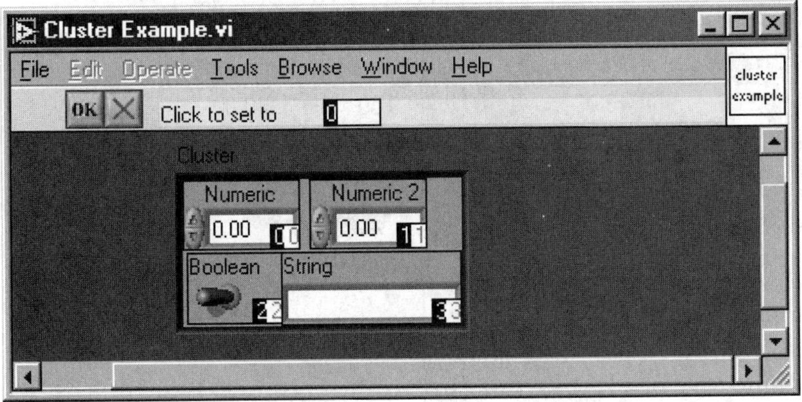

Figure 1–143
This screen is seen when a cluster's elements are reordered.

A similar ordering can be done to control the <Tab> order of front panel objects via the ***Edit»Set Tabbing Order. . .*** *menu item.*

Next, go to the block diagram. Drop the Unbundle function from the **Functions»Cluster** palette. Wire the cluster to the left half of the Unbundle function. Notice how the Unbundle function automatically grows itself to four elements. It knows to do this by following the wire to the cluster of four elements (see Figure 1–144).

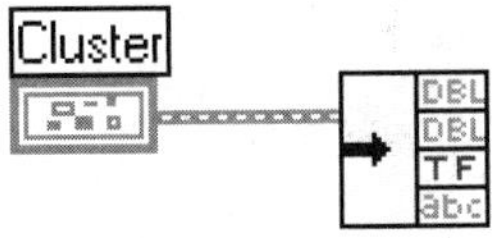

Figure 1–144
The Unbundle function gives you access to the elements of a cluster.

Now, let's use the Unbundle function's outputs. On your front panel, horizontally clone your cluster (<Ctrl-Shift>-drag an *edge* of `Cluster` with the Positioning tool) to the right of the original `Cluster`, then make the new `Cluster 2` an indicator. Now, type the numbers shown in Figure 1–145 into `Cluster` (still a control).

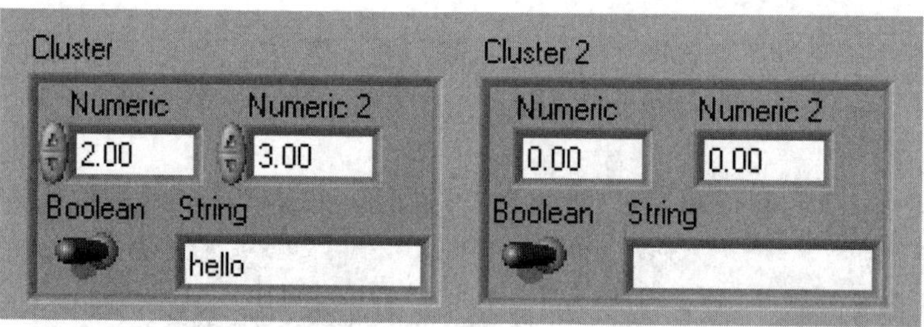

Figure 1–145
The numbers to be typed into `Cluster`.

Build the block diagram (shown in Figure 1–146) from this front panel, using these tips:

1. The Bundle function is in the **Functions»Cluster** palette.
2. The far right wire must be connected to the Bundle's rightmost terminal, not the center one.
3. The Not function is in the **Functions»Boolean** palette.

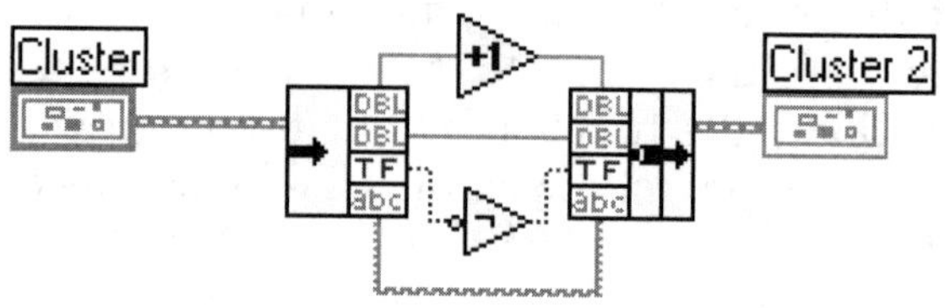

Figure 1–146
The elements of a cluster are manipulated with the Bundle and Unbundle functions.

Run the VI, and you should get the image in Figure 1–147.

Notice how cluster element 0 `Numeric` was incremented, and the Boolean value was toggled, which should make sense as you look at the block diagram. What's that center terminal on the Bundle function doing? Nothing, in this case, but you could clean up the block diagram a bit by wiring it as in Figure 1–148.

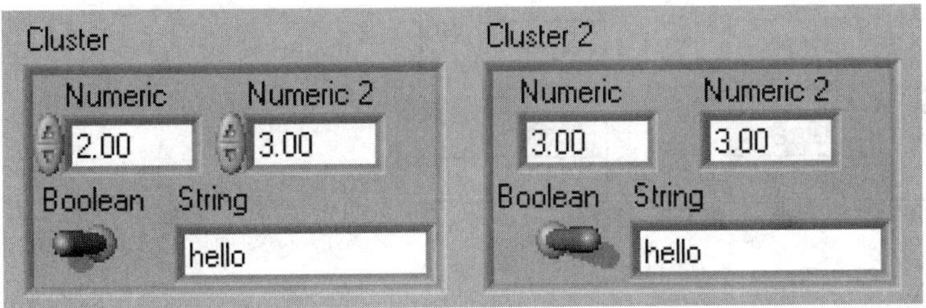

Figure 1–147
This is the result of manipulation as shown in Figure 1–146.

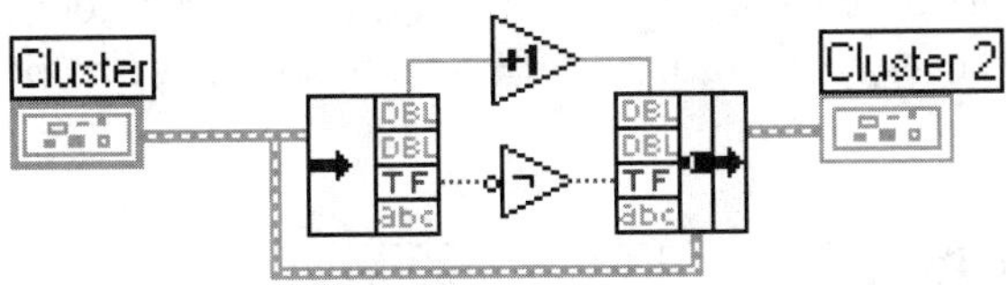

Figure 1–148
The middle terminal of the Bundle function is used to produce the effect of Figure 1–146.

The center terminal of a Bundle function allows you to leave any number of its other inputs unwired by providing the Bundle function with default values (and data types) of the incoming cluster's unwired elements.

In the beginning, there was the Unbundle function with its corresponding Bundle function. But the LabVIEW developers thought, "Hey, let's get fancy." This happens quite a bit. So along came the fancy Unbundle By Name function and its corresponding Bundle By Name function. These functions use cluster elements' labels (names) to identify them. We'll now see how to use these "By Name" alternatives, as they're very helpful. Pop up on the Unbundle function, select **Replace**, and replace it with the Unbundle By Name function. Similarly, replace the Bundle with its corresponding Bundle By Name function. After the replacements, you will need to move things around quite a bit with the Positioning tool, and you will likely need to delete lots of wires (if not all of them) then rewire everything, so you get the image in Figure 1–149.

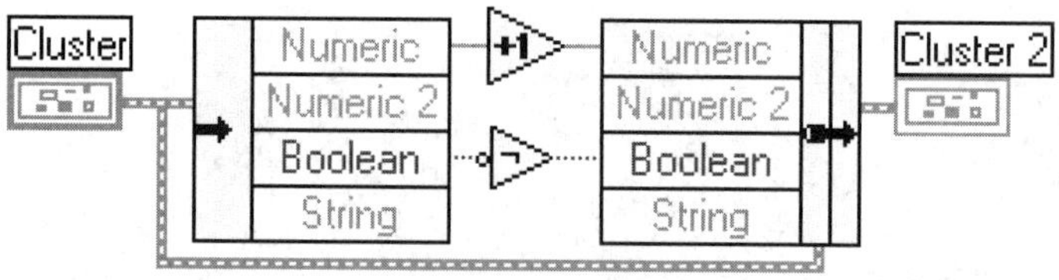

Figure 1–149
The Unbundle By Name and Bundle By Name functions produce the effect of Figure 1–146.

Notice how we have a broken **Run** button at this point. This can be remedied by removing all unwired elements from our new bundling and unbundling functions (pop up on each unwired element, and select **Remove Element**), so you get the image in Figure 1–150 (after some repositioning of wires and other items).

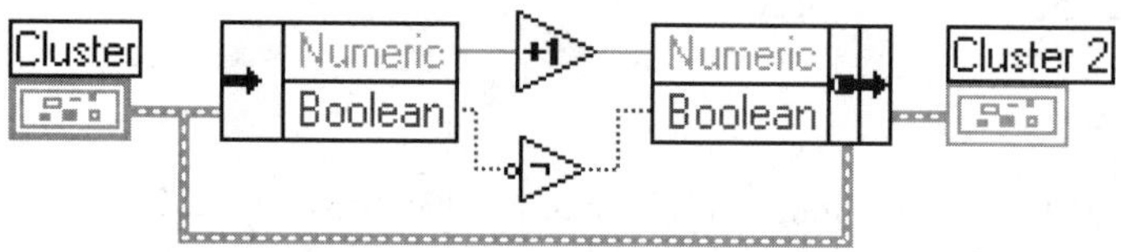

Figure 1–150
The middle terminal of the Bundle By Name function produces the effect of Figure 1–146.

This should work just fine, if there are no broken wires hiding anywhere. Notice you can pop up on the elements of the Bundle By Name and Unbundle By Name functions and change them to a different element quite easily (such as `Numeric` to `Numeric 2`). Left-clicking with the Operating Tool also does this.

The last point to make about the "By Name" functions is this: The Bundle By Name function requires that you wire its center terminal, whereas the regular Bundle did not require this, provided you had *all* its inputs wired.

1.12.2 Other Cluster Uses

Clusters are very handy for passing complex data structures to charts and graphs. Back in the section on graphs, we didn't go into the XY Graph because we had yet to discuss clusters in depth. Build the VI shown in Figure 1–151, using these tips:

1. First, drop a Digital Control, shrink it a bit horizontally, and label it x. Drag the label to the left, so that it looks like this:

2. Clone this control below itself, label the clone y, drop a cluster, drag the two Digital Controls into the cluster, and select the cluster's **AutoSizing»Size to Fit** menu item, so that it looks like this:

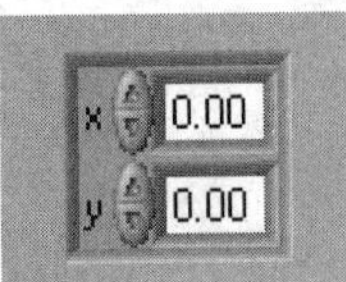

3. Drop an array, drag the cluster into the array, grow the array as shown in Figure 1–151, then enter the values shown. This is an array of clusters of two floating-point numbers. We can pass this particular data type to an XY Graph. The NaN stands for Not a Number, and causes the corresponding point to not be drawn on a graph or chart.
4. Since you've gone through the trouble of entering this data, pop up on the array (not an element of the array), and using the **Data Operations** menu item, make its current value default.
5. On the front panel, create an XY Graph. Any other type of graph will not accept the data type we've created, so make sure you get the right graph. Pop up on the graph's Plot Legend, and change the line width to its widest. Pop up again on the graph and disable autoscaling for *both* x- and y-axes. Add the free label, using the Labeling tool, saying `autoscaling off for both axes`. Make sure the axes are both ranging from 0 to 10.
6. Wire the two terminals together on the block diagram. Your VI should now look like Figure 1–151 (see Table 1.4 to see why the color of your wire is brown).

Now, go back to the front panel and run the VI so you can see the power of XY Graphs. Save your VI as `XY Graph Example.vi`.

One final point about clusters—they are quite handy for passing information into subVIs. If you find yourself with too many inputs and/or outputs on a subVI, and don't want a connector pane that looks something like this or this, group some inputs and/or outputs into a cluster. This technique can lead to sloppiness, unless the clusters you create can be used throughout your VIs and subVIs, in which case this technique can lead to efficiency.

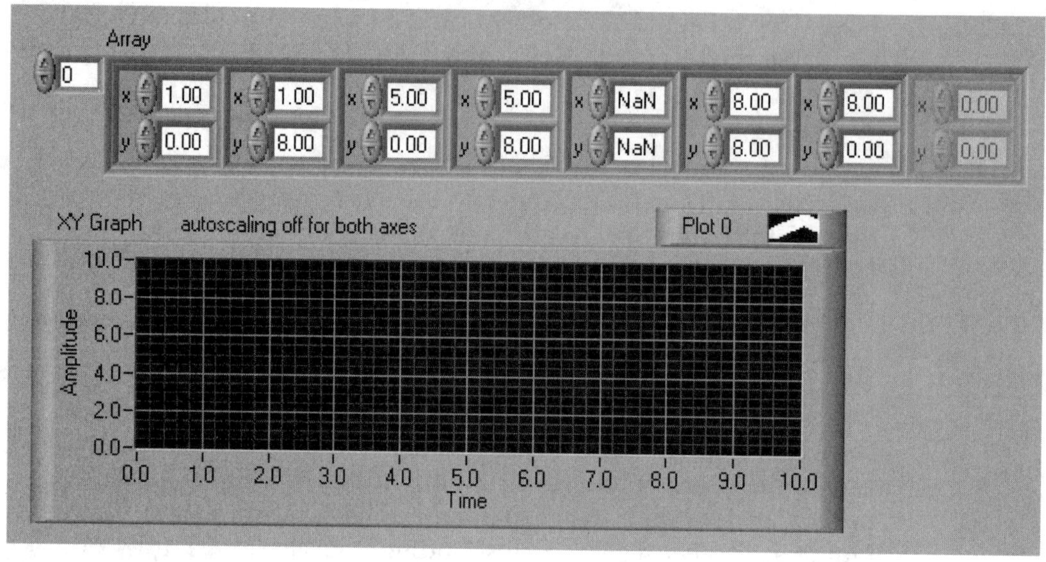

Figure 1–151
This VI illustrates a data format for the XY graph.

When I worked at NI on the LabVIEW team, those two crowded connector patterns in the previous paragraph were created to please a particular NI employee named Monnie, thus were internally known as the *Monnie Pleaser* and the *Super Monnie Pleaser*, respectively.

1.13 File I/O

Please make sure you understand the material in Appendix A before reading any of this section.

File I/O means file input and/or output, where a *file* is a group of bytes residing on some form of nonvolatile memory—*nonvolatile* means that the data is retained when the computer's power is turned off. As of the writing of this book, this nonvolatile memory usually comes in the form of rotating magnetic disks, which are typically called hard drives, hard disks, or floppy disks. But the important point is not where a file resides, but the fact that it stores nonvolatile data.

1.13.1 High Level File I/O

Let's go ahead and use LabVIEW's high-level file I/O VIs (used like functions, but are actually subVIs) to read and write files. Interruption: Please read Appendix C now if you have not already done so; it applies to many file I/O VIs if you've changed versions of LabVIEW on your computer. Descriptions of the two most useful high-level file I/O VIs, copied directly from LabVIEW's Help window, are shown in Figure 1–152.

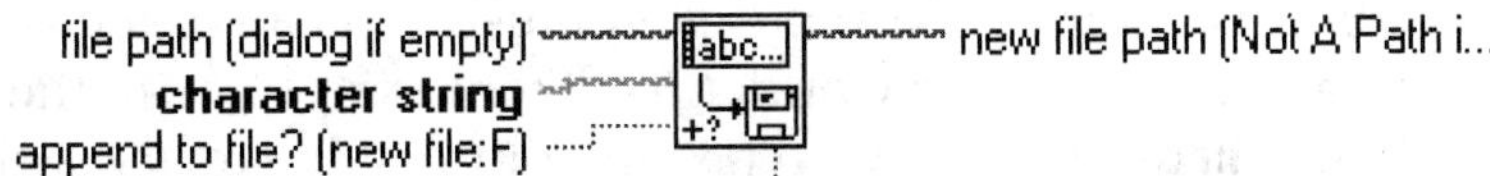

Write Characters To File.vi

Writes a character string to a new byte stream file or appends the string to an existing file. The VI opens or creates the file beforehand and closes it afterwards.

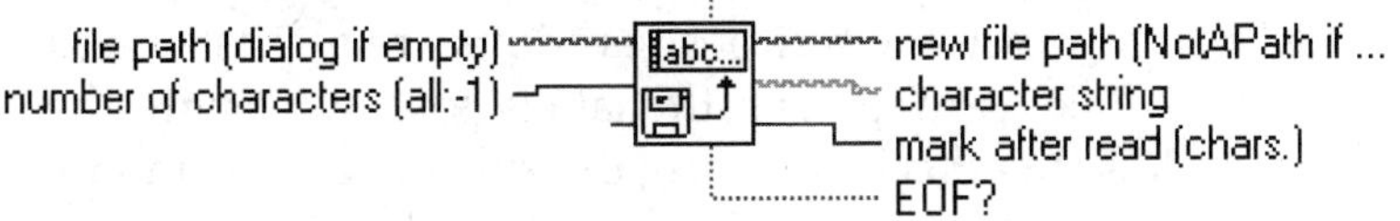

Read Characters From File.vi

Reads a specified number of characters from a byte stream file beginning at a specified character offset. The VI opens the file beforehand and closes it afterwards.

Figure 1–152
The two most useful high-level file I/O VIs (from LabVIEW's Help documentation).

The **file path** wire shown above has a *path* data type that we have not yet discussed. The path is very similar to the string data type, except it is only used to specify the location of a file. On a PC, path data will look like `C:\Folder 1\Folder 2\Filename.txt`. On other platforms, the delimiters are often characters other than the backslash. Many characters, such as :, /, \, <, >, ?, *, |, and " are not allowed in file names.

The **character string** wire shown in Figure 1–152 is the data being written to (or read from) a file. It is a group of bytes, but not necessarily readable text.

All other inputs and outputs may be studied at your leisure, as the objective here is to have a quick tutorial. In this section, we will only be reading or writing the entire contents of a file. Later, we will see how to read and write parts of a file.

Create a new VI with the front panel and block diagram shown in Figure 1–153, using these tips:

1. For the front panel strings, the left one is a control, and the right one is an indicator. Create one of the strings first, set its font and appearance, then clone it to create the other string. To do this, drop a string, grow it, then show its scrollbar. Use the text ring in the tool bar 13pt Application Font to set the font inside the string controls. In order to do this, you must select all of the string's text (if any), and set it to Courier New font, size 14, with the text ring. Since there's probably no text in your newly-created strings, simply place the cursor inside the strings when setting their fonts. Courier New (and Courier) are *monospace* fonts, meaning that all characters have the same width.
2. If this were not an example VI, you might want to hide the buttons' labels on your front panel, but show them on your block diagram, since the buttons' captions are almost the same as their labels. Take care to keep the labels consistent with the captions if you do so, as this is breaking my rule of "no redundant code!" But I suppose redundancy in a *very* localized context is sometimes helpful. The front panel labels are shown here for clarity.
3. Be careful not to confuse your Path control with your String control. Most functions with a terminal expecting a String wire will not accept a Path wire, and vice versa. LabVIEW text usually says "File Path" where this book says "Path."
4. The `Path` control has been grown a bit to accommodate long paths. Your path should indicate a real directory on your hard drive (likely not the one you see below). It should not initially specify a real file name; otherwise, you'll be overwriting one of your own files! Path controls in LabVIEW support drag-and-drop from the operating system.
5. The subVIs shown on the block diagram are the two described above, found in the **Functions»File I/O** palette.

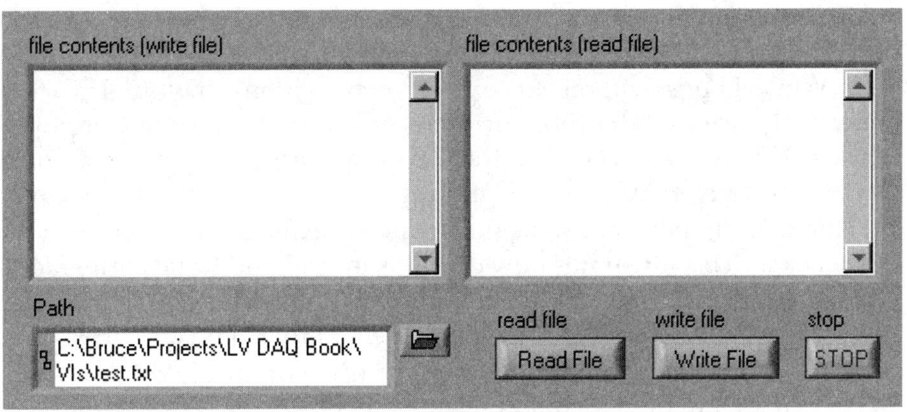

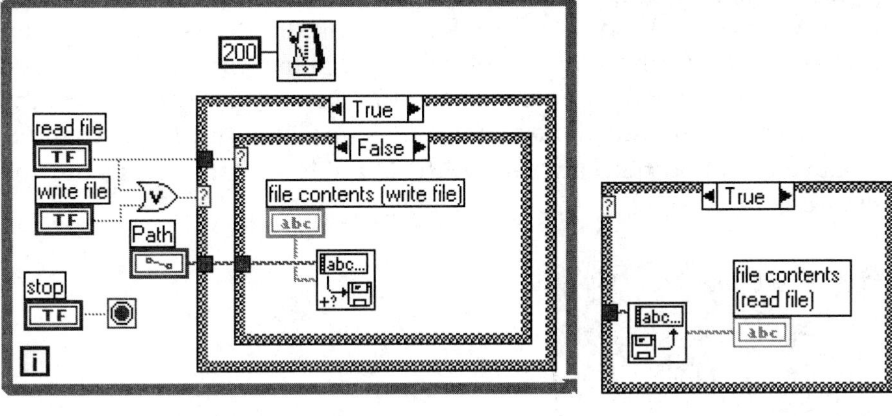

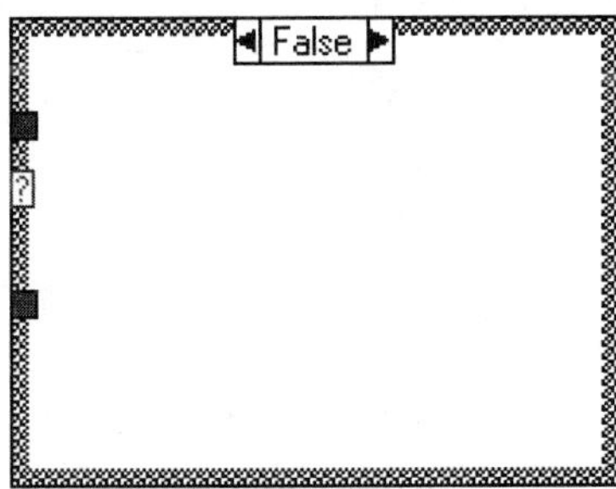

Figure 1–153
A simple, but flexible VI illustrates high-level File I/O.

First, notice the 200 ms delay in the loop from the Wait Until Next ms Multiple function. This is a politeness issue. I will often sneak such a delay into our While Loops without an explicit explanation for you. This technique effectively leaves more time for other programs to use the computer's processor. This 200 ms delay also translates to a response time of about 200 ms, so you may prefer 100 ms or 50 ms, if you notice this tiny delay. I don't.

Run this VI; first write a file (make up a *new* file name), then read the same file back in. This illustrates how easy it is to read and write entire files in LabVIEW. If you end your file name with the `.txt` extension, you could easily use any text editor to read the file independently from LabVIEW.

We'll do a quick overview of some of the more useful high level file I/O VIs, then go into more detail in the next section.

Save this VI as `High Level File IO.vi`, as we will be using it later when we discuss local variables. So that you don't lose your path information you have on the front panel, pop up on that path and make its current value the default value, then save the VI again. If you want this path to stay useful, leave that path alone on your disk until you finish this chapter.

Have a look at the **Functions»File I/O»File Constants** palette, shown in Figure 1–154.

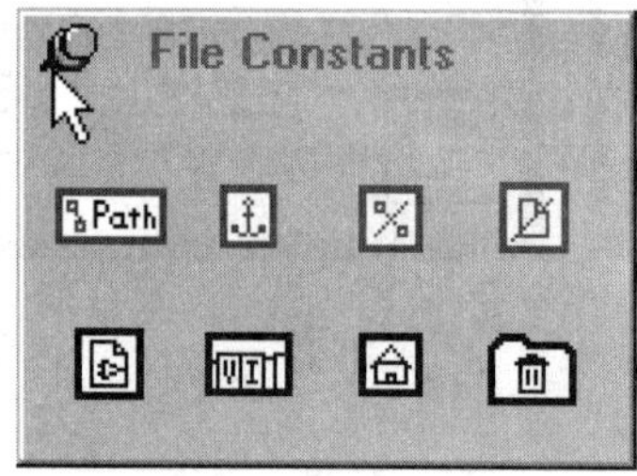

Figure 1–154
The **Functions»File I/O»File Constants** palette.

There is a Path Constant in the upper left, allowing you to specify your path on the block diagram, just like constants of other data types. The Current VI's Path constant is very useful, as it allows you to programmatically access data in the same folder as your VI. The Not A Path constant is often used as an error condition to be returned from some file I/O functions.

Let's move up one palette level and look at the **Functions»File I/O** palette, shown in Figure 1–155.

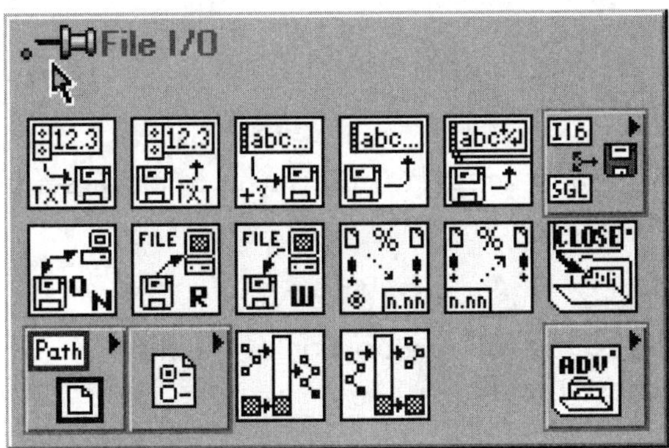

Figure 1–155
The **Functions»File I/O** palette.

To programmatically build paths, such as this sequence

```
C:\
C:\Folder 1
C:\Folder 1\Folder 2
C:\Folder 1\Folder 2\Filename.txt
```

indicates you can use the Build Path function. Conversely, to strip names from the ends of paths, you can use the nearby Strip Path function

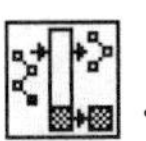 .

Other items in this palette are built for spreadsheet files (only ASCII spreadsheet files, such as tab-delimited or comma-separated files, in spreadsheet lingo). Still other items work only with binary files, where bytes are interpreted as the I16 or SGL data types. Such files would appear as garbage if you opened them with a text editor.

1.13.2 Low Level File I/O

Internally, whenever any file read or write takes place on any computer, a *file reference number* is created. The basic LabVIEW data type for this is the Byte Stream File RefNum, which I'll simply be calling *RefNum* in this chapter, even though LabVIEW has many types of RefNums. Here's what it looks like on a front panel:

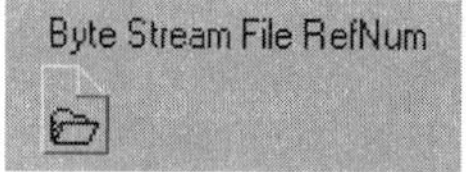

On any computer running any program, the basic internal sequence of events for reading a file is this:

1. Open the file in *read mode*. This creates a RefNum.
2. Using this RefNum, read the file or a portion of the file.
3. Using this RefNum, close the file.

Writing a file is very similar—just substitute the word *write* for *read* in the sequence above.

In general, if a file is opened for writing, it cannot be opened again for writing, neither by LabVIEW nor by any other program; the operating system enforces this rule. However, the same file can be opened simultaneously for reading.

When a file is opened, and we have a RefNum, there is also a *file position* associated with this file. LabVIEW's low-level file-reading function allows you to set this position when doing reads. Suppose we want to read the middle four bytes of a file containing the text `Hello!`. Figure 1–156 how this is done in LabVIEW, using a valid path I have on my computer:

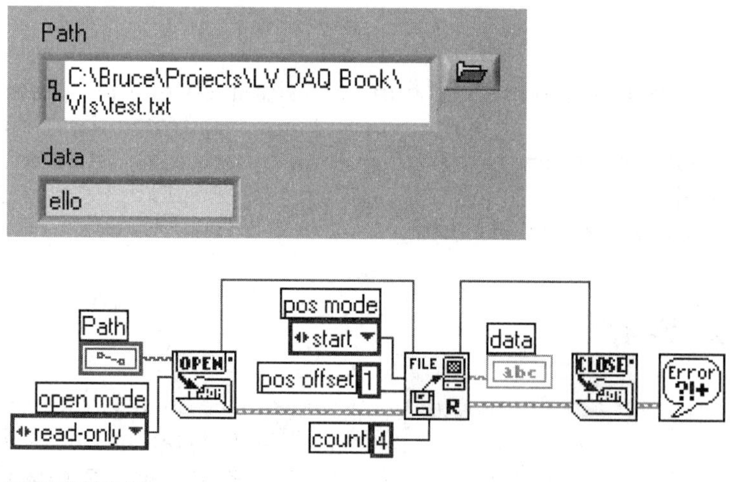

Figure 1–156
An example of low-level File I/O.

See how the four file functions are strung together by the wire on the bottom? This wire is the error cluster, designed to be used in this daisy-chaining fashion. It's often used whenever a group of error-prone functions are called in sequence, so that whenever an error occurs somewhere along the line, it's passed all the way to the error-handling function at the end of the daisy-chain. This helps to pinpoint the source of the error. Use this error cluster technique in your programs wherever possible!

Trying not to bog you down with details, the picture in Figure 1–156 shows the low-level file-opening, file-reading, and file-closing functions (to be named later in a table). A `4` is passed to the `count` parameter, thus instructing this function to only read four bytes. A `1` is passed to a the `pos offset` parameter of the file-reading function, so we start with the letter `e`. A `0` into `pos offset` would have started reading from the first byte, thus giving us `Hell` (I swear that was accidental) instead of `ello`.

LabVIEW offers you many other powerful file I/O functions listed in Table 1.10. Anytime you see "string" in the description column below, realize you can convert data from *any* format to a string, and vice versa, so this "string" can handle *any* of your file I/O needs. Such conversions will be detailed in Chapter 6 (you will read about the Type Cast function and string "flattening" functions). Table 1.10 provides a summary of all file I/O functions that I've actually used in my numerous LabVIEW projects, where the "functions" that are actually VIs (higher level "functions") are shown in `monospace` font.

Table 1.10 *A Summary of File I/O Functions*

Name	Icon	Description
`Write To Spreadsheet File.vi`	12.3 TXT	Writes a string to a spreadsheet file (only ASCII files, such as tab-delimited or comma-separated files, in spreadsheet lingo).
`Read From Spreadsheet File.vi`	12.3 TXT	Reads a string from a spreadsheet file (only ASCII files, such as tab-delimited or comma-separated files, in spreadsheet lingo).
`Write Characters To File.vi`	abc... +?	Writes a string of characters to any file.
`Read Characters From File.vi`	abc...	Reads a string of characters from any file.

Table 1.10 A Summary of File I/O Functions (continued)

Name	Icon	Description
`Open/Create/Replace File.vi`		Opens, creates, or replaces a file, and returns a RefNum. The RefNum may be used for reading, writing, and other low-level file operations.
Read File		Given a RefNum, reads any number of bytes from anywhere in a file; returns a string (or user-specified data type).
Write File		Given a RefNum and a string (or user-specified data type), writes any number of bytes to anywhere in a file.
New File		Given a path, creates a new file and opens it, returning a RefNum.
Open File		Given a path, opens a file and returns a RefNum.
Close file		Given a RefNum, closes a file. You should close all files you have opened when you're finished with them.
Build Path	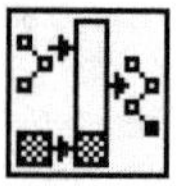	Given a path and a string, appends the string to the path, returning a larger path.
Strip Path		Given a path, strips the last string from the end the path, returning a smaller path.
`Read From I16 File.vi`		Given a file composed of I16s, reads them to an array of I16s without explicitly using the string data type.
`Write To I16 File.vi`		Given an array of I16s, writes them to a file without explicitly using the string data type.
`Read From SGL File.vi`		Given a file composed of SGLs, reads them to an array of SGLs without explicitly using the string data type.

Table 1.10 A Summary of File I/O Functions (continued)

Name	Icon	Description
`Write To SGL File.vi`		Given an array of SGLs, writes them to a file without explicitly using the string data type.
File Dialog		Pops up a file dialog box, allowing you to select a file or folder. Returns a path along with some other information.
Flush File		Given an open file that has some data buffered in memory due to previous file writes, this physcially moves the data to the disk.
File/Directory Info		Given a path, returns information about a file or folder, such as the file size, the modification date, whether the path refers to a file or folder, etc. This can be used to determine whether a path refers to an existing file.
Volume Info		Given a path, returns information on the size of the volume (the floppy disk, the hard drive, etc.), and how much free drive space exists. This is useful for monitoring disk space, so you'll know when you're about to run out.
Move		Given two paths, moves a file or a folder and its contents from one location to another.
Copy		Given two paths, copies a file or a folder and its contents from one location to another.
Delete		Given a path, deletes a file. The file cannot be open when you attempt this deletion.
List Directory		Given a path to a folder (a directory), lists the files and folders within.
New Directory		Given a path, creates a folder (a directory).
Path To Array Of Strings		Given a path, returns an array of strings containing the volume name, folder names, and ending file name. All of these items are optional in a path.

Table 1.10 A Summary of File I/O Functions (continued)

Name	Icon	Description
Array Of Strings To Path		Opposite of the Path To Array Of Strings function, above.
Path To String		Given a path, converts its data type to a string.
String To Path		Given a string, converts its data type to a path.

1.13.3 File Formats: ASCII Versus Binary

Technically, data formatting isn't really a part of any file manager, but you should understand at least the basics of data formatting if you want to program with file I/O. Remember the simple file containing the text `Hello!`? We could think of the file as illustrated in Figure 1–157, where each box represents one byte.

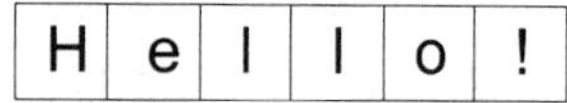

Figure 1–157
ASCII data illustrated as one byte per rectangle.

This data is saved in ASCII (sometimes called *text*) form. Often in LabVIEW, the need arises to store numeric data to files, then read it back later. Suppose we had a spreadsheet full of data. To simplify this example, our spreadsheet will contain only two rows and two columns:

1.234, 2.345

10.678, 11.555

Figure 1–158 illustrates how your spreadsheet application might save its numeric data in CSV (comma-separated value) form, which is also ASCII; thus, any text editor can read this data. CSV file names should end with

`.csv`, so your spreadsheet application can automatically recognize it. Each box represents one byte:

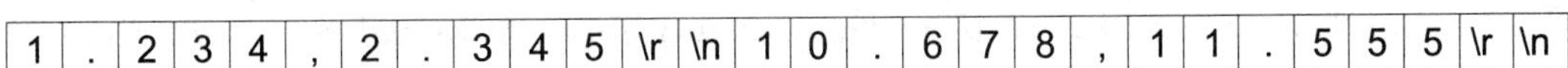

Figure 1–158
28 bytes of ASCII data.

Notice the `\r` and `\n` characters here; together, they correspond to the *end of line* (or *return*) character on a PC, which is what you get when you hit the <Return> (or <Enter>) key on your keyboard in a text editor. Oddly, it takes two bytes (on a PC) to represent this character. Non-PC platforms may use just one of these two bytes to represent the *end of line* character. Table 1.11 lists some technical details on these characters:

Table 1.11 *End of Line Characters*

Name	Icon	Backslash Symbol	Hexadecimal Value	Decimal Value
carriage return		\r	0x0d	13
line feed		\n	0x0a	10

Suppose we were really concerned about disk space or numeric precision, and wanted to save these same four numbers more efficiently. We could use the SGL type (32-bit floating-point number). On disk, we could save the file as in Figure 1–159, where each box represents four bytes (32 bits).

1.234	2.345	10.678	11.555

Figure 1–159
16 bytes of binary data as four groups of four bytes.

This data is saved in *binary* form. Unlike the previous two examples of ASCII data, each byte does not correspond to a human-readable character. A text editor could make no sense of this data. If you're curious, the internal bit patterns of these 32-bit numbers are loosely explained in Appendix A.

Reading and writing data in either ASCII or binary form is conceptually straightforward, but can be tricky to implement in real life, regardless of your programming language. So, which is better for file I/O, ASCII or binary? ASCII files are easier to understand for most people, because any text editor can read them, and people can more easily visualize the individual bytes as characters. If ASCII files are put in the right form, many spreadsheets can read them as well. Binary files are more efficient in terms of disk space, numeric precision, and processing speed. In other words, binary files produce smaller, more precise data files that can be processed by the computer more quickly.

1.13.4 Formatting Data for File I/O

This section assumes that you'll be reading or writing a 1D or 2D array of numbers, as this is the situation in most LabVIEW applications. I almost always find it easiest to convert my data into string format if it's not already, then use this string with LabVIEW's file I/O functions. Some of LabVIEW's high-level file I/O functions hide the string data type for you, by showing you only numeric data types like the I16 and SGL data types. With the low-level file I/O functions, you can manipulate data in non-string formats by specifying a different data type in the Open File function's **datalog type** input. This trick will not be discussed further in this book, but you may prefer this approach.

I will first cover in detail the Format Into String function (in the **Functions»String** palette), as it is often useful. Suppose you wire it, as shown in Figure 1–160, to a String Indicator on your front panel.

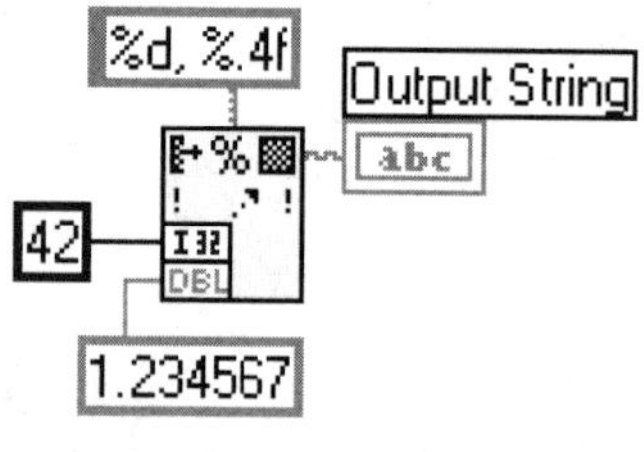

Figure 1–160
The Format Into String function is *very* useful.

When you run this, you get the following text in your front panel String Indicator: Output String 42, 1.2346

Here's how the Format Into String function works: The percent character is special, and the letter after the % tells the function to format either a number, string, or path in a particular way. In our example, the %**d** tells the function to format an integer (**d** = **d**ecimal), and the %**.4f** tells the function to format a floating-point number with four digits of precision (notice the rounding of the 5 up to 6 in the number displayed, 1.2346). This weird syntax is a result of the C programming language's `printf` function. For a technical description of what can follow the %, get any reference book on the C programming language and look up the `printf` function. Some minimal Web searching on "printf" is likely to give you enough useful information so you don't really need a book; look for examples, as they are extremely helpful.

In the example shown in Figure 1–160, if you were writing this string to a file and you wanted to add an *end of line* character to the end of this line, you would want to use the Concatenate Strings function to add an End of Line constant, as shown in Figure 1–161.

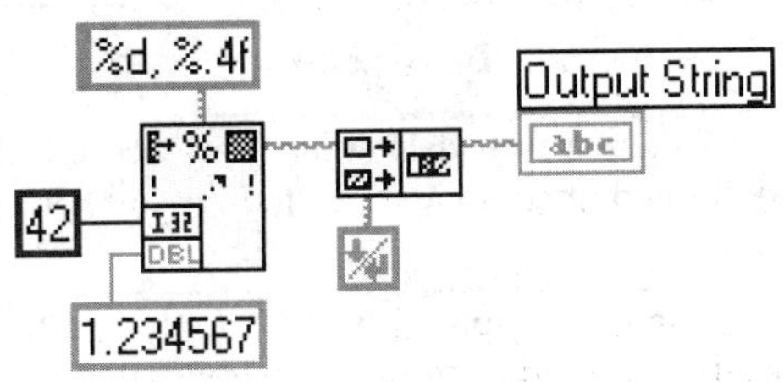

Figure 1–161
The Format Into String function uses a platform-independent End of Line constant.

If you're really clever, you can do the above without the Concatenate Strings function! This End of Line constant translates into the carriage return and/or line feed character, depending on the platform. On the PC, it translates to the carriage return followed by the line feed. On other platforms, it translates to just one of these two characters. If you knew for a fact that your program would never be run on non-PC platforms, you could have added the *end of line* characters directly to the "format string" constant, and skipped the string concatenation function, as in Figure 1–162.

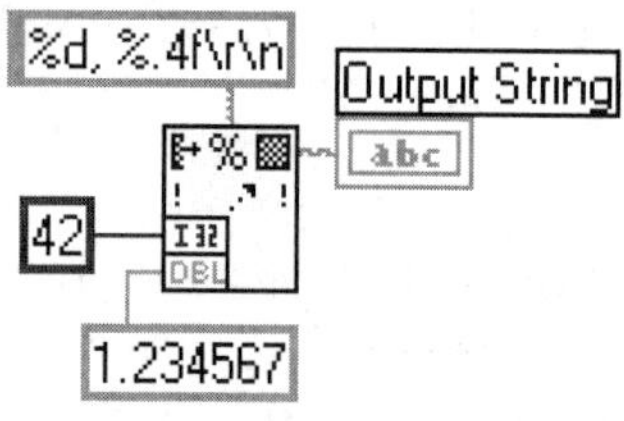

Figure 1–162
The Format Into String function now uses a carriage return and line feed regardless of the platform.

Here is an overview of LabVIEW's arsenal of data formatting functions useful for file I/O (and other things, of course). There's a bit of overlap with Table 1.10. The functions described in Table 1.12 are found at various locations under the **Functions»File I/O**, **Functions»Strings**, and **Functions»Advanced»Data Manipulations** palettes.

Table 1.12 *Data Formatting Functions: Useful for File I/O*

Name	Icon	Description
`Write To Spreadsheet File.vi`		Writes a string to a spreadsheet file (only ASCII files, such as tab-delimited or comma-separated files, in spreadsheet lingo).
`Read From Spread-sheet File.vi`		Reads a string from a spreadsheet file (only ASCII files, such as tab-delimited or comma-separated files, in spread-sheet lingo).
`Write To I16 File.vi`		You can use this, described in Table 1.10, or the Flatten To String function (below) instead, as it's more generic.
`Read From I16 File.vi`		You can use this, described in Table 1.10, or the Unflatten From String function (below) instead, as it's more generic.
`Read From SGL File.vi`		See `Write To I16 File.vi`, above.
`Read From SGL File.vi`		See `Write To I16 File.vi`, above.

Table 1.12 Data Formatting Functions: Useful for File I/O (continued)

Name	Icon	Description
Format Date/Time String		Good for creating human-readable time stamps in ASCII files.
Array To Spreadsheet String	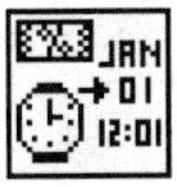	Changes a numeric array to a spreadsheet string, 1D or 2D.
Spreadsheet String To Array	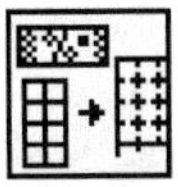	Changes a spreadsheet string to a numeric array, 1D or 2D.
Format Into String	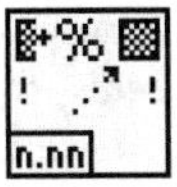	**VERY USEFUL:** Changes many different types of numbers, strings, and paths into a single string over which you have much formatting control.
Scan From String		Converts a string with numbers and text into separate numbers and pieces of text on different LabVIEW wires; something like the opposite to the Format Into String function.
Match Pattern		**VERY USEFUL:** Searches through a string for a specified pattern. A bit difficult to learn, but worth it.
Pick Line		**VERY USEFUL:** Picks a line from a multiline string. If you have ASCII data with multiple lines of text, this is one way to pick out the individual lines.
Scan String For Tokens	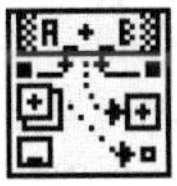	Searches through a string for *tokens* (special delimiters).
Additional String To Number Functions (This is a subpalette.)		Self-explanatory—the subpalette looks like this: 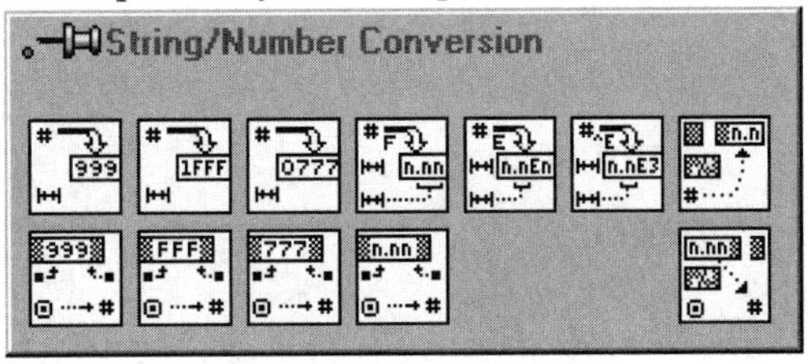 String/Number Conversion

Table 1.12 Data Formatting Functions: Useful for File I/O (continued)

Name	Icon	Description
Type Cast		**VERY USEFUL:** (advanced) Useful for converting binary data to ASCII, and vice versa; works only with data types of limited complexity.
Flatten To String		**VERY USEFUL:** Changes any data type into a string, no matter how complex.
Unflatten From String		**VERY USEFUL:** Undoes what the Flatten To String function did.
Swap Bytes		Some equipment or computers use data with byte-swapped 16-bit integers—this will swap bytes inside each integer. For example, a Macintosh computer stores 16-bit integers with the most significant byte first, while a PC stores the least significant byte first.
Swap Words		Same idea as the Swap Bytes function, only 16-bit integers within 32-bit integers are swapped.

Exercise 1.2

Write a VI that generates and saves to disk a pair of floating-point random numbers once every 500 ms for 10 seconds (exact 500 ms timing is not really possible unless you have a real-time setup, so just get close). The disk file should be specified by a Path Control on the front panel and named `fake-data.dat`. The disk format should be CSV, like this:

```
0.0037,1.5194
0.4615,1.7076
0.7643,1.3877
```

The first random number of each pair should be in the range 0 – 1, and the second in the range 1 – 2. The numbers should have four digits of precision.

All of these hints apply to the solution shown, and may not be relevant should you build this VI another way.

1. You can base this VI on a For Loop with the constant 20 wired to its count terminal.
2. Create a 1D array within each iteration of the For Loop, and pass it through a tunnel to create a 2D array coming out of the For Loop.
3. Pass the 2D array to `Write To Spreadsheet File.vi`. Be sure to wire proper constants to this subVI.

Solution

Figure 1–163 shows one of many ways to implement this VI.

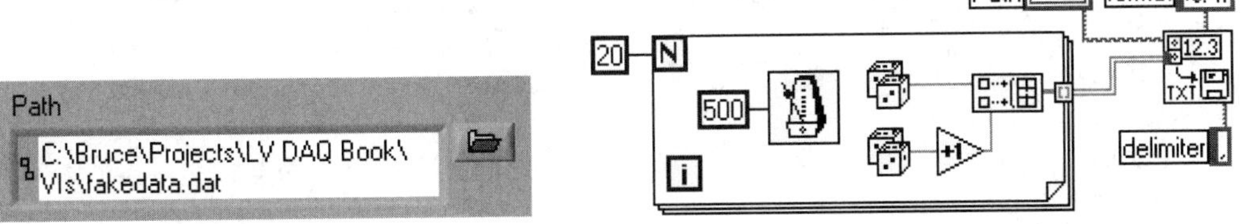

Figure 1–163
One way to implement the VI in Exercise 1.2

1.14 Deviations from Dataflow: Local and Global Variables

Rule #1 for locals and global variables: *Avoid them.* The only times I still get confused in LabVIEW are when a local or global variable is involved. LabVIEW existed for quite a long time without them, and most people were quite happy with it (except for the fact that LabVIEW had no Undo capability back then). Local and global variables will often be called locals and globals in this section, for the sake of brevity.

1.14.1 Local Variables

A local variable is simply a copy of the data in a front panel control or indicator. Here is an example of why you might want to use a local. Suppose your VI must have two While Loops running simultaneously, and you wanted to stop them both with the push of one button. Before reading past the end of this sentence, see if you can explain why this VI won't work as described. See Figure 1–164.

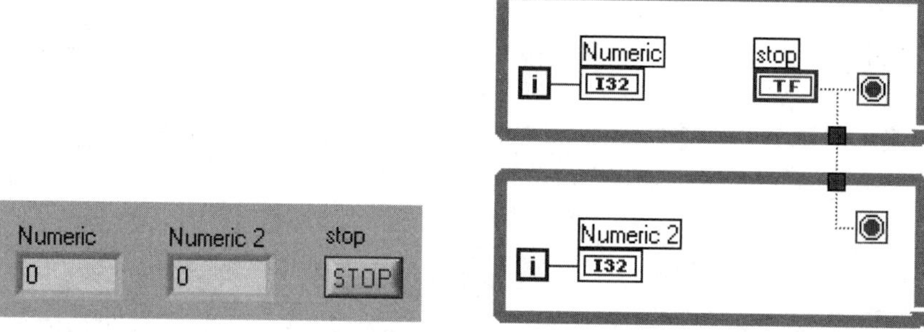

Figure 1–164
A tricky VI.

Why does this not work as described? The top While Loop never finishes until the `stop` button is pressed. The rules of dataflow view both loops as big nodes. Data cannot flow from the top loop to the bottom loop until the top loop has stopped, so the loops are never running simultaneously. Can you figure out a way to have the `stop` button stop both running loops?

Unless you are experienced with LabVIEW's locals or globals already, or are another Einstein, you probably could not figure this one out. One slick way to do what we want would be to use a local that refers to the `stop` button. Build the VI in Figure 1–164, saving it as `Locals Example.vi`, and verify that it does not work as described. Now, pop up on the `stop` button's terminal on the block diagram, and select **Create»Local Variable**. Clone this local so you have two locals. Make one a *Write Local*, and leave the other one as a *Read Local* (this is analogous to controls vs. indicators), by popping up on the local and selecting the **Change To Write** or **Change To Read** menu item if needed. Wire your diagram as in Figure 1–165.

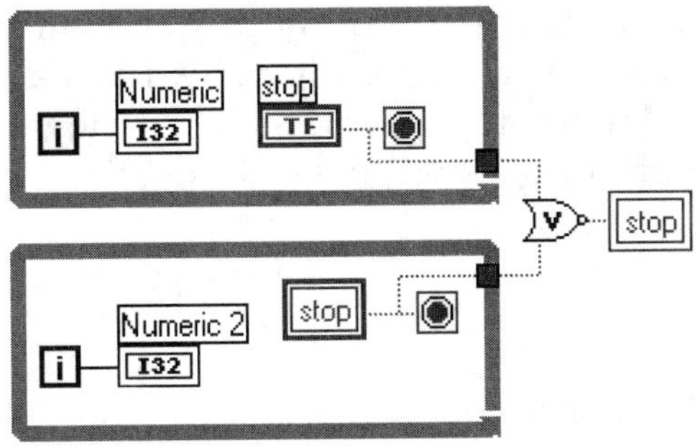

Figure 1–165
Another tricky VI with locals.

Notice the broken **Run** button. You will need to pop up on the `stop` button on the front panel and change its mechanical action to something that's non-latching. Since buttons with latching cannot have locals, we must pop the front panel button up ourselves with that far right local in the block diagram in Figure 1–165. We are writing a False to the `stop` button there, which pops it up.

Open `Password Caller.vi`, and run it twice with the correct password, `foo`. Notice that the second time the password dialog box pops up, your password is showing! This is not a very good feature, especially from a security perspective, so let's fix it. Go to the block diagram, double-click `Password.vi` (with the Positioning tool or Operating tool), then open its block diagram. Pop up on the `Enter your password:` string's terminal, and select the **Create»Local Variable** option. Make sure it's a Write Local (its border should be one pixel thick, not two), as we'll be writing data to it.

Modify `Password.vi` as in Figure 1–166 (that new box is an Empty String constant).

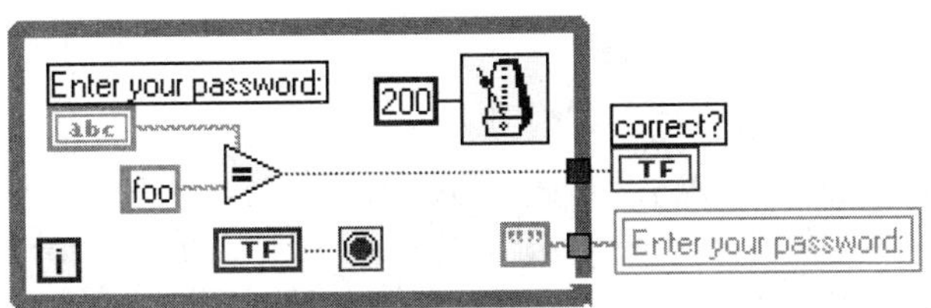

Figure 1–166
Locals make our previously created password subVI more useful.

This will automatically empty the `Enter your password:` string after the VI has closed. Save `Password.vi`, and close it. Now, run `Password Caller.vi` a couple of times, and notice that your password string is empty when the password dialog box pops up *after* the first run.

Exercise 1.3

Open `High Level File IO.vi` that you created earlier, and delete the large String Indicator on the right, as well as any broken wires. Change the remaining string's label to `file contents`. Use this String Control both for writing data to your file and reading data back. Use a local to do the latter. Your front panel can be simplified as in Figure 1–167.

Figure 1–167
The VI shown in Figure 1–153 can be simplified to this.

Solution

Simply create a local variable (a Write Local) for the `file contents` string control, then wire this local to where your deleted String Indicator used to be, as shown in Figure 1–168.

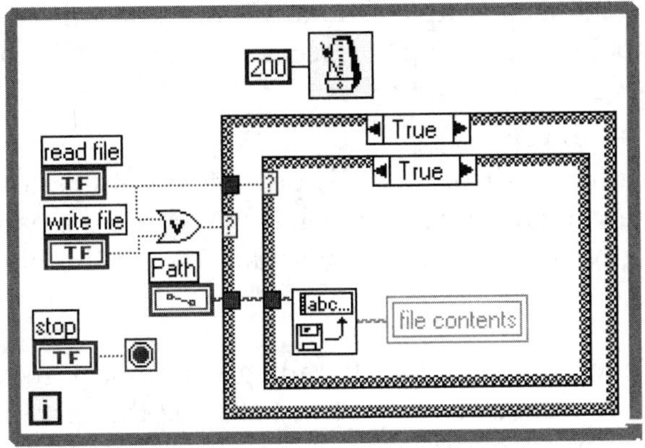

Figure 1–168
The VI shown in Figure 1–153 can be simplified to this by using a local.

Save this new, improved VI again.

Bonus: You can modify this VI further to read and write binary files, since the string is capable of viewing hex data, should you ever have the need.

1.14.2 Global Variables

Global variables (globals, for short) are somewhat similar to locals, in that the same piece of data can be referenced from multiple places within a block diagram. However, globals exist on their own special front panel, unlike locals, which always require a preexisting front panel control or indicator. More importantly, any global can be used in any number of VIs, unlike locals, which are restricted to just one VI. If you really must use the global, you can find it in the **Functions»Structures** palette. Here is a step-by-step guide on how to create and use a global:

1. Drop a Global Variable on the block diagram from the **Functions»Structures** palette. The empty global looks like this: [?]
2. Double-click this global (with the Positioning tool or Operating tool). You are now looking at your new global's very own front panel. A global's front panel is special because it has no block diagram.

3. Drop any control, such as a simple Digital Control, on this front panel. Be sure to label it (for example, `global 1`); otherwise, you cannot use it.
4. Save the global as `Globals.vi`.
5. Back on the block diagram from step 1, pop up on your global and select the name you chose for its label. If you chose the name `global 1`, your global should now look like this on the block diagram: [global 1]

I find it convenient to have all the globals for any given project in the same front panel. I always call it `Globals.vi` out of habit, as this habit always makes it easy for me to find my projects' globals. Be careful not to use globals from this `Globals.vi` in other projects, though. If you copy VIs from one project to another that references these globals, you should also copy your `Globals.vi`.

Time for the mandatory warning.

1.14.3 Why You Should Avoid Locals and Globals

Locals and globals are dangerous! From the lingo of digital hardware designers comes the term *race condition*, which means two events will occur in random order, and the proper behavior of the system requires that one of these events occurs first—therefore the system is unreliable.

It is very easy to accidentally produce a race condition whenever you're using globals. The example in Figure 1–169 may work perfectly, in which one loop generates 50 points of data at 10 Hz, and the other loop charts those 50 points.

However, this VI is unreliable; there is no guarantee that the loops will stay synchronized! The bottom loop could easily miss some data.

In general, locals and globals are safe to use whenever you can guarantee that their data is valid whenever they're being read. For example, I will write a value to a global only once, before I ever read it—then I'll read it at multiple points in my program. I must take extra care *not* to read it before I write to it the first time! Things can get confusing when your locals or globals are inside a loop, where one loop reads and the other writes to the same global (as shown Figure 1–169), so don't call me if you get stuck with locals *or* globals!

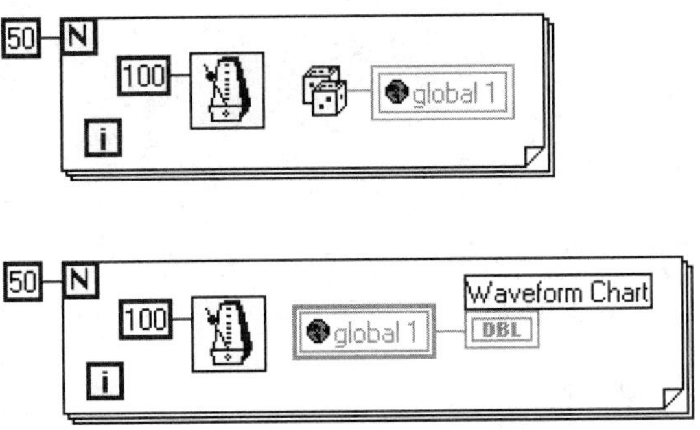

Figure 1–169
Global variables used incorrectly.

The other downside to locals and globals is that an extra copy of their data is generated for every Read Local or Read Global (*read* here means data is read from them by the block diagram, like a control). This can degrade performance if the local or global data is a large array or other type of large data (containing many bytes).

1.15 Property Nodes

Property Nodes, formerly called Attribute Nodes, are one of my favorite features in LabVIEW. With them, you can programmatically make a front panel object invisible, change its color, move it around, and so on. Different types of front panel objects have different types of properties you can change. Most objects allow you to change their visibility, and most objects, like the graph, have properties that are specific only to their object's type. For example, a graph might allow you to change a plot's color, but a numeric would not, since it has no plots.

Create the front panel objects shown in Figure 1–170 on a new VI (a Boolean and Numeric control).

To create a Property Node, simply go to the front panel object, or better yet, its block diagram terminal, pop up on it and select **Create»Property Node**. Do this now with the Numeric control, and find the new Property Node on the block diagram. To begin with, the Property Node should look like the item on the left in Figure 1–171.

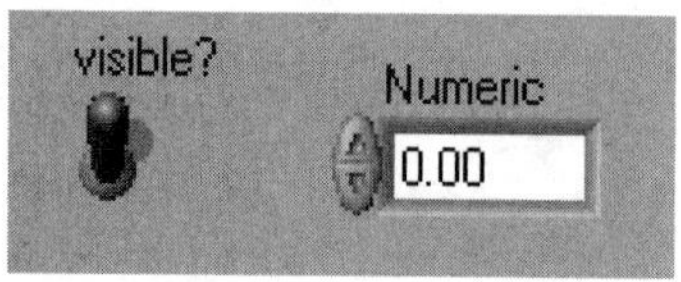

Figure 1–170
A Boolean and Numeric control.

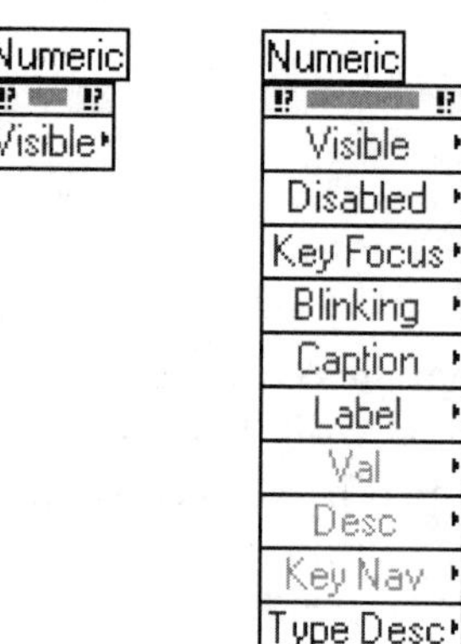

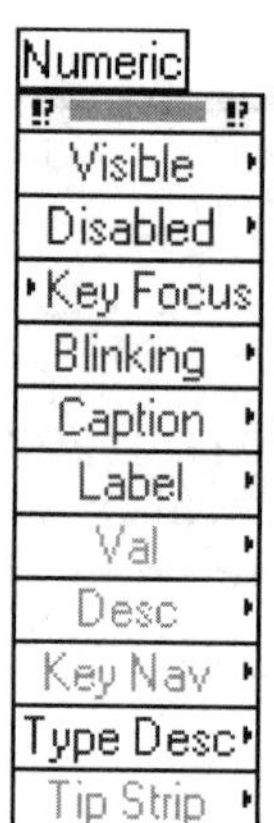

Figure 1–171
These are Property Nodes showing various properties of Boolean and Numeric controls that you can read or write.

The middle Property Node in Figure 1–171 is a result of growing the left Property Node, thus showing more properties. When a Property Node is showing multiple properties, it executes them from top to bottom.

The right Property Node in the figure is a result of popping up on the **Key Focus** property and selecting the **Change To Write** menu item. Property Nodes let you know whether data is going in or coming out by means of the little black arrows, shown in the figure. This is unlike many other one-terminal block diagram nodes, like front panel objects' terminals and local variables, that let you know this with a one-pixel or two-pixel thick border.

You can pop up on any Property Node to change the property shown. Take a look at a few different properties by using the Operating tool and left-clicking the Property Node with the Help window showing—the Help window will describe each property. Notice that different properties have different data types.

As an example, let's make a Digital Control invisible. Wire your block diagram as in Figure 1–172, then run the VI a few times with the `visible?` control in different states.

Figure 1–172
A simple illustration of the Property Node.

Open `Password Caller.vi` again, then open `Password.vi` from its block diagram. Open the block diagram of `Password.vi`. Pop up on the `Enter your password:` string's terminal, and select the **Create»Property Node** option. Modify `Password.vi` as in Figure 1–173.

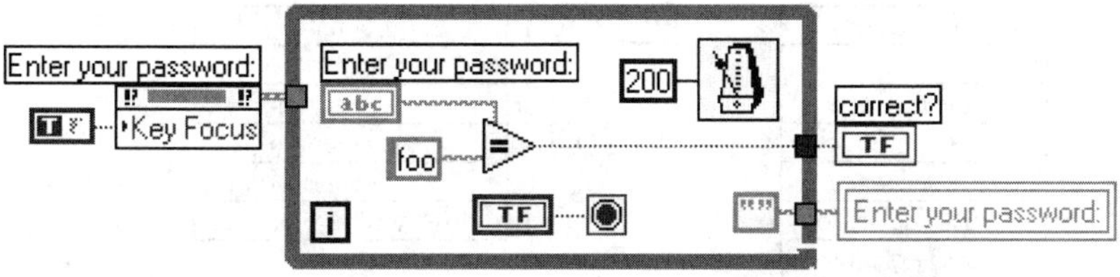

Figure 1–173
Our previously-created password VI is refined once again.

Save `Password.vi`, close it, and try it out now from `Password Caller.vi`. Notice that you can now immediately start typing your password when the dialog box pops up, because the Key Focus property places your cursor on the `Enter your password:` string.

1.16 Printing

Printing in LabVIEW can be very tricky. What works fine on one printer might not work so well on another. By now, most of these issues have been resolved, but be aware of this potential problem. When I worked on the LabVIEW development team, I was very glad that printing was not my department!

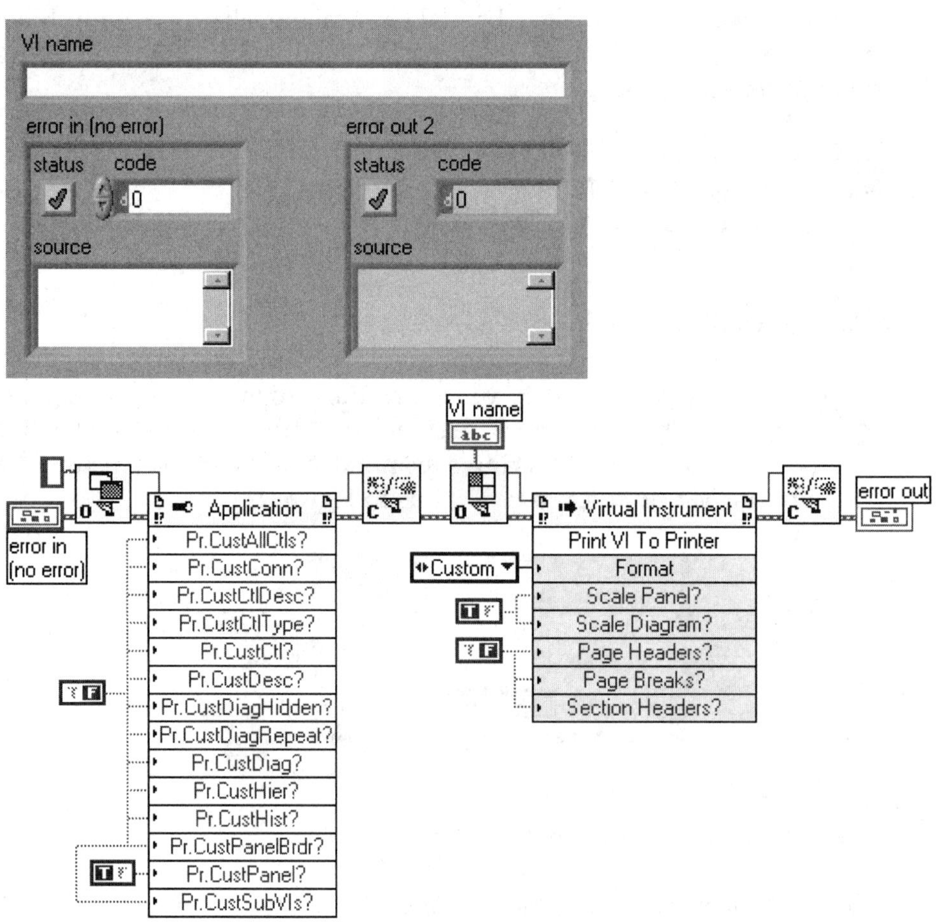

Figure 1–174
Property Nodes are useful for customizing printing.

Probably the simplest way to print your results in LabVIEW is to send your results to a subVI that is dedicated to printing. This subVI will be called your *print VI* hereafter. As you're developing your application, you can save time (and paper) by making this VI a dialog box that pops up and displays your results rather than actually printing them. Then, when you're ready to complete your application, change the print VI so that it no longer pops up, but so that it has its **Operate »Print at Completion** property set. Other handy printing tips:

1. Color the background (of the front panel) of your print VI white. This speeds up the printing, as the printer doesn't print any ink where it sees white.
2. If you have objects on the front panel of your print VI that you don't want printed, which you probably will in order to get data to your print VI, use the Property Node to make them invisible just before printing, or simply hide them with their terminals' **Hide Control** menu item.

If you want to get rid of the header information printed across the top of your page, you might need to get fancy and tweak the printing properties, using some advanced functions, as in Figure 1–174. The new functions shown here can be found in the **Functions»Application Control** palette. The two large nodes are the Property Node (different than our previously discussed Property Node) and the Invoke Node.

Finally, LabVIEW has some built-in report-generating functions found in the **Functions»Report Generation** palette.

1.17 Finding Objects in LabVIEW

LabVIEW has a very nice *find* feature. Hit <Ctrl-F>, and up pops a box like the one in Figure 1–175 (possibly with something other than the Add function shown).

This powerful utility lets you find objects, like functions or subVIs, or even text. In general, keep the **Search Scope** ring to **All VIs in Memory**, as shown. Sometimes you will want the **Include VIs in vi.lib** box checked, other times not. `Vi.lib` is a folder in LabVIEW's directory that contains many of LabVIEW's built-in VIs.

When **Search for:** is set to **Objects**, as it is shown in Figure 1–175, you can quickly find any VI in memory by popping up on whatever **Select Object:** icon is showing and selecting the **VIs by Name...** option.

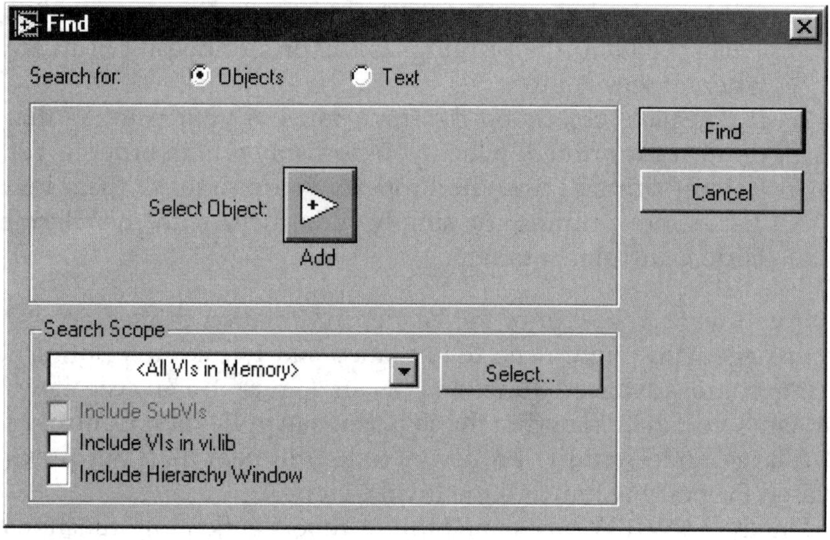

Figure 1–175
LabVIEW's find feature.

1.18 Learning More

As you might have guessed by now, you cannot learn everything about LabVIEW in one chapter of one book. This was just to prepare you for doing some useful LabVIEW/DAQ work—if you've thoroughly learned the material in this chapter, you're ready.

For more detail on LabVIEW in general, read a more in-depth LabVIEW book, such as the excellent *LabVIEW for Everyone*, described in Section 1.1.

After you have a good working knowledge of all information in this chapter, you may want to take a LabVIEW class, sponsored by NI. I've enjoyed teaching dozens of LabVIEW classes when I worked there (and afterwards). Unless you're a genius, you had better know some LabVIEW basics before you set foot in any of these classes! Becoming familiar with this chapter and working through the examples will adequately prepare you for the basic LabVIEW class.

Close any VIs you might have open with the **File»Close All** menu item, as we don't want our computer screen to become too cluttered.

Signals and DAQ

2

DAQ (data acquisition) hardware will be described in this section, as the hardware basics should be understood before attempting DAQ software. First, we'll need to get a firm grasp on many relevant hardware concepts and terms. The term *signal* means an electrical signal that corresponds to some real-world value, such as a voltage corresponding to a temperature. DAQ devices allow the computer to read or write such signals. This chapter covers the following diverse topics, all of which are needed to prepare you for DAQ:

1. Ohm's Law
2. Categories of Signals
3. Signal Conditioning
4. DMA and FIFO
5. NI's DAQ Products

2.1 Ohm's Law

If you are familiar with Ohm's Law already, skip to Section 2.2. Otherwise, you should understand these electrical terms: *voltage, current,* and *resistance.* All matter, including you, is composed of incredibly tiny atoms, which have electrons circling around them. These electrons normally stick to their atoms, unless there is some sort of voltage (electrical force) in the area, in which case electrons might flow from atom to atom. Electricity is electron flow.

Electricity can flow very easily through *conductors,* such as metal wire, yet it cannot flow through *insulators,* such as glass, rubber, dry wood, most plastics, and other such materials. *Resistors* are somewhere between conductors and insulators, as electricity can flow through them, but the substance of a resistor offers some resistance to the electron flow.

Imagine electricity flowing through a resistor, and I'll explain a few key electrical concepts by making an analogy to water flowing through a pipe. **Voltage** on the resistor corresponds to **water pressure**, and is measured in volts. Two other terms for voltage are *potential* and *EMF* (electromotive force). **Current** through the resistor corresponds to the **water's flow rate,** like gallons per minute, and is measured in amps (A), or *amperes*. In the world of DAQ, currents are often so small that milliamps (mA) are used (1,000th of an amp). **Resistance** of the resistor corresponds roughly to the **water pipe's resistance to flow**, which would be inversely related to the square of the pipe's diameter. Electrical resistance is measured in Ohms (Ω). In the DAQ world, resistances are so large that kOhms (kΩ) are often used (1,000 Ohms). An amp is defined as a certain number of electrons flowing per second. I'm too lazy too look up that number, and it's not really relevant, but it's a big number, because electrons are unbelievably small.

Ohm's Law defines a relationship between voltage (V), current (I), and resistance (R), as follows:

$$V = IR$$

In Figure 2–1, Figure 2–2, and Figure 2–3, we show three systems that are analogous to one another.

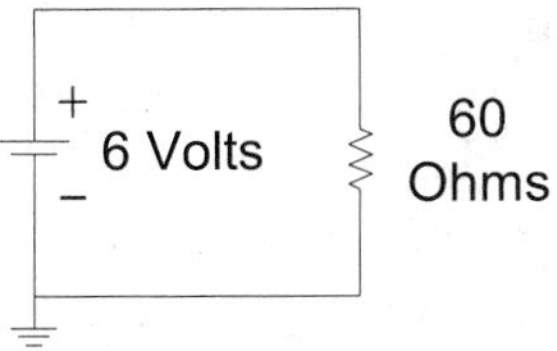

Figure 2–1
System 1—an electrical schematic.

Figure 2–1 is an electrical schematic, in which the lines represent wires. The item labeled "6 Volts" to the left is a voltage source, such as a battery. It causes electrons to flow around the loop through the resistor, such as a light bulb, on the right. Ohm's Law applies to circuits like this, even if they're part of a more complex circuit. We could plug V and R into the equation to figure out that the current here, I, is 0.1 A, or 100 mA.

The little symbol at the bottom left of the circuit represents the circuit's *ground*, which is any part of the circuit at zero volts. All portions of the wire connected to the ground symbol, up to the bottoms of the battery and the resistor, are at ground voltage (or very close).

Figure 2–2
System 2—an implementation of System 1.

Figure 2–3
System 3—can be compared to Systems 1 and 2, believe it or not!

A flashlight (Figure 2–2) is a real-life implementation of System 1. The 6-volt battery in the type of flashlight shown in System 2 corresponds to the voltage source, and the light bulb corresponds to the resistor.

Water flowing through your showerhead (Figure 2–3) corresponds to electricity flowing through the resistor (or light bulb) in the previous cases. The water pressure is fairly constant (provided you don't flush the toilet), like the voltage was. Like electricity, the water's flow rate is *inversely* proportional to the resistance it encounters while flowing. If you plugged up half the holes in the showerhead, this would be like doubling your resistor's resistance in System 1, so the water would flow about half as fast.

In this showerhead analogy, imagine all the water going down the drain and getting pumped back through the showerhead (hopefully it gets filtered!). That is more like the two electrical circuit examples discussed above, in that they are closed systems (the same electrons are cycled around a loop).

2.2 Categories of Signals

In the world of DAQ, a signal is typically a voltage (described in the last section), although sometimes people use current as their signal instead. With DAQ, you implicitly have a computer that is reading signals (input), generating signals (output), or both. Also, signals can conveniently be divided into two categories—analog and digital. An analog signal is one that can vary continuously between two values, such as the temperature on a thermometer, as shown in Figure 2–4.

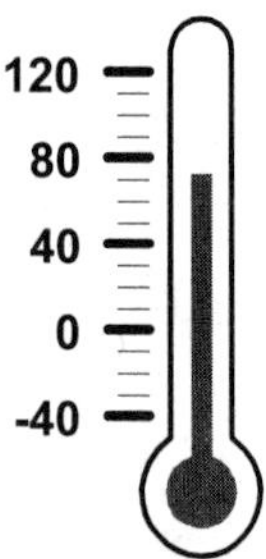

Figure 2–4
A thermometer illustrates an analog value.

Your temperature might be 20 degrees, 40 degrees, 22.3443 degrees, or any number of values; such is the nature of analog signals.

A digital signal is one that can have only two values, such as a warning light on a car's dashboard. These two values are often called *on* or *off*, or *True* or *False*. More loosely defined, a digital signal can be composed of *any number* of these *on/off* signals, so the resulting digital signal could take on a larger number of values, in the same manner that many bits could form a byte.

The objective in DAQ is either to read some real-world value, such a temperature, into the computer (input), or to change some real-world value with the computer (output). A device that converts the real-world value to and from an electrical signal, such as a voltage, is called a *transducer*. This electrical signal moves into or out of the computer by means of a DAQ device.

For example, a computer's sound card is a DAQ device. The transducers involved are the speakers (for output) and a microphone (for input); they convert electrical signals to sound, and vice versa.

In this book, DAQ signals have been divided into five general groups:

1. analog input
2. analog output
3. digital I/O
4. counter
5. other signals: bus-based and serial

NI *usually* only includes the first four of these five categories when talking about DAQ, so we will focus upon these throughout the book. We will mention the fifth where appropriate.

2.2.1 Analog Input Signals

Analog signals are used to measure physical phenomena that may continuously vary between two values, such as temperature, pressure, and flow rate. Simply put, analog input is where the computer reads an electrical signal that can vary between two values. There are a number of terms we must use in order to properly discuss analog signals:

1. channels
2. samples
3. scans

4. multiplexer
5. grounding
6. differential vs. single-ended
7. source impedance
8. gain
9. range
10. resolution
11. scan rate
12. waveforms
13. triggering

Before we go into all these details, Figure 2–5 provides a very fundamental view of what's happening inside any analog input DAQ device, including any ordinary sound card.

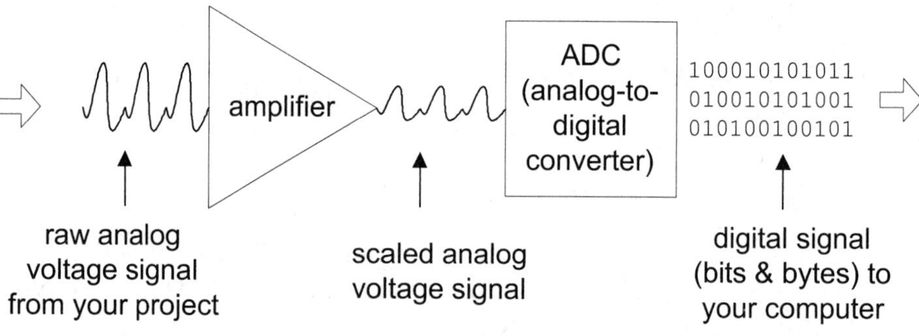

Figure 2–5
The basics of an analog input DAQ device are shown.

Figure 2–6 provides a close-up view of that amplifier, swiped from one of NI's documents.

A **channel** is the electrical connection to a signal. A **sample** is the conversion of an electrical signal to a digital value, also called a *data point* in this book. A **scan** is the collection of one or more samples, across one or more channels, at the same point in time (or nearly the same point in time). Some DAQ devices may have a few microseconds' delay between different channels' samples in a scan. To reduce this delay to a nanosecond or less, use DAQ devices with *simultaneous* sampling—for most applications, this is not an issue.

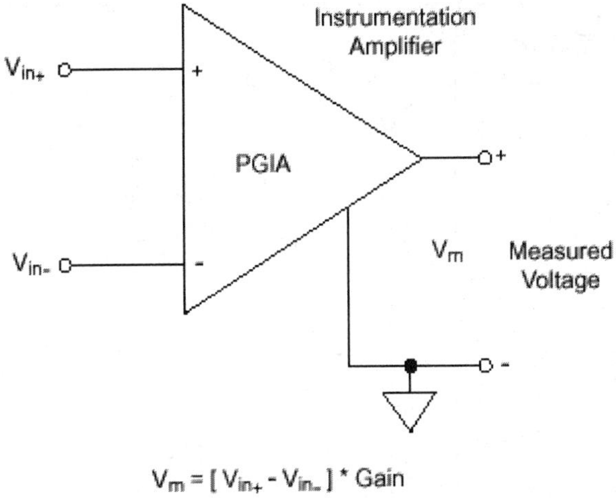

Figure 2–6
A PGIA (programmable gain instrumentation amplifier) multiplies a voltage by a user-programmed factor.

Multiplexers are used so that only one amplifier measures many analog input signals (thus saving money and parts). Figure 2–7 shows how a multiplexer might work when configured in *differential* mode, to be described soon, where one signal is wired to ACH0 and ACH8, the next to ACH1 and ACH9, the next to ACH2 and ACH10, and so on.

Grounding affects how the DAQ device is connected to the signal you're measuring. The signal you're measuring will either be *grounded,* meaning its ground is electrically connected to your DAQ device's ground, or *floating,* meaning it is not. As an example, a thermocouple is just two wires connected together that produce a voltage from which you can determine the temperature. Figure 2–8 shows a simple thermocouple with no shielding.

The thermocouple is a floating device, as it is not electrically connected to ground, so you would need to connect it to your DAQ device in a special manner involving extra resistors (discussed later in this section). If instead, you were measuring the voltage on a grounded instrument (for example, with a three-prong power plug such that the third prong was ultimately connected to your DAQ device's ground), your connection scheme would be different, as illustrated later in Figure 2–10.

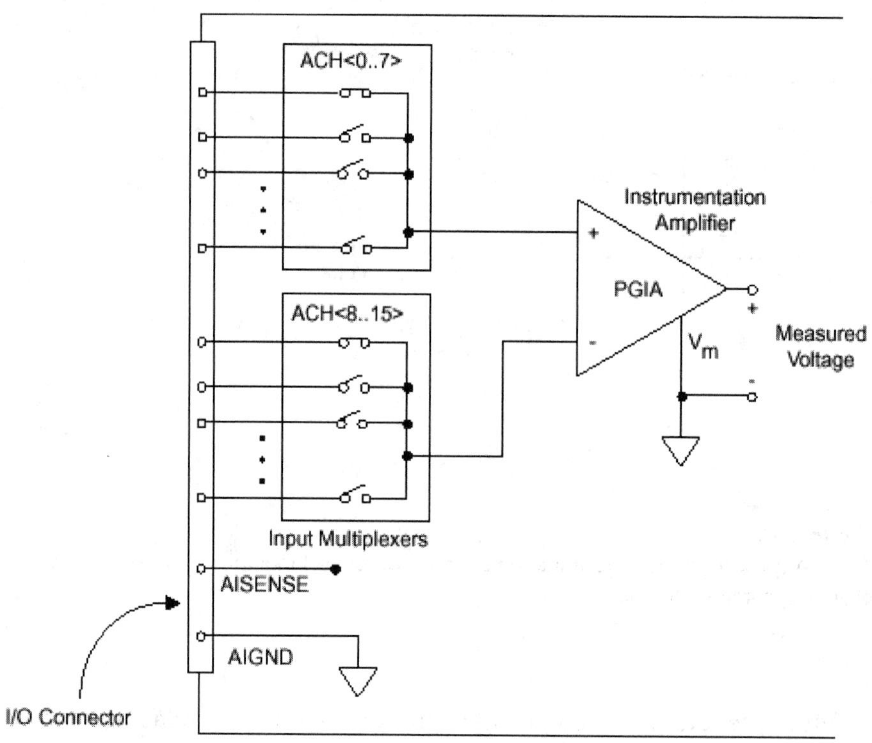

Figure 2–7
The I/O Connector shown in Figure 2–7 is where you connect the wire pairs whose voltages you want to measure.

Figure 2–8
A simple thermocouple.

Ground-Checking Trick: If you don't know whether a signal is grounded, here is an effective trick to determine this. First, get a voltmeter, such as a Fluke handheld meter (I like the 7x series), set this meter to measure resistance, then measure the resistance between the signal's ground and the DAQ device's ground. Often, the DAQ device's ground is the same as the computer's ground. If your meter indicates any resistance (not an open circuit), you are grounded. If your meter indicates an open circuit, *reverse the connection of the meter probes*, and if it still indicates an open circuit, the signal is floating. If the circuit is not open in either direction, it is grounded.

The ground of the signal you're measuring may be at a different voltage than the ground of your DAQ device. This could lead to inaccuracies in the voltage you're measuring, in that an unwanted current, called a *ground loop*, can exist between the two grounds. Figures 2–10 and 2–20 illustrate ground loops. The worst case I've heard of was in a power plant, where one supposed ground was actually just over 10,000 volts. This could lead to death if you're not careful.

Source impedance is effectively how strong your signal's voltage-driving power is. In electrical engineering terms, this would be the resistance in the *Thevenin Equivalence* model of your source. The lower the impedance, the stronger the voltage source. Keep the ratio of the input source impedance to the DAQ device's impedance very low, much less than one, for more accurate measurements. This is almost never a problem—typical signal sources such as RTDs, strain gauges, thermocouples, and thermistors all have very low source impedances compared to DAQ devices, which are designed to have very high impedances, usually in the GΩ range, on their inputs.

Differential and **single-ended** connections are different ways of connecting an analog input signal to your DAQ device. A differential connection is one in which the DAQ device's analog input signal has its own reference signal (or signal return path), which means two wires per signal. A single-ended connection is one in which multiple analog input signals share a common reference signal, with only one unique wire per signal.

Following are NI's recommendations on when you should use differential signals (*noise* means electrical noise, not sound—see Section 2.3.1 for details).

You should use differential input connections for any channel that meets any of the following conditions:

- The input signal is low level (less than 1 V).
- The leads connecting the signal to the DAQ device are greater than 10 ft (3 meters).

- The input signal requires a separate ground-reference point or return signal.
- The signal leads travel through noisy environments.

Figure 2–9 is a diagram comparing the two basic connection methods with multiple signals.

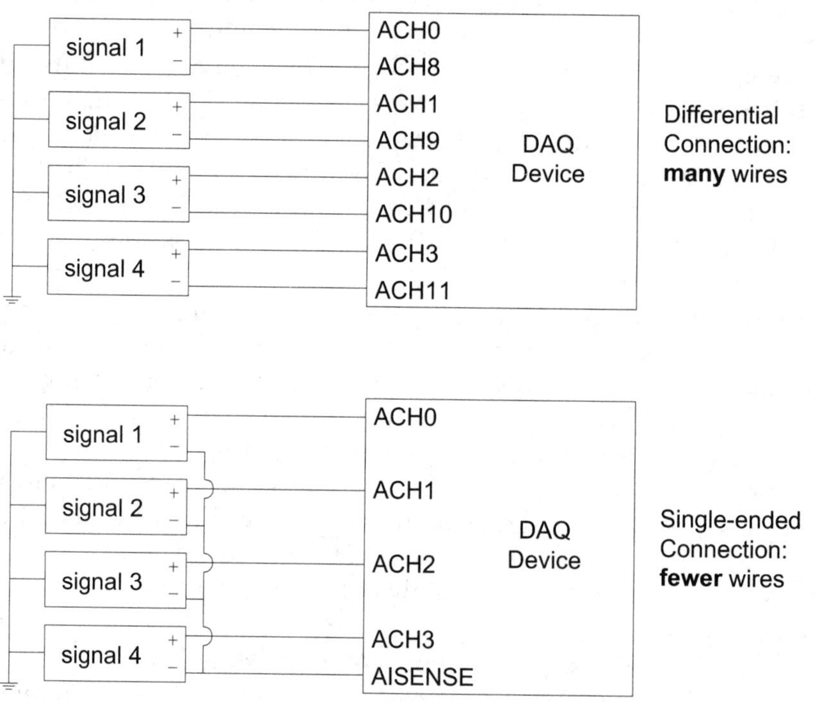

Figure 2–9
This is the simplest way to illustrate the difference between differential and single-ended connections. There is a ground associated with the DAQ device, as well, as you will see in Figure 2–10.

There are two different types of single-ended connections: RSE (referenced single-ended) and NRSE (non-referenced single-ended). RSE means the signal's negative terminal is connected to the computer's ground, while NRSE means it's not. On DAQ connector blocks, there is an AISENSE connector that is used in different ways as illustrated in NI's immensely helpful DAQ Connections Table, shown in Figure 2–10.

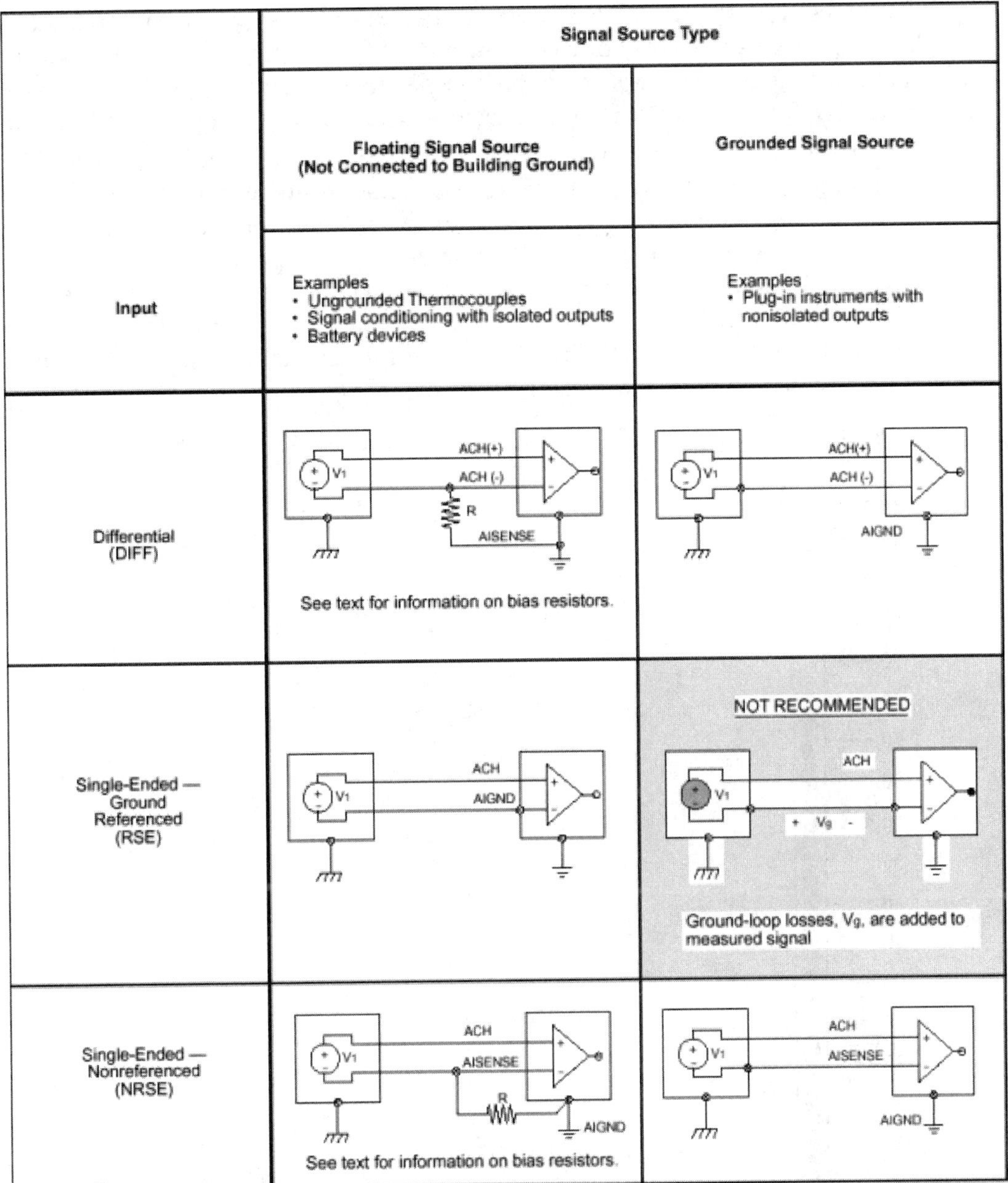

Input	Signal Source Type	
	Floating Signal Source (Not Connected to Building Ground)	Grounded Signal Source
	Examples • Ungrounded Thermocouples • Signal conditioning with isolated outputs • Battery devices	Examples • Plug-in instruments with nonisolated outputs
Differential (DIFF)	V_1, ACH(+), ACH (-), R, AISENSE See text for information on bias resistors.	V_1, ACH(+), ACH (-), AIGND
Single-Ended — Ground Referenced (RSE)	V_1, ACH, AIGND	NOT RECOMMENDED V_1, ACH, + V_g - Ground-loop losses, V_g, are added to measured signal
Single-Ended — Nonreferenced (NRSE)	V_1, ACH, AISENSE, R, AIGND See text for information on bias resistors.	V_1, ACH, AISENSE, AIGND

Figure 2–10
DAQ Connections Table.

The resistors shown are *bias resistors;* they keep the signal from floating too far away from your DAQ device's ground. Their resistance should be at least 100 times the resistance of the source impedance so you'll have less than 1% gain error. Typical values for bias resistors are anywhere from 100 kΩ to 1 MΩ.

Gain is the factor by which the DAQ device's amplifier multiplies the incoming voltage. **Range** refers to a signal's maximum and minimum values. Range is closely related to gain. The range of the thermometers shown in Figure 2–11 is about –40 degrees to 120 degrees. Let's suppose we're measuring voltage. We must set up our DAQ equipment to measure within a certain voltage range, such as from 0 to 10 volts, -5 to +5 volts, 0 to 100 mV, and so on. Most DAQ devices contain a built-in amplifier that can adjust its gain (a PGIA), which obviously affects the range.

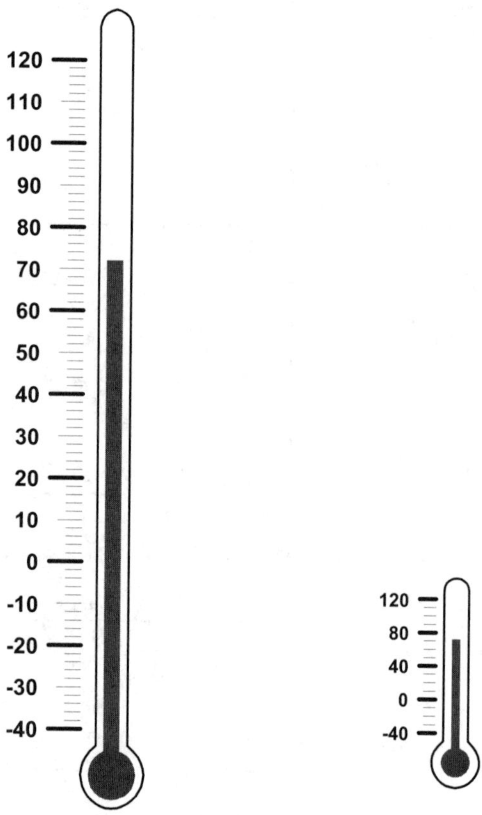

Figure 2–11
The thermometer on the left has a higher resolution, allowing you to distinguish between smaller values.

Resolution is the smallest difference in a value that a measurement system can detect. For example, look at the two thermometers in Figure 2–11.

To your eye, the thermometer on the left has a higher resolution, since you can distinguish between smaller values in temperature. Such is the case with the tick marks, as well; the tick marks on the left correspond to smaller differences in temperature.

With DAQ, this same concept applies to measuring voltages. Analog DAQ measurement involves an ADC (analog-to-digital converter) in the DAQ device, which converts an analog signal's value to an unsigned N-bit integer. Your ADC might have 12-bit or 16-bit resolution, meaning the resulting integer has 12 or 16 bits. A larger number here corresponds to greater resolution. Why? A 12-bit integer can take on 4,096 different values (2^{12}) ranging from 0 to 4095. A 16-bit integer can take on 65,536 different values (2^{16}) ranging from 0 to 65,535. This is exactly like having more tick marks on your thermometer. Supposing your signal's voltage range is 0 to 5 volts, the 12-bit ADC would be able to distinguish voltages every (5 / 4,096) volts, or 1.221 mV. The 16-bit ADC would of course be much more sensitive, distinguishing voltages every (5 / 65,536) volts, or 0.0763 mV.

Scan rate is simply the number of times per second the ADC measures the value of all of its signals. For any given signal, it is also called *sampling rate*. You should be very careful to sample *at least* twice the rate of the highest frequency of the signal you're measuring (5 to 10 times is better); otherwise, you will get inaccurate data. To illustrate, suppose you are sampling an 8 Hz sine wave at a rate of 100 Hz (see Figure 2–12).

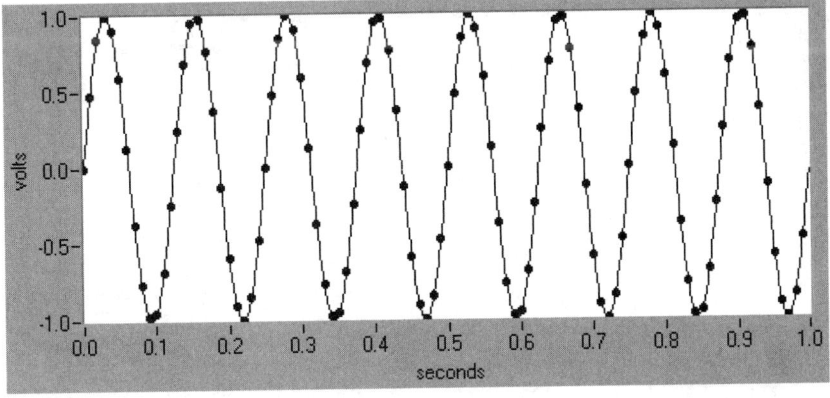

Figure 2–12
Samples of this waveform are taken where the black dots are shown.

The only information actually seen by the computer is represented by the black dots in Figure 2–12 (spaced at 100 Hz, or 0.01 seconds), which is a good representation of the underlying sine wave.

On the other hand, suppose you are sampling that same 8 Hz sine wave at 10 Hz, as shown in Figure 2–13.

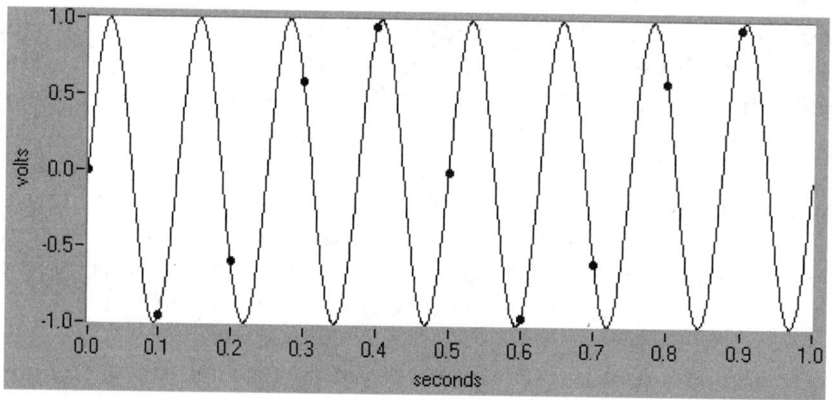

Figure 2–13
Samples of this waveform are taken where the black dots are shown—they are spaced too widely.

The computer sees only the black dots, which appear to represent a completely different waveform, shown in Figure 2–14.

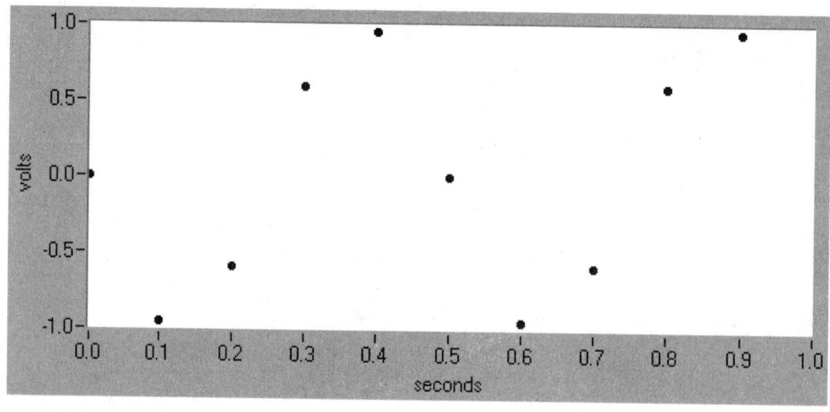

Figure 2–14
Aliasing is illustrated, since these samples do not represent the underlying waveform in Figure 2–13.

This effect is called *aliasing*, and the result is a false signal of a lower frequency than the actual signal, as was just illustrated. The only way to prevent aliasing is to sample at least twice as fast as the highest incoming frequency. Of course, you cannot always control what the highest frequency is—in this case, you must either sample faster or filter out the high frequency data using analog circuitry before it hits the ADC. Analog filters do this; there is no way to correct the aliasing problem using a digital filter with your computer after the signal has passed through the ADC. NI has some DAQ devices with built-in anti-aliasing filters.

When performing frequency-based calculations, you should generally sample at least four times the rate of the highest incoming frequency for better accuracy, even though you are required to sample at only twice the rate.

If you are interested in the time-domain characteristics of the incoming signal, you will likely want to sample even faster than you would for measuring the frequency-domain characteristics. You may have a better idea of how fast you need to sample after you see graphs of the incoming signals.

Waveforms are simply sequences of analog or digital input data (or output data) that have been sampled (or updated, for output) at a regular rate. The set of samples taken at 100 Hz, shown in Figure 2–12, is a good example of a waveform. NI's DAQ devices have built-in hardware mechanisms (to be discussed in Section 2.4 of this chapter) to ensure that your samples are evenly spaced in time, even when you're sampling at a very fast rate.

Triggering simply means starting your DAQ by means of some special change in your signal. Some DAQ devices have hardware triggering mechanisms; others don't. Triggering tells your DAQ device to start sampling whenever the incoming voltage exceeds, or falls below, a voltage that you specify—you can usually adjust some other parameters concerning just how fast and how far the incoming voltage must change in order to trigger.

2.2.2 Analog Output Signals

Most concepts concerning analog output signals are the same as they are for the analog input signals, except

1. Sampling is called updating.
2. Instead of voltage signals being changed into digital signals on your computer by an ADC, digital signals on your computer are

changed to voltage signals by a DAC (digital to analog converter, of course).

3. DACs are more expensive than ADCs.

Take care not to attempt to drive too much current during analog output! This could destroy your DAQ device. Your DAQ device is likely rated at a certain number of mA, which you should not exceed.

Some DAQ devices cannot produce accurate waveforms on their analog outputs, due to the lack of DMA or FIFO associated with their analog output (discussed in Section 2.4).

2.2.3 Digital I/O Signals

Digital signals differ from analog signals in that they can take on only two values instead of many. Compare an analog signal graph to a digital signal graph, shown in Figure 2–15.

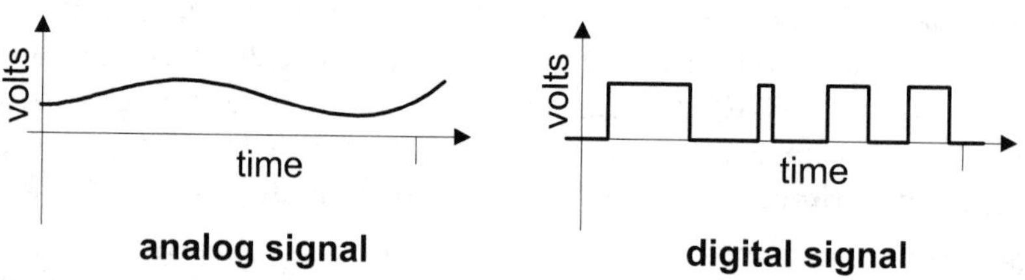

Figure 2–15
The two fundamental signal types are compared.

Notice that our digital signal is one voltage or the other—nothing else. Of course, a little slop is part of any digital signal specification, since it's tough to have an *exact* voltage, but it's convenient to think of them as only one voltage or the other.

On NI's DAQ devices, digital signals are typically either zero volts or five volts. These types of signals are sometimes called *TTL* (transistor-to-transistor logic). Other voltage pairs used are 0 – 24 VDC, 3 – 32 VDC, 0 – 120 VAC, and 0 – 240 VAC. RS-232 and RS-485 serial ports have different voltage values for their digital signals, but we will discuss them in their own section.

Digital input signals always measure electrical signals that have exactly two states. Some examples are a valve being opened or closed, a light being on or off, a pressure exceeding a certain value or not, etc.

Digital output signals are quite similar to digital input signals, except for the obvious fact that the signals are going out of the computer rather than coming in. Common uses for digital output signals are to open or close relays, to open or close valves, to turn lights on or off, or to activate audible or visible alarms. Most built-in digital signals in NI's DAQ devices cannot produce enough voltage to drive a relay (an electrically activated switch for large voltages or currents). To control relays, extra circuitry (to be discussed later) is needed between many DAQ devices' digital I/O lines and the relay. If you intend to use relays, you may build your own relay interface (discussed in Chapter 8). However, it is more time-efficient to buy a relay interface—NI has many DAQ products with solid-state or mechanical relays built in.

The DAQ device seller usually forgets to mention that with many DAQ devices, particularly those that drive relays, your application might require that the "power on" state (state when the power is first turned on) of the digital output is either high or low. Not all DAQ devices give you an option of choosing this power on state.

2.2.4 Counter Signals

Inside certain DAQ devices are special circuits called counters, or sometimes counter/timers—we will simply call them counters in this book. These chips handle only digital signals at the TTL level, and they are especially designed to count or time such signals by counting every high-to-low (or low-to-high) transition of the digital signal. You could have a counter counting transitions of this signal for one second (see Figure 2–16).

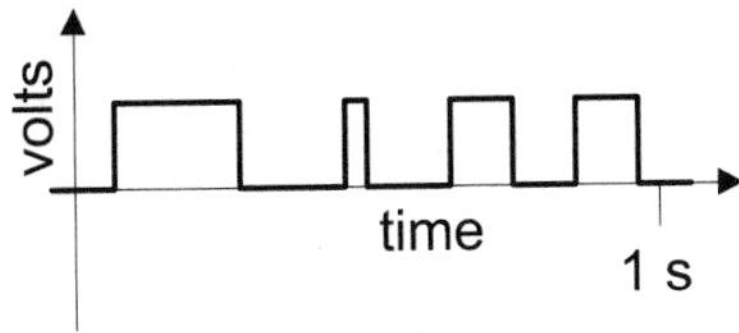

Figure 2–16
A digital signal with four pulses.

The counter could have been counting rising *or* falling transitions; either way, it would have counted four. Inside the counter is an N-bit digital integer, or *register*, which keeps track of the count. If your counters have N bits, and you think your count will exceed 2^N, you can *chain* M counters together so you effectively have one counter with NM bits that can count to $2^{(NM)}$. Some DAQ devices require that you add a physical wire to chain counters, while others do not.

The *timer* part of the counter (remember, it's also called a counter/timer) expects a regularly changing signal, and it can be used to tell you the frequency of the digital signal. This is based upon how much time elapses between consecutive low-to-high transitions (or vice versa) of a digital signal. You could build a VI that reads the signal shown in Figure 2–16 for one second, and it would think it sees a 4 Hz signal because there are four transitions within one second.

2.2.5 Other Signals: Bus-Based and Serial

The term *bus* in this book means "a group of digital signals," not a school bus. Technically, serial signals like RS-232 are also bus-based, but the word *serial* implies that only one wire is being used to transmit the data in a particular direction.

Busses that ship with many computers as of this writing are the serial port, the parallel port, and the USB port. The serial port, shown in Figure 2–17, is the very common 9-pin male connector on the back of most computers.

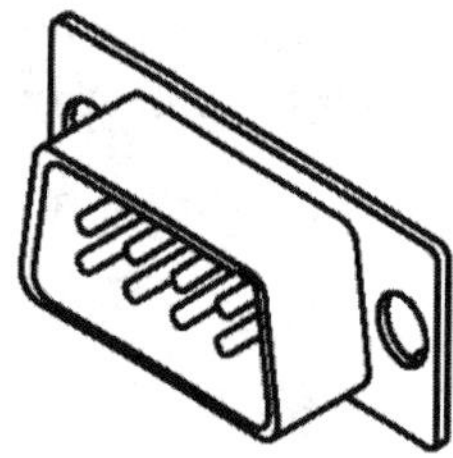

Figure 2–17
A 9-pin serial port connection can be found on most computers.

Older equipment usually has 25-pin connectors as their serial port. The 25-pin parallel port on a computer looks similar to Figure 2–17, but is the female version. RS-232 and RS-485 are both forms of serial communication, although the port commonly built into computers handles RS-232 only. A simple converter can convert RS-232 signals to RS-485 signals, and vice versa.

For laboratory equipment and scientific instruments, a common bus is the 25-pin GPIB bus, meaning General Purpose Interface Bus. GPIB is also called IEEE-488 or HPIB.

NI's term DAQ does not typically apply to busses, but rather to simple analog and digital signals. In my opinion, bus-based signals, including serial signals, really are DAQ signals. They just have a special connector, and a special handshaking interface (specific timing, voltage, and signal patterns) controlled by hardware as well as software.

2.3 Signal Conditioning

Most ADC chips buried within your DAQ device expect to read a voltage in the range of 0 to 5 volts. But transducers produce many different types of electrical signals, hence this section. We will discuss (1) noise, (2) low-level and high-level signals, (3) isolation, and (4) common-mode voltages.

NI has a number of excellent application notes available for you to download from their Web site. After you read this section, which is just an introduction of the more important signal conditioning concepts, you may want to download the following two application notes from *www.ni.com* if you are interested.

Application Note 025: Field Wiring and Noise Considerations for Analog Signals

Application Note 048: Signal Conditioning Fundamentals for PC-Based Data Acquisition Systems

2.3.1 Noise

Many DAQ systems need to take noise into consideration. By "noise," I don't mean irritating sounds, like rap music; I mean unwanted electrical signals that your DAQ device might mistake for a real signal. If you've ever heard static on a radio station, on your phone line or cell phone, or seen it on a television screen, you've experienced electrical noise. Observe a pure sine wave next to a noisy sine wave, in Figure 2–18.

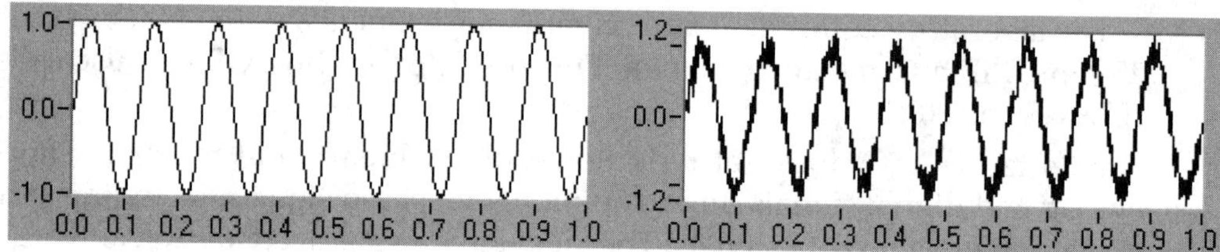

Figure 2–18
A pure sine wave on the left and a noisy sine wave on the right.

Any time you have an electrical loop of wire, as all DAQ systems have in some form, it picks up at least some electrical noise because it's acting like a radio antenna. The longer the wire, or more accurately, the bigger the area in the wire loop, the more noise it will pick up. Here are a few steps you can take to reduce the effects of noise in an analog signal:

1. Minimize the length of wire between your transducer and the ADC; once digitized, a signal is much less susceptible to noise.
2. Use *twisted pair* wires (thus minimizing the area of the wire loop). Better yet, use shielded wires and/or coaxial cables.
3. If possible, use a larger voltage signal on the length of wire between your transducer and the ADC.
4. If desperate, use a current signal (usually 4–20 mA or 0–20 mA). Current signals can run many hundreds of feet without collecting noticeable noise!
5. Use differential signals rather than single-ended.
6. Finally, you can filter the signal, although the previous items are better because they help prevent the noise from getting into your signal in the first place. Hardware filtering (NI sells this) is inherently more effective than software filtering.

To see how much noise is on a voltage-measuring DAQ system, simply set up the measurement software (you'll do this in a later chapter) as if you were measuring a voltage, then measure zero volts. To get zero volts, simply remove the voltage-sensing wires from your transducer, then temporarily connect them to one another. Once you're supposed to be measuring nothing, and you measure something, then you know that "something" is pure noise.

If you cannot get rid of the noise on your signal by any of the above means, and you're measuring a slowly changing signal, you may be able to filter it with a low-pass filter, which does not let high-frequency signals through. Many DAQ devices have these filters built in. This doesn't do much good if you're measuring a high frequency signal in the same frequency range as the noise, of course—you would be likely to filter out your real signal along with the noise.

Your last resort for reducing noise is to handle it in software. You can use a sophisticated software filter, or simply average a number of samples together to reduce the effective noise. If you average many samples to produce one sample, your noise is reduced in proportion to the square root of the inverse of the number of samples used in such an average—see Chapter 6, Section 6.5 for details.

2.3.2 Low-Level and High-Level Signals

Low and *high* in this section referr to a signal's voltage levels. For example, a high-level signal is likely powering your clothes dryer, if it's an electric one.

For low-level signals, noise will be more apparent relative to the size of your signal, so you may need to somehow increase the size of your signal and/or use the other noise-reducing tricks mentioned above. At some point before a low-level signal hits the ADC, it will need to be amplified so that it's in the amplifier's voltage-measuring range, usually around 0 to 5 volts. Most DAQ devices have a built-in amplifier designed to do this.

For high-level signals, some DAQ devices have an amplifier with a fractional gain, and many are especially geared towards high-level signals. If gain accuracy isn't that important, you could build a resistor-divider as in Figure 2–19.

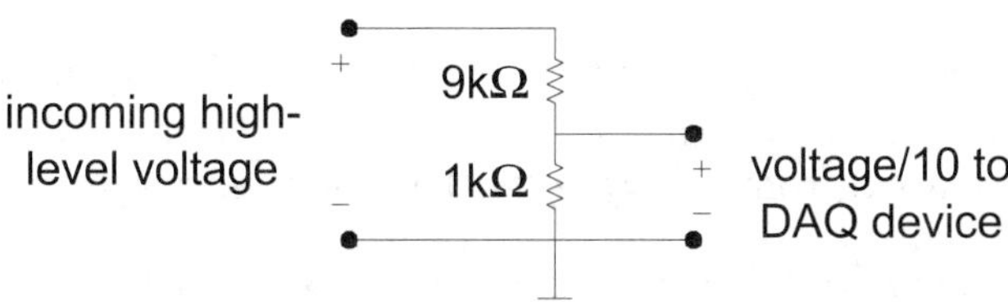

Figure 2–19
A voltage divider.

I just randomly picked these resistor values, but you would need to pick them to keep the voltage in the optimal range, depending on your input signal and on your DAQ device. Most resistors have 5% accuracy, although they can be found in 1% and 0.1% accuracies. Resistors' thermal characteristics are sometimes more important then their "percentage" accuracy, so pay attention to this when designing your system.

If there's a possibility that the incoming signal might exceed the DAQ device's safety range, thus frying it, you may want to consider isolation.

2.3.3 Isolation

Isolation means that there is no direct electrical connection between two parts of a circuit. For example, an optically isolated electrical connection, sometimes called an *optocoupler,* has three basic sections. One section converts the voltage to light, the second section consists of light shining through an electrically insulating piece of transparent material (no electricity flow), and finally, the third section converts the light back to a voltage. The two electrical sections are not connected electrically in any way, so if a dangerously high voltage occurs in section one, which may be vaporized, it won't *electrically* hurt section three. Isolation will cure ground loop problems as well as protect against excessive voltages.

Take care when soldering optocouplers; they're often very sensitive to heat.

2.3.4 Common-Mode Voltages

I cannot describe common-mode voltages as well as NI's Application Note 048, *Signal Conditioning Fundamentals for PC-Based Data Acquisition Systems,* which you can download from *www.ni.com*. The following is an excerpt.

> When you connect your sensor or equipment ground to your DAQ system, you will see any potential difference in the grounds on both inputs to your DAQ system. This voltage is referred to as *common-mode voltage*. If you are using a single-ended measurement system(*as shown in Figure 2–20*) the measured voltage includes the voltage from the desired signal, V_S, as well as this common-mode voltage from the additional ground currents in the system, V_G.

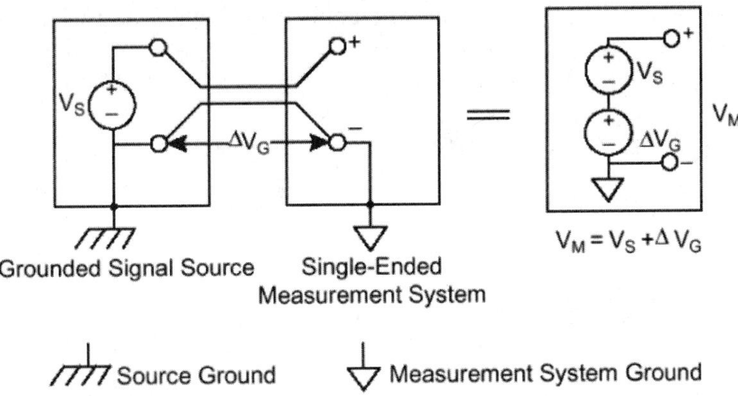

Figure 2–20
Common-mode voltage is illustrated.

All DAQ devices have limits in terms of voltages that can appear on their inputs. If you exceed these limits, you could destroy the DAQ device. You must include the common-mode voltage when considering these limits! For example, if you're measuring a signal that ranges from 0 to 5 volts, and your common-mode voltage is 9 volts, your signal will range from 9 to 14 volts from your DAQ device's point of view. This will exceed the limits of some DAQ devices.

If your DAQ device can measure in differential mode, this can reject a certain amount of common-mode voltage.

2.4 Keeping Time with DMA and FIFO

Whether you are doing analog input or output, or digital I/O, there are times you'll need to sample data or generate data at a very consistent rate. For instance, if you're acquiring an analog input waveform for frequency analysis, you need your samples to be evenly spaced in time. DMA (Direct Memory Access) and FIFO (first-in, first-out) buffers are two hardware mechanisms that make this possible. Both mechanisms buffer DAQ data into high-speed memory before the computer processes it so that no real-time data is lost. FIFO's memory resides on the DAQ device, but DMA uses your computer's normal memory.

For example, suppose we have a perfectly timed system, and we sample a 2 Hz sine wave at 100 Hz for one second. Figure 2–21 shows the perfect 2 Hz sine wave, superimposed on the locations of the 100 Hz samples.

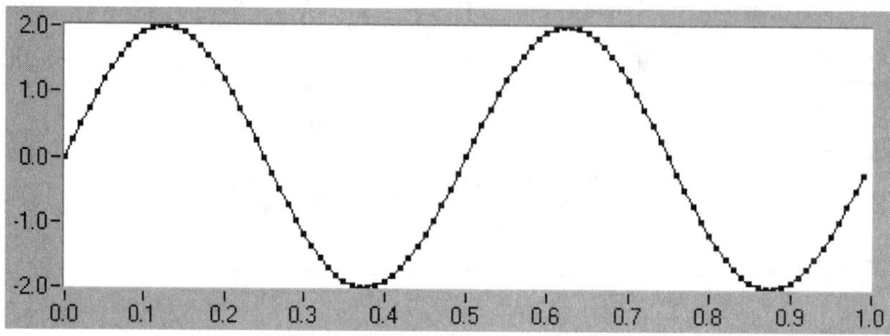

Figure 2–21
A sine wave is adequately sampled.

Next, suppose we are using software timing to determine when each of these 100 samples is taken, with no DMA or FIFO. Also, suppose our software is based upon a typical operating system that is not real-time, and the processor gets busy with something else during the time interval between 0.3 seconds and 0.5 seconds in the graph shown in Figure 2–21. The section of the sine wave shown in Figure 2–22 would not be sampled.

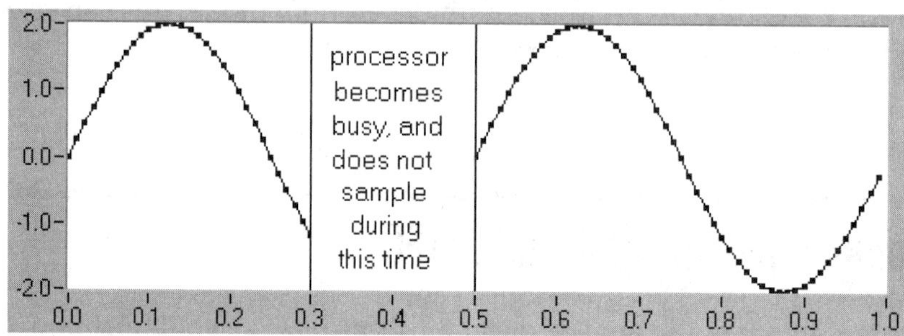

Figure 2–22
The sine wave from Figure 2–21 loses several samples.

Your software would not be aware that these data points were missing, so you and your computer would think that the signal looks like the signal in Figure 2–23.

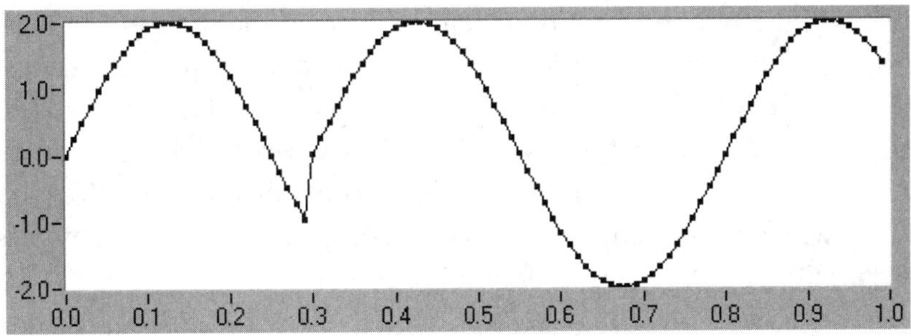

Figure 2–23
The computer would interpret the sine wave from Figure 2–22 as such, unaware that the timing has been corrupted.

DMA and FIFO can compensate for the microprocessor's absence for a length of time proportional to their buffers' lengths; typically, DMA has a much longer buffer length. Figure 2–24 shows how they relate to the DAQ device and the computer

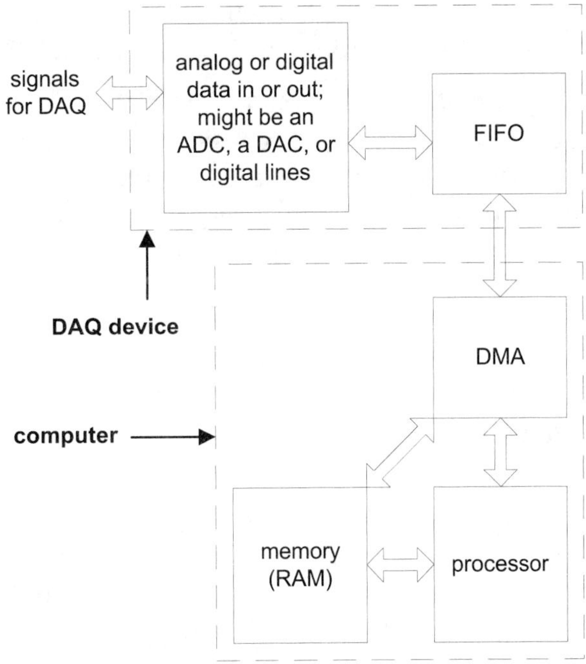

Figure 2–24
DMA is illustrated in a DAQ system.

To describe where they got their names, look at the diagram in Figure 2–24 and imagine data coming into the computer and being stacked up in the FIFO before the DMA can read it. The *first* piece of data *in* needs to be the *first* piece of data *out*, otherwise data will be sent to the computer out of order. FIFO means first-in, first–out. Next, notice how the DMA hardware has a two-way arrow going *directly* to the computer's memory, bypassing the processor? DMA means Direct Memory Access.

If you have both DMA and FIFO, you may typically ignore the FIFO's buffer size, because any data bottlenecks will usually occur in the larger DMA buffer. DMA and FIFO always involve buffers of finite length. The FIFO buffer is a fixed size, and is dependent on your DAQ device. However, you may control the size of the DMA buffer, as is described in Chapter 3 when *continuous* analog input is discussed. A good rule of thumb is to allocate at least 10 seconds worth of DMA buffer when reading or writing waveforms continuously; this allows your computer's processor to completely ignore your DAQ hardware and software for up to 10 seconds.

On PCs, DMA is built-in, but on other platforms, such as the Macintosh, DMA is not; you may be forced to buy a separate DMA board if you need it.

DAQ devices are more likely to have DMA on their analog input than on their analog output, so check for DMA on the analog output if you need to generate high-speed, accurate analog waveforms. FIFO is often not enough. NI's arbitrary waveform generator and function generator products are designed for this type of analog output.

2.5 NI's DAQ Products

This section provides a review of NI's DAQ products, but you may want to get your hands on NI's physical product catalog (recommended), or browse their Web site to see the latest DAQ products. Although I have high-speed Internet access, I find that the physical catalog is currently the best way to get an overview of NI's DAQ products.

Networked and distributed systems will be covered in Chapter 10.

2.5.1 NI's DAQ Product Line Overview

NI's DAQ section of its product catalog is beginning to make the catalog quite thick. DAQ devices vary widely in their function and form; some go inside the computer, some do not, and some are used at great distances from the computer. There are many ways to categorize the DAQ products. Possibly the most basic way would be to divide them into internal and external devices. Internal devices are used inside the computer, like those shown in Figure 2–25.

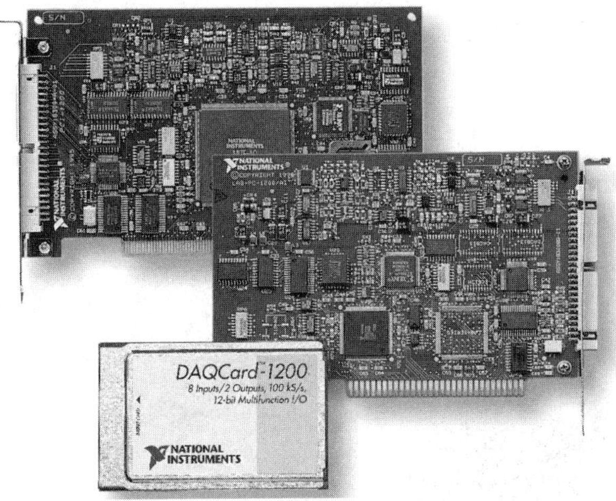

Figure 2–25
Examples of NI's internal DAQ equipment.

External devices (modules, chassis, etc.) are used outside the computer; some examples are shown in Figure 2–26.

Many of NI's DAQ devices use a cable to connect to a connector block, or terminal block, with screw terminals. Pictured in Figure 2–27 are examples of connector blocks.

Figure 2–28 shows examples of cables, shielded and not shielded, that connect a variety of DAQ devices to connector blocks.

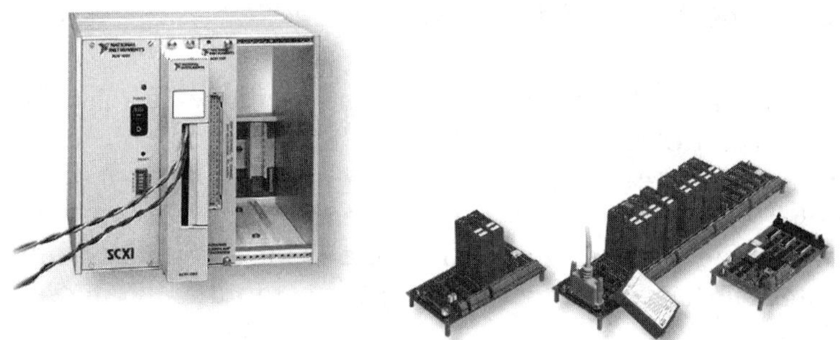

Figure 2–26
Examples of NI's external DAQ equipment.

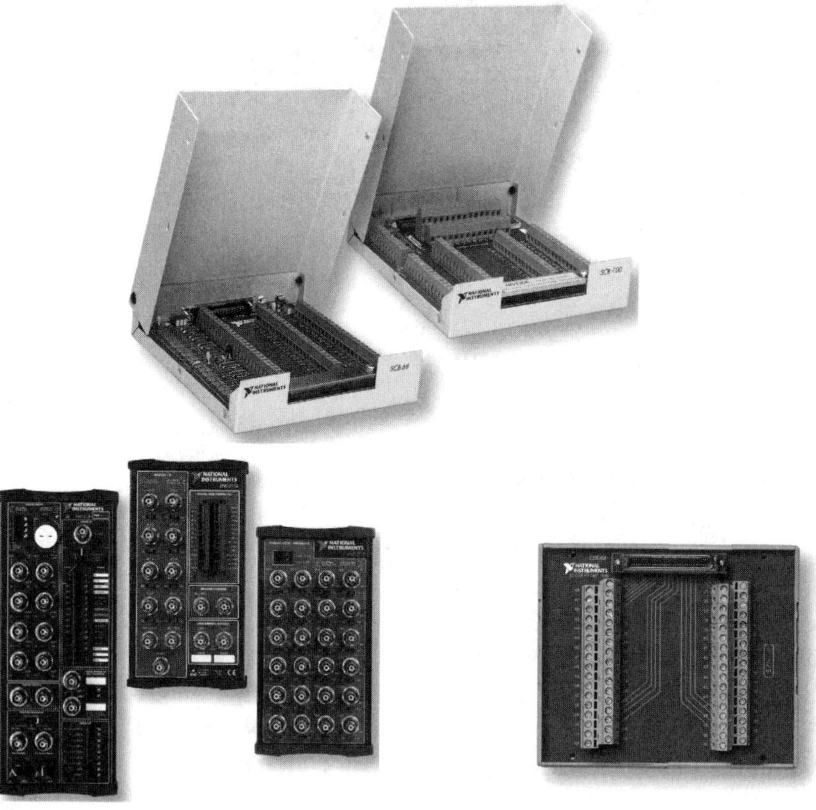

Figure 2–27
Examples of NI's connector blocks.

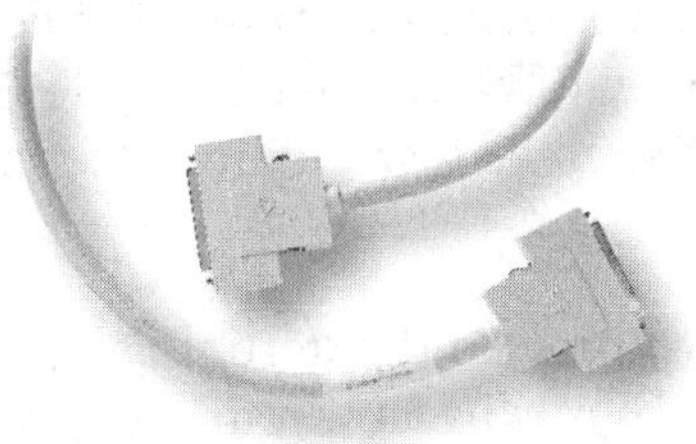

Figure 2–28
Examples of NI's cables.

I could make up any number of schemes to categorize the DAQ devices, but let's stay consistent with the NI product catalog's scheme.

- **High-Performance (E Series).** E Series devices are high performance boards, many of which are multifunction (analog, digital, and counter functionality all on one board). Most of these devices are internal, but some are external.
- **Real-Time.** Real-time devices, currently all internal, have their own processor in the device and can achieve true real-time, deterministic performance. The devices also have basic DAQ functionality, such as analog, digital, and counter. See Chapter 9 for further details on real-time issues.
- **Dynamic Signal Acquisition.** Dynamic signal acquisition devices are designed for rapidly changing signals; the signals are sampled at a high rate and with high resolution. Some of these devices have anti-aliasing filters. They are ideal for audio, speech, sonar, or vibration applications. DMA is always used.
- **Analog Output.** NI has two categories of analog output devices—static single-point updates and high-speed waveform generation. The high-speed devices utilize DMA to produce accurately timed waveforms with no glitches or hiccups.
- **Timing I/O.** The timing I/O products have counters built in. They can be used for event counting, pulse and frequency measurements, pulse and pulse-train generation, buffered measurements, and quadrature encoder measurements.

- **Low-Cost.** Some DAQ devices are called "low cost," possibly because they are less expensive than the majority of NI's other DAQ devices. They also have reduced functionality, such as fewer gain selections, fewer channels, and so on.
- **Portable Measurements.** PCMCIA products and external DAQ devices (with RS-485, USB, IEEE 1394 [FireWire] and parallel port interfaces) fall into this category.

2.5.2 Low-Cost Signal Conditioning—SCC and 5B

I really like SCC devices, as they can offer low-cost signal conditioning. An SCC system consists of shielded carriers, signal conditioning modules, a DAQ device, cables, and panelettes. A panelette, in SCC terms, is a small panel for the SCC shielded carrier that may have special connectors. Shown in Figure 2–29 are (1) an assembled SCC system, (2) a lone SCC module, and (3) a partially disassembled SCC shielded carrier with three SCC modules on its left.

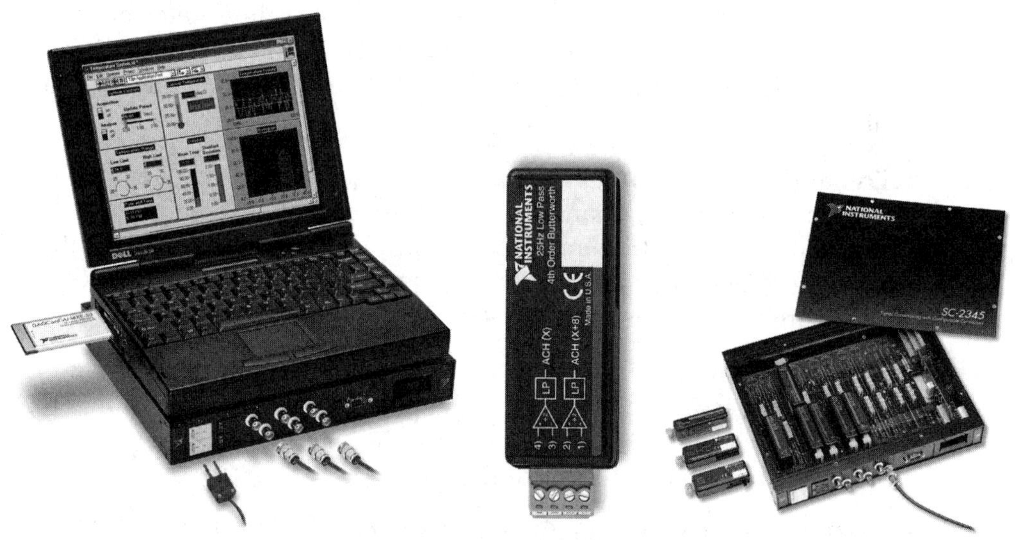

Figure 2–29
NI's portable SCC DAQ equipment.

The SCC modules are currently designed for thermocouples, strain gauges, isolation of analog or digital input and digital output, current measurement, and voltage attenuation.

The 5B Series devices offer a different look, shown in Figure 2–30.

Figure 2–30
5B Series devices, on a rack and in a LapDAQ enclosure.

2.5.3 Relay Control Devices

NI makes a wide range of relay control devices; some are solid state, some are mechanical. Read Section 2.2.3, particularly the note at the end, if you're considering relays. NI's lowest cost relay control devices use mechanical relays, the ER-8 and ER-16, shown in Figure 2–31.

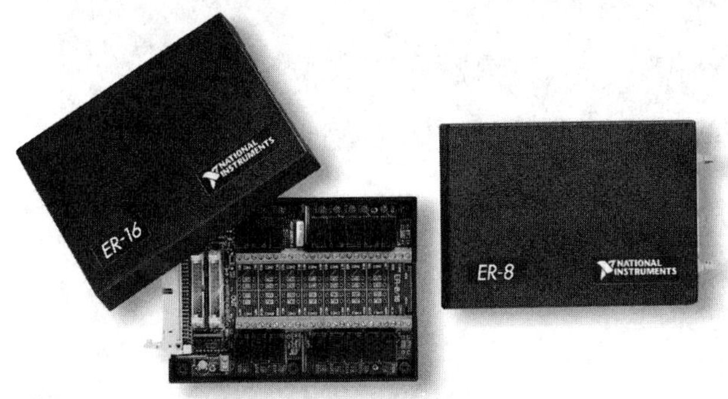

Figure 2–31
NI's ER-8 and ER-16 relay control devices.

2.5.4 PXI and VXI

A normal PC sometimes cannot tolerate rigorous industrial environments. One solution is to buy an industrial computer chassis and use industrial DAQ components, such as SCXI. Another solution is to use PXI, which combines PC technology with industrial packaging, integrated triggering, and LabVIEW. CompactPCI is the core architecture used by PXI to offer superior mechanical integrity and more slots than a normal PC. Figure 2–32 shows some PXI racks and PXI modules.

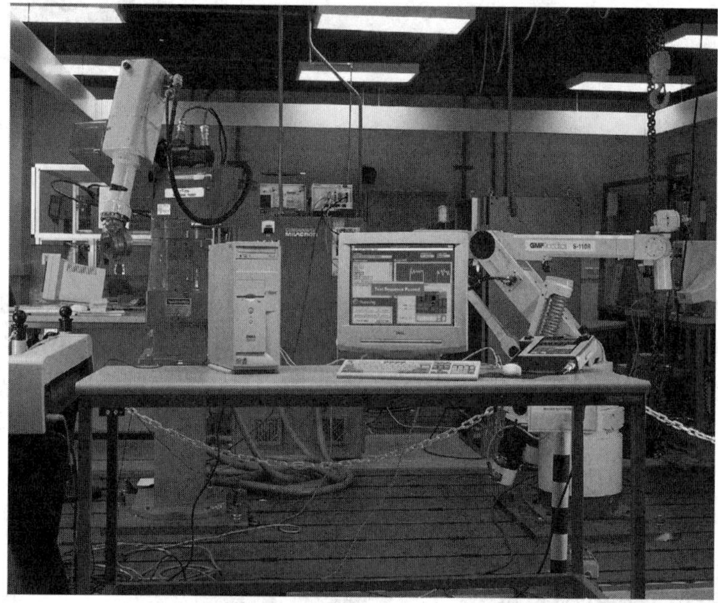

Figure 2–32
PXI equipment.

VXI is the highest-end platform for industrial DAQ applications. However, it is usually the most expensive solution possible—unless you build your hardware out of solid gold.

2.5.5 Systems with Many Signals

For systems with many signals, it may be more economical to use SCXI or PXI, or perhaps the AMUX-64T analog input multiplexer. When you need to synchronize two or more DAQ boards inside one PC or PXI chassis, many of NI's internal DAQ devices may be connected with the RTSI bus cable, which allows such synchronization.

Basic DAQ Programming Using LabVIEW

3

This section assumes you have a real DAQ device with analog input and either analog output or digital I/O. If not, see *www.LCtechnology.com/noai.htm* for suggestions. If you have skipped the preceding chapters and you just want to watch LabVIEW work with your DAQ device right away, continue through Section 3.1.5. Here's the background information you need:

1. A VI (pronounced vee-eye) is a LabVIEW program, consisting of two windows—a front panel and a block diagram. By default, the front panel is gray, and the block diagram is white.
2. Through Section 3.1.5, you will need to be in just one mode in LabVIEW. Once LabVIEW is up and one of its windows is active, hit your <Tab> key until the cursor looks like this:

LabVIEW provides many "easy-to-use" VIs for performing DAQ. These are all based upon *NI-DAQ*, which is the underlying free, downloadable software driver for NI's vast array of DAQ hardware. All of NI's programming environments (not just LabVIEW) use NI-DAQ. As we'll see later in Chapter 11, NI-DAQ can also be used with non-NI software programming environments. If you already know how to program in another language, and have a rather simple DAQ application in mind, you might not want to use LabVIEW. But if you're new to programming, or if you want to build robust DAQ applications quickly, use LabVIEW. In my opinion, LabVIEW provides the fastest and most robust way to build most DAQ applications with NI's DAQ products.

Before you can do anything with DAQ from LabVIEW, or any language, you must assign a number to each DAQ device so LabVIEW can know what it's communicating with. That is the primary purpose of the MAX (Measurement & Automation Explorer), to be discussed first.

Here's a breakdown of the contents of this chapter:

1. Assign Device Numbers to DAQ Devices
2. LabVIEW DAQ Functions: An Overview
3. Using the Basic LabVIEW DAQ Functions to Display Data
4. LabVIEW File I/O with DAQ

This section will be written as if you have a DAQ device with at least two analog inputs and at least two analog (or digital) outputs. If so, you should follow along in this chapter while actually building the software in LabVIEW and using your DAQ device. If not, just pretend ☺. I highly recommend that you work through this chapter with a *real* DAQ device and a *real* LabVIEW on a *real* computer. Why? You will be shown herein many of the tricky tricks that you'll need to know to do your basic LabVIEW/DAQ work; these tricks won't sink in so well unless you actually use them.

Word of warning: Whenever you use real equipment, rather than this idealistic book, you will almost certainly run into problems of some sort. That's where NI's tech support department comes in handy! You might buy a ribbon cable (see Figure 3–1) from NI for your DAQ device for $30 (before tax and shipping!) that clearly contains about $3 worth of material.

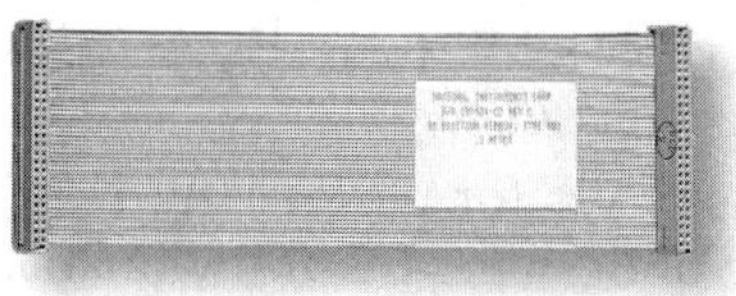

Figure 3–1
Three dollars worth of material worth $30? (Hmmm...)

Why is this? NI has high quality products with excellent support. That support somehow manages to answer even very difficult questions without a tech support fee! When you buy NI products, you are also buying their services as well, to a degree.

3.1 Assign Device Numbers to DAQ Devices

Before we can have LabVIEW using any DAQ device, we must assign a number to each DAQ device so LabVIEW can identify it. Figure 3–2 shows an intentionally simple VI that reads two analog input voltages, but it won't work until you run MAX with two NI devices, both having analog input ability—then you must assign the devices *device numbers* 1 and 2. Don't bother building this if you're following along with a DAQ device, as it requires two devices as well as information we have not yet covered. One value comes from device 1, and the other comes from device 2.

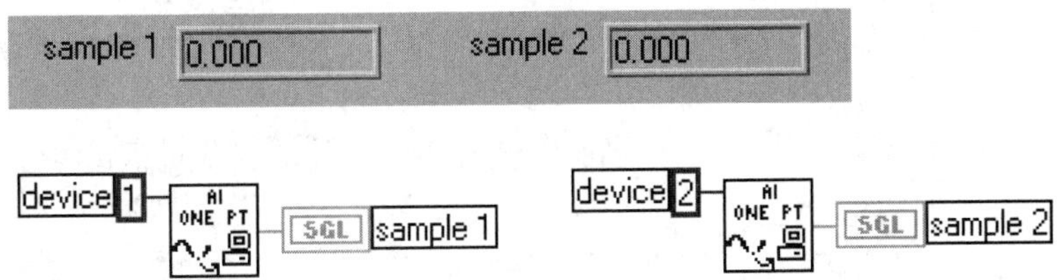

Figure 3–2
An unrealistically simple DAQ example, illustrating device numbers (won't work until you run MAX).

As NI-DAQ evolves, the utility that assigns device numbers to DAQ devices changes rapidly. I hope it's still called MAX by the time you're reading this book, but if not, the basic concepts will still apply to whatever new software assigns your device numbers. MAX not only assigns device numbers to DAQ devices—it also configures other important parameters for DAQ devices. Figure 3–3 shows MAX's current interface, which changed even as I was writing this book.

As mentioned earlier, MAX may be changed by the time you're reading this, so keep in mind that the screens you are using could differ slightly—or drastically—from the ones shown in this book. And MAX is different for platforms other than the PC.

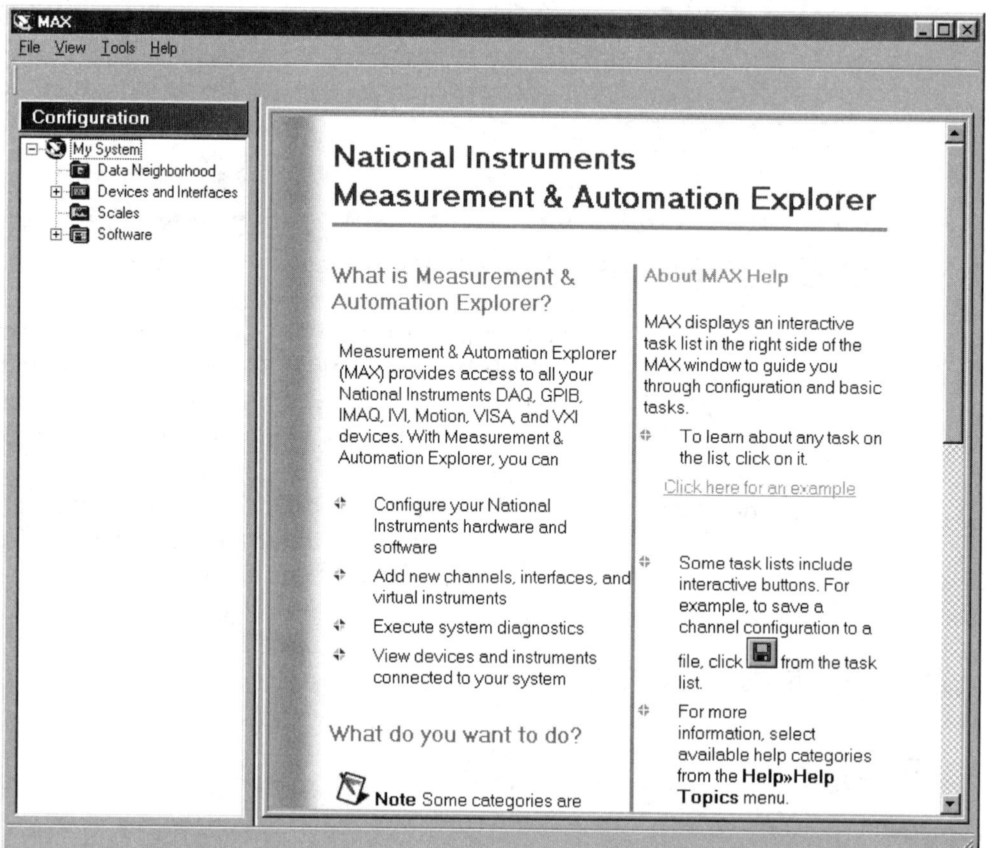

Figure 3–3
The MAX interface.

Use MAX to figure out how to install your DAQ device. MAX currently does a great job of walking you through this process, step by step, and the process varies for different types of DAQ devices. In the **Help** menu of my MAX, the relevant help is in **Help»Help Topics»NI DAQ»Contents»DAQ Devices»Adding DAQ Devices**. Follow the steps therein, cross your fingers, and you're likely to get your device configured and a device number assigned.

MAX works with DAQ devices and much of the recent NI hardware, including General-Purpose Interface Bus (GPIB) and motion control devices. It can do much more than simply assign a device number to each DAQ device. MAX can detect devices, especially if the devices are "plug and play" (in a cross-platform sense, this means: plug it in, and theoretically it works without jumping through hoops to install it—also known as "plug and pray"). Through its test panels, you can verify the operation of your hardware and make simple measurements. You can perform simple analog or digital input and/or output. You can operate counters. Like MAX's predecessors, it allows you to set voltage ranges on the analog I/O and to electrically set analog input to differential or single-ended (RSE or NRSE). If your device allows mixing differential and single-ended analog input signals, MAX can configure them. It also allows you to assign names to incoming signals, so you can use these names, rather than numbers, in your LabVIEW program. You will probably find that using names for the signals will simplify your projects. You can make up meaningful names that you can easily remember—this is a great feature from NI, and it very easy to use. The list of MAX's features goes on and on; it is a truly powerful and useful utility (not to mention necessary for DAQ with LabVIEW).

3.1.1 Preparation for DAQ

I will be using an AT-MIO-16DE-10, which looks remotely like the middle board in Figure 2–25, while stepping through the basic DAQ operations. I chose this device because it has analog input and output, digital I/O, and counters. This allows me to illustrate all of the major categories of the so-called "Easy I/O" VIs. If your DAQ device doesn't have some of these options, just use it where you can. If your device has no analog output, as many don't, you can use its digital output instead in this chapter.

If you have an older DAQ device, they are not so clever about installation—they often require a computer without lots of gadgets already attached, as they can only work with certain computer resources, such as interrupts. In general, newer DAQ devices approximate plug and play much better.

Step 1: *Install LabVIEW* (if it is not already installed). Remember your installation folder for later use. You may be using a later version of LabVIEW, but the basic information in this book will remain useful for years to come.

Step 2: *Install NI-DAQ* (if it is not already installed). Unlike LabVIEW, you can download this from NI's Web site, *www.ni.com,* for free.

Step 3: *Install a DAQ Device* (if it is not already installed). Follow the instructions in MAX's **Help** menu to install your device. If your device is not plug and play, you may be forced to jump through extra hoops to make it work. For my E Series device, it was truly a matter of plug and play; MAX automatically gave me a device number of 1. This number is your primary link between your DAQ device and LabVIEW. I can see the device number by expanding the **Devices and Interfaces** icon in the left screen of MAX (Figure 3–3). Even more device-specific information becomes available when you double-click the device's name, as shown in Chapter 1, Figure 1–31.

See the left pane in Figure 1–31? You may have any number of assorted NI devices here, such as the system shown in Figure 3–4 that I was actually working on recently, with two motion control devices (able to control four stepper motors apiece) and two DAQ devices.

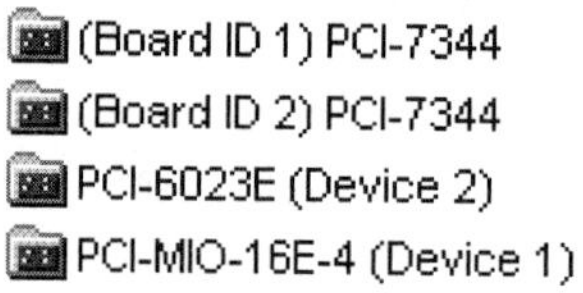

(Board ID 1) PCI-7344
(Board ID 2) PCI-7344
PCI-6023E (Device 2)
PCI-MIO-16E-4 (Device 1)

Figure 3–4
DAQ devices as MAX sees them.

Step 4: *Configure your DAQ Device.* Since I am connecting an analog input to an analog output on the *same* device, I know their grounds are internally connected. To keep the wiring simple, I choose a single-ended connection. Thus, our **DAQ Connections Table** (Chapter 2, Figure 2–10) indicates that I should use NRSE; in fact, it warns against using RSE for grounded connections. However, I know we will be seeing negligible ground-loop losses because all of our grounds are connected on the same DAQ device. Therefore, I also know I can safely break the rules of that **DAQ Connections Table** and go with RSE. This will save us one wire (AISENSE to AIGND).

If I click the **Properties** button Properties... with my DAQ device showing, as seen in Chapter 1, Figure 1–31, I see the window shown in Figure 3–5. Unless your DAQ device is an AT-MIO-16DE-10, you will likely have different tabs and different resources shown.

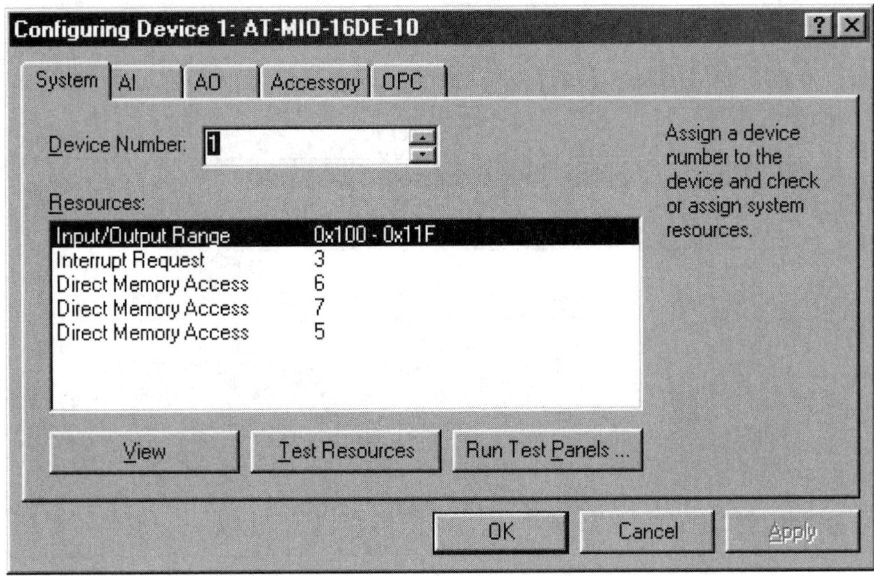

Figure 3–5
Device properties in MAX.

This shows me again that my device number is 1, which I could change here if I wanted. This screen also shows me which computer resources my device is using, such as the interrupt number, the I/O address, and related information. If you don't see some of these resources, your DAQ device may not be able to perform some of the functions mentioned in this chapter, like sampling multiple channels (**Interrupt Request** may be needed) or continuous analog input/output (**Direct Memory Access** may be needed). I'm happy to see these resources have been granted—they usually are, especially with more modern equipment. Ideally, you shouldn't be concerned about these resources, but with some of the older devices, you must set them up yourself. These annoying older DAQ devices often have little physical switches and/or jumpers that may require some tweaking ☹.

Since we will be using analog input and output, we must make sure the configuration is set up right by using the **AI** and **AO** tabs. First, click your **AI** tab, and you should see something like Figure 3–6.

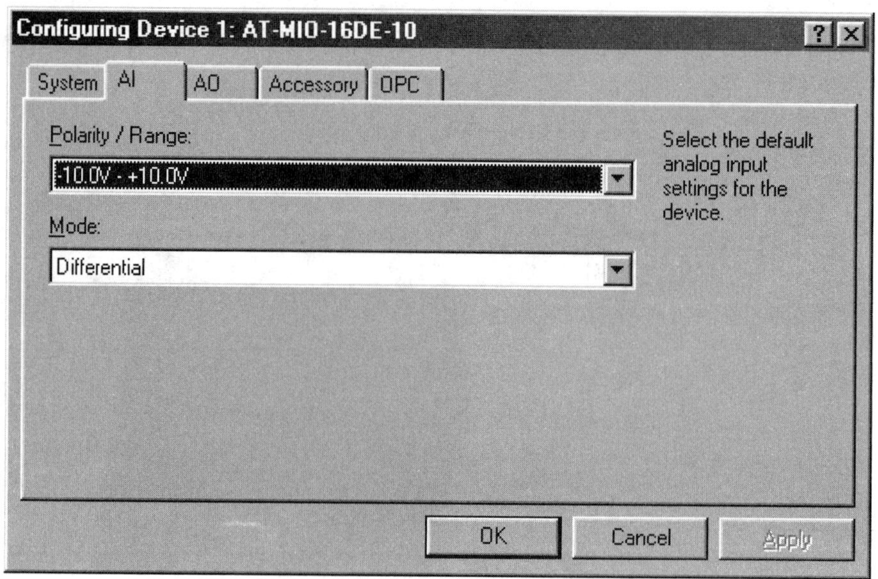

Figure 3–6
Analog input properties can be changed in MAX.

Select +/- 10V (or +/- 5V, for many devices) for **Polarity / Range**, and "Referenced Single Ended" instead of "Differential" for **Mode**, in this **AI** (analog input) tab. Since you have this screen open, you may as well see the other options under each of the pull-down menus, but be sure you finish with the settings as described in this paragraph.

If you have no **AO** (analog output) tab as shown in Figure 3–7, skip the next paragraph, as you can use digital lines instead.

Make sure you have **Bipolar** selected, not because you're manic-depressive, but since we're reading both positive and negative values, consistent with our previous analog input configuration. If you have made any changes under the **AO** or **AI** tab (or any tab), be sure to click the **Apply** button or the **OK** button instead of the **Cancel** button, or else your changes will not take effect. This rule applies to all Windows-compliant dialog boxes. **Apply** and **OK** are identical, but **OK** closes the window afterwards.

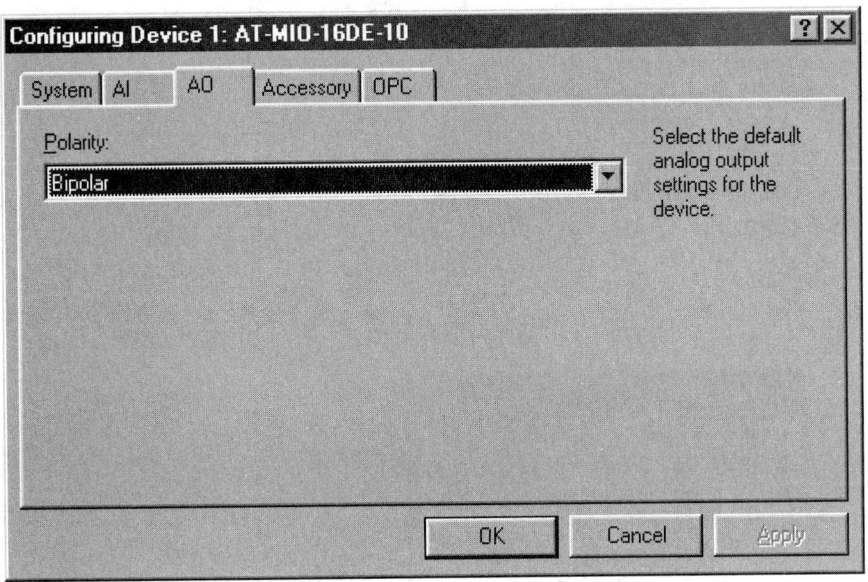

Figure 3–7
Analog output properties can be changed in MAX.

Step 5 (optional): *Assign names to your signals.* In the left **Configuration** pane of MAX, as shown in Figure 3–3, expand the **Data Neighborhood** icon, and see whether your DAQ device already has its channels named. If so, you may not want to rename them if it will interfere with ongoing DAQ work, so skip this step, but at least peruse the channels' properties without changing them.

If you are allowed to create new channels, pop up on the **Data Neighborhood** icon, select the **Create New…** menu item, select the **Virtual Channel** icon, then click the **Finish** button. You will now be guided though a series of panels in which you may describe your signal in detail. As you go through these panels, accept all the defaults, but set the following fields:

Field	***Value***
Channel Name	voltage 0
Channel Description	raw voltage
Units	V
Which analog input mode will be used?	Referenced Single Ended

One of my favorite features with MAX is the ability to easily duplicate channels while automatically numbering them. If you have created your "voltage 0" channel as described, you can now see a **voltage 0** icon when you expand your **Data Neighborhood** icon. Pop up on that **voltage 0** icon, select the **Duplicate...** menu item, and see how easy it is to create a **voltage 1** channel. Go ahead and create a **voltage 2** channel by popping up on the **voltage 1** icon. In MAX, your **Configuration** pane should have your newly created channels, looking something like Figure 3–8.

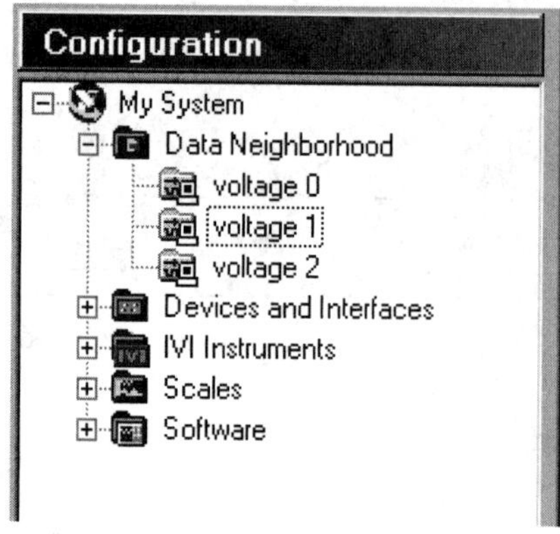

Figure 3–8
MAX allows you to name your own channels.

Figure 3–9
Don't fry your DAQ device.

We will not be using channel names in this book, in order to keep things simple and generic, but I highly recommend their use and will point out where they could be used. For example, they are not just restricted to analog input signals—they are designed for use with all types of signals.

Step 6: Test your DAQ device through MAX. DAQ devices are sensitive to electrical damage, whether from common static electricity or from wiring errors.

In this section, you will connect real wires to your real (expensive) DAQ device. To be safe, you should completely wire your terminal block and double-check it before you connect it to your DAQ device—connect the cable to your DAQ device only after you're really sure about your terminal block wiring. Whenever you wire a DAQ device, you must take extra care to wire to the right terminals. For example, if you connect two outputs together, even for a split second, you can destroy your board. Wiring +5V to ground will likely blow a fuse on your DAQ device, if it's lucky enough to have a fuse—otherwise you have a fried DAQ device (Figure 3–9).

This step verifies that the basic connection between your computer and your DAQ device is working. Usually, I use MAX only to assign a device number, to verify that I'm getting a valid reading on an analog input, and to ensure that the counters are working. But since we're covering DAQ basics here, we'll do some extensive MAX testing that will involve connecting our DAQ device's terminals in certain ways. At this point, you should refer to your DAQ device's *pinout*. My AT-MIO-16DE-10's pinout looks like Figure 3–10.

Concerning digital I/O ports, their numbers vary widely among DAQ devices. From MAX's point of view, the ports on my particular DAQ device (the AT-MIO-16DE-10) are numbered 1 through 4, though other devices' ports may start with a different number. Therefore, I am going to refer to your digital ports as the "first [digital I/O] port," "second port," "third port," and so on, throughout this chapter, which may have no relationship to the ports' actual numbers. Once you're in LabVIEW, as opposed to MAX, the ports may start with 0, regardless of what MAX thinks! Like digital I/O ports, counters are also numbered in crazy ways on various DAQ devices. For multiple counters or multiple digital I/O ports, you cannot always count on consecutive numbering (sometimes, numbers are skipped).

Signal	Pin	Pin	Signal
AIGND	1	51	PC7
AIGND	2	52	GND
ACH0	3	53	PC6
ACH8	4	54	GND
ACH1	5	55	PC5
ACH9	6	56	GND
ACH2	7	57	PC4
ACH10	8	58	GND
ACH3	9	59	PC3
ACH11	10	60	GND
ACH4	11	61	PC2
ACH12	12	62	GND
ACH5	13	63	PC1
ACH13	14	64	GND
ACH6	15	65	PC0
ACH14	16	66	GND
ACH7	17	67	PB7
ACH15	18	68	GND
AISENSE	19	69	PB6
DAC0OUT	20	70	GND
DAC1OUT	21	71	PB5
EXTREF	22	72	GND
AOGND	23	73	PB4
DGND	24	74	GND
DIO0	25	75	PB3
DIO4	26	76	GND
DIO1	27	77	PB2
DIO5	28	78	GND
DIO2	29	79	PB1
DIO6	30	80	GND
DIO3	31	81	PB0
DIO7	32	82	GND
DGND	33	83	PA7
+5 V	34	84	GND
+5 V	35	85	PA6
SCANCLK	36	86	GND
EXTSTROBE*	37	87	PA5
PFI0/TRIG1	38	88	GND
PFI1/TRIG2	39	89	PA4
PFI2/CONVERT*	40	90	GND
PFI3/GPCTR1_SOURCE	41	91	PA3
PFI4/GPCTR1_GATE	42	92	GND
GPCTR1_OUT	43	93	PA2
PFI5/UPDATE*	44	94	GND
PFI6/WFTRIG	45	95	PA1
PFI7/STARTSCAN	46	96	GND
PFI8/GPCTR0_SOURCE	47	97	PA0
PFI9/GPCTR0_GATE	48	98	GND
GPCTR0_OUT	49	99	+5 V
FREQ_OUT	50	100	GND

Figure 3–10
AT-MIO-16DE-10 pinout (your device will likely have a different pinout).

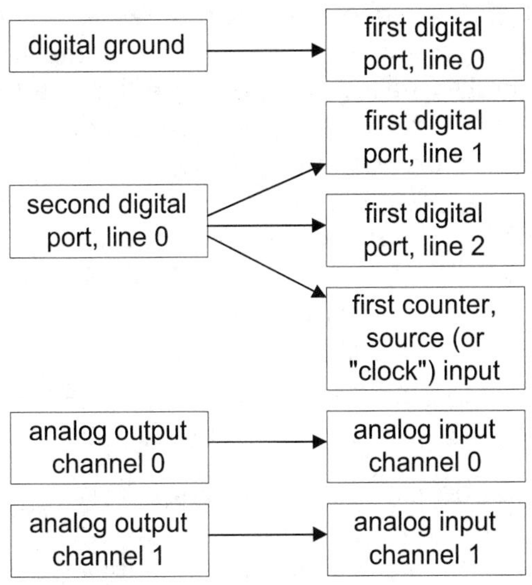

Figure 3–11a
Wiring diagram for your DAQ device *with* analog outputs.

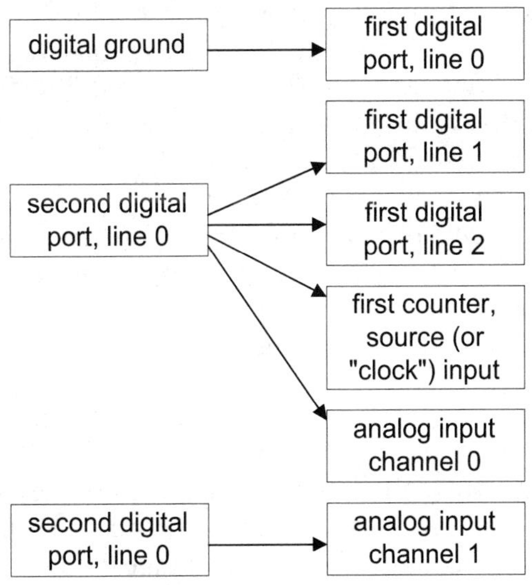

Figure 3–11b
Wiring diagram for your DAQ device *without* analog outputs but *with* digital I/O.

Before making any connections, reboot your computer so you can make sure that your DAQ device's digital I/O ports all come up in the input mode. Or, if you're already comfortable with DAQ, make sure that the first digital port is entirely set to input mode. Otherwise, if your DAQ device happens to have its first and second digital ports simultaneously set to output mode at this point, you could destroy your DAQ device.

If you have analog outputs on your DAQ device, wire your device as in Figure 3–11a.

If you don't have analog outputs, substitute digital outputs as shown in Figure 3–11b.

Go ahead and wire as shown in Figure 3–11a or 3–11b, using these instructions:

1. Connect analog input channels 0 and 1 to analog output signals 0 and 1, respectively. In the pinout of Figure 3–10, with single-ended analog input, this would mean connecting ACH0 and ACH1 to DAC0OUT and DAC1OUT. If your device has no analog output, use your *second* digital I/O port, lines 0 and 1. In the pinout, this would be PA0 and PA1 (the first digital I/O port, in this highly weird case of the AT-MIO-16DE-10, is DIO**X**—the second is PA**X**, the third PB**X**, and the fourth PC**X**).
2. Connect line 0 of your second digital I/O port to lines 1 and 2 (not line 0) of your first digital I/O port. In the pinout, this would mean connecting PA0 to DIO1 and DIO2. Next, connect line 0 of your first digital I/O port to digital ground. In the pinout, this would mean connecting either DGND terminal to DIO0.
3. Connect line 0 of your second digital I/O port (the same line previously used) to the source input of your first counter. In the pinout, this would mean connecting PA0 to PFI8/GPCTR0_SOURCE—your first counter might be numbered 0 or 1 or whatever; it depends on your DAQ device.

On my AT-MIO-16DE-10, I've connected the following terminals to produce the connections above: 3-20, 5–21, 24-25, 27-29, 29-97, and 47-97. Unless you happen to have an AT-MIO-16DE-10, your wiring will probably be completely different, as different DAQ devices have different pinouts. Figure 3–12 is a picture of the actual connections on a CB-50 connector block when I performed this experiment with a completely different DAQ device, the Lab-PC+ (fewer terminals produce a better picture).

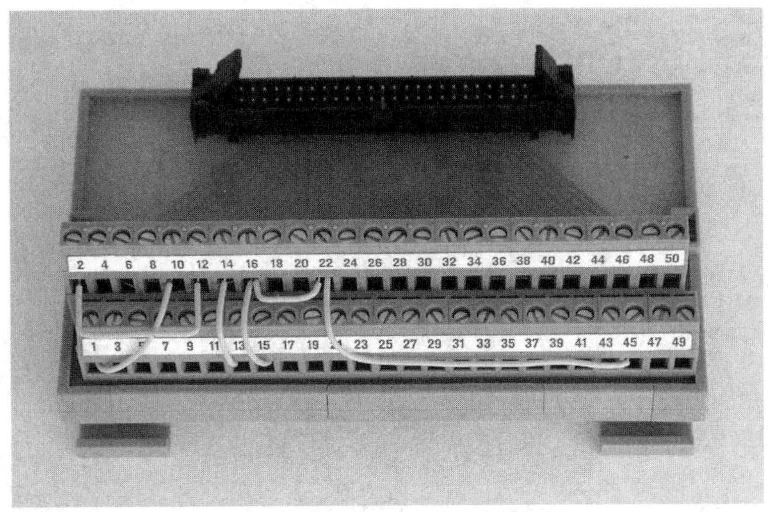

Figure 3–12
Actual working connections using the Lab-PC+, not the AT-MIO-16DE-10.

Currently, the Lab-PC+ and all DAQ devices with the number 1200 in their titles, *and* with analog output capability, would be wired according to Figure 3–12, as they have the same pinouts.

3.1.2 Analog I/O Test Panels

Now, lets look at the handy test panels, built into MAX. At this point, you should be on MAX's screen, where you have your device's **Device Number** in the area on the right, as shown in Figure 1–31. You should also see a **Test Panel** button like this, Test Panel , or a smaller version of that button, in the row of buttons across the top of your screen. Click that **Test Panel** button—a test screen will pop up. If it does not, that means your computer is not reading your DAQ device, so you'll need to figure out why. If you or anybody you know cannot figure this out, contact NI for help (see preface).

By default, the **Analog Input** tab's screen pops up for those devices with analog input (you may see a different set of tabs, depending on your device), as seen in Figure 3–13.

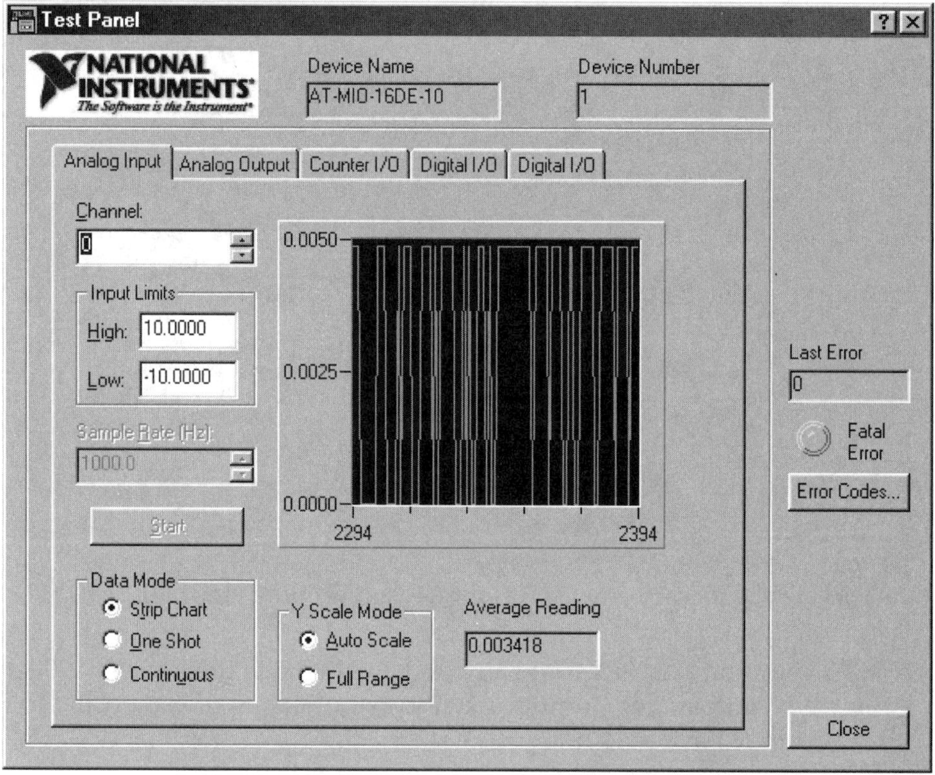

Figure 3–13
MAX allows you to run a basic test on analog input channels.

*If we had named our signals **voltage 0**, **voltage 1**, and **voltage 2**, we could be using these names in the **Channel** box shown in Figure 3–13.*

Since I have wired my analog input channel 0 to my analog output channel 0, and my analog output channel 0 happens to have 0 volts on it, my analog input channel is reading roughly 0 volts, as shown above. That bumpy signal shown above is normal whenever any DAQ device reads a constant voltage. Real signals always have a tiny amount of noise on them, and the autoscaling feature of this chart display tends to magnify such noise.

This is a perfect time to discuss the resolution of your ADC. Since the ADC on my DAQ device is 12 bits, and since my input range is 20 volts (-10V to +10V), I can calculate the "sample interval," or resolution, from one ADC value to the next. Twelve bits effectively divides your 20 volt range by 2^{12} = 4,096, and 20 / 4,096 = 0.00488. This means the actual voltage we're reading is just a tiny bit above 0 volts, and has a little noise on it, as it is fluctuating between 0 and +0.00488 volts. It could have just as easily been a solid 0 (unlikely, as real devices have noise), or perhaps it could have been fluctuating between 0 and -0.00488 volts; this is all within the noise range of this DAQ device.

Let's actually use the DAQ device now. If your device does not have analog output, skip ahead and read the digital I/O information in Section 3.1.3, then come back here and pretend that your digital outputs are analog outputs. If you must do this, line 0 of your second digital output port will correspond to what we will call analog output channel 0. Line 1 of that same digital output port will correspond to our analog output line 1. You will only have the choice of 0 volts or 5 volts when you use digital output, but that's better than no choice at all!

For those of you with analog output, you should have your terminals wired as described earlier, so that when you place a voltage on your analog output channel 0, it will show up on your analog input channel 0. Also, analog output channel 1 should show up on analog input channel 1, if you've wired correctly. Click on your **Analog Output** tab, and you should a screen like the one in Figure 3–14.

Change the DC voltage to 1, click the **Update Channel** button, then immediately click the **Analog Input** tab so you can catch the transition from 0 volts to 1 volt (if your **Analog Input** tab is showing a strip chart), as in Figure 3–15.

If this works, you have verified your wiring between analog input channel 0 and analog output channel 0. Now enter a 1 into the **Channel** item of your **Analog Input** tab's screen. Using the **Analog Output** tab as before, first change its channel to 1, then change your voltage to 2 (not 1), then click the **Update Channel** button immediately followed by clicking the **Analog Input** tab as before. You should see analog input channel 1 change from 0 to 2 volts.

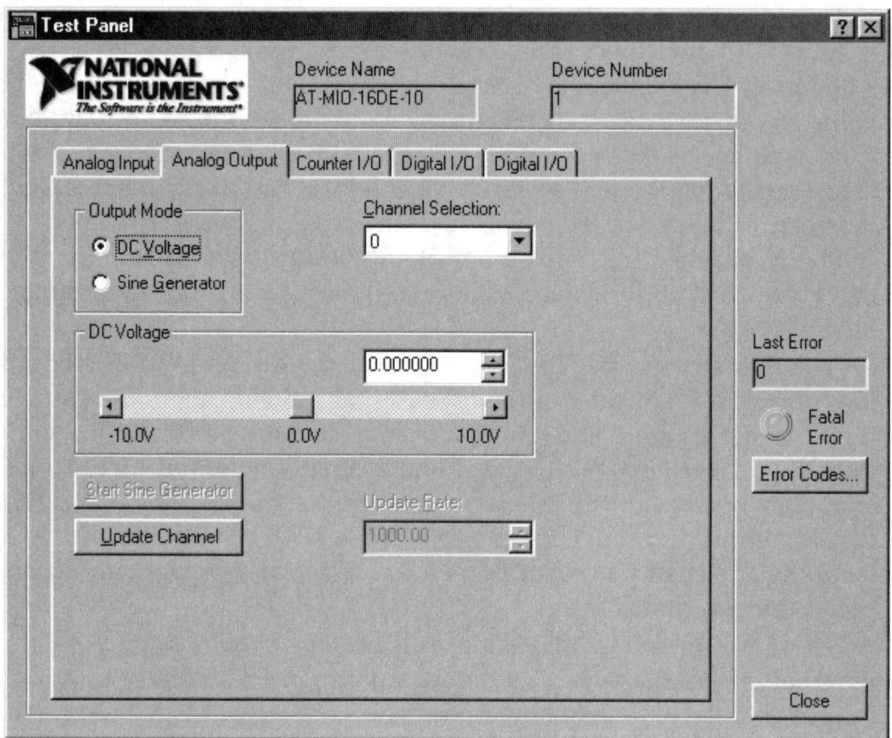

Figure 3–14
MAX allows you to run a basic test on analog output channels.

3.1.3 Digital I/O Test Panels

Word of warning: Since digital I/O lines can be set to inputs or outputs via LabVIEW, take extra care not to set any two digital lines to outputs if they are wired together, which they should be at this point. You can destroy your DAQ device this way, through software alone!

Select the **Digital I/O** tab of your test panel (or the one on the left, if you have multiple **Digital I/O** tabs like I do here). You should see something like the panel in Figure 3–16.

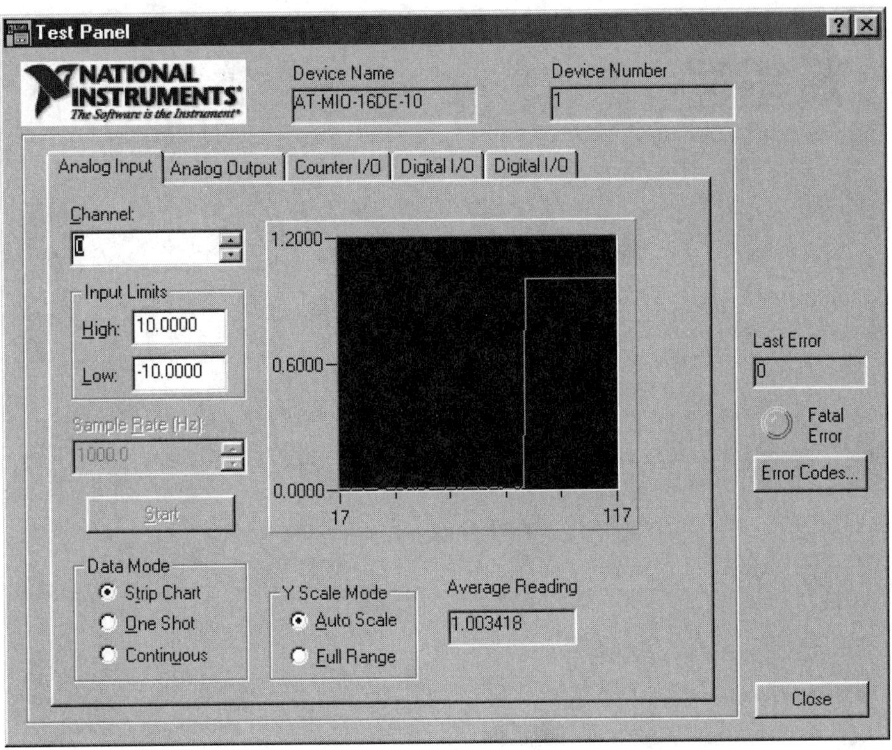

Figure 3–15
Analog input channel 0 has a voltage jump.

Your device will probably have only one **Digital I/O** tab. On my crazy DAQ device, Figure 3–16 shows that my first digital I/O port is accessed through the left **Digital I/O** tab. I must access my other three ports by clicking the right **Digital I/O** tab. In Figure 3–17, I've selected this tab, set my **Output Port** to 2 (which is my device's *second* digital I/O port), then checked the rightmost check box. The way I've wired the ports allows me to use this check box, followed by the **Write Output** button, to toggle lines 1 and 2 on my *first* digital I/O port, shown in Figure 3–16.

If your DAQ device has just one of the above **Digital I/O** tabs, you should be able to set up your ports so you can see lines 1 and 2 on your first port changing when you toggle line 0 of your second port, all on the same screen.

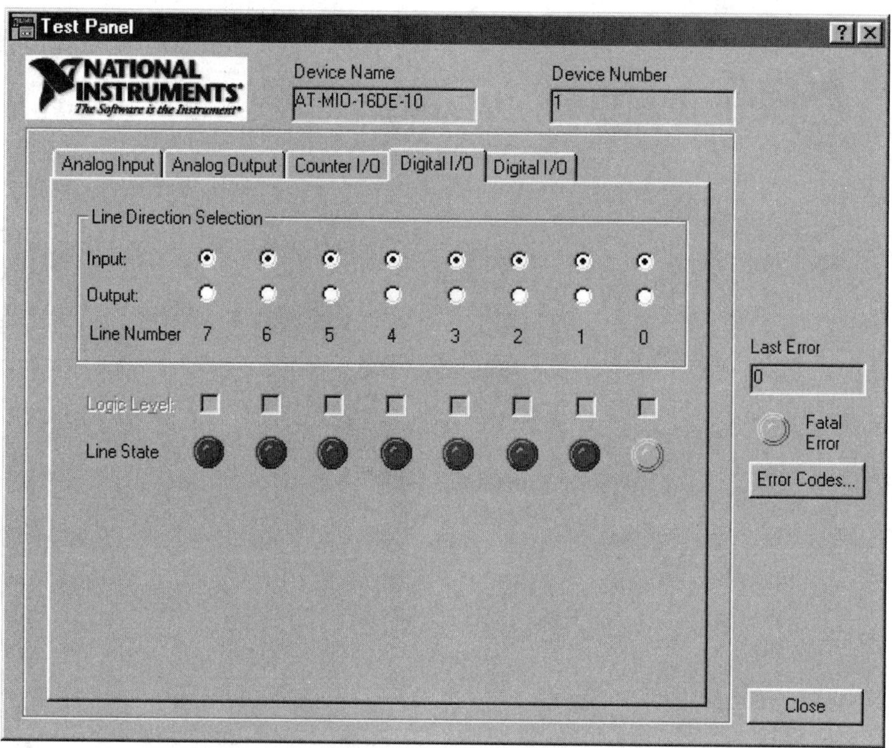

Figure 3–16
My first digital I/O port is an input.

3.1.4 Counter Test Panels

Click the **Counter I/O** tab, and you should see a panel like the one in Figure 3–18.

Counter Mode will vary per DAQ device. This is going to sound very unscientific, but select any counter mode, push the **Start** button, and see if **Counter Value** (or the like) begins counting. If it does, your counter works! If not, select another counter mode, push the **Start** button, as see if **Counter Value** now begins counting. Continue this admittedly unscientific procedure until you see the **Counter Value** rapidly changing; then you know your counter is working! Why this unscientific procedure? There are too many counting modes to cover here, and there are no modes common to all DAQ devices, so experience has taught me that this is the best way to proceed. For my AT-MIO-16DE-10, the mode shown in Figure 3–18 is the only one that changed **Counter Value**. This **Counter I/O** screen is a good way to verify that your DAQ device is indeed talking to the computer just after installation.

Keep your DAQ device wired as such for the remainder of this chapter. After this chapter, we'll no longer use a physical DAQ device.

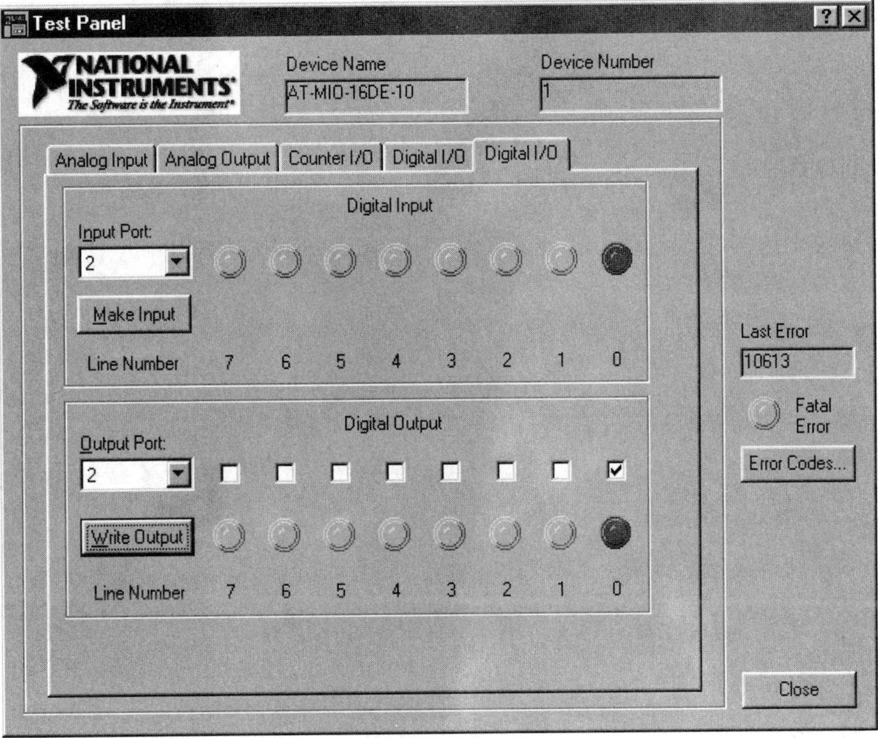

Figure 3–17
My second digital I/O port is an output.

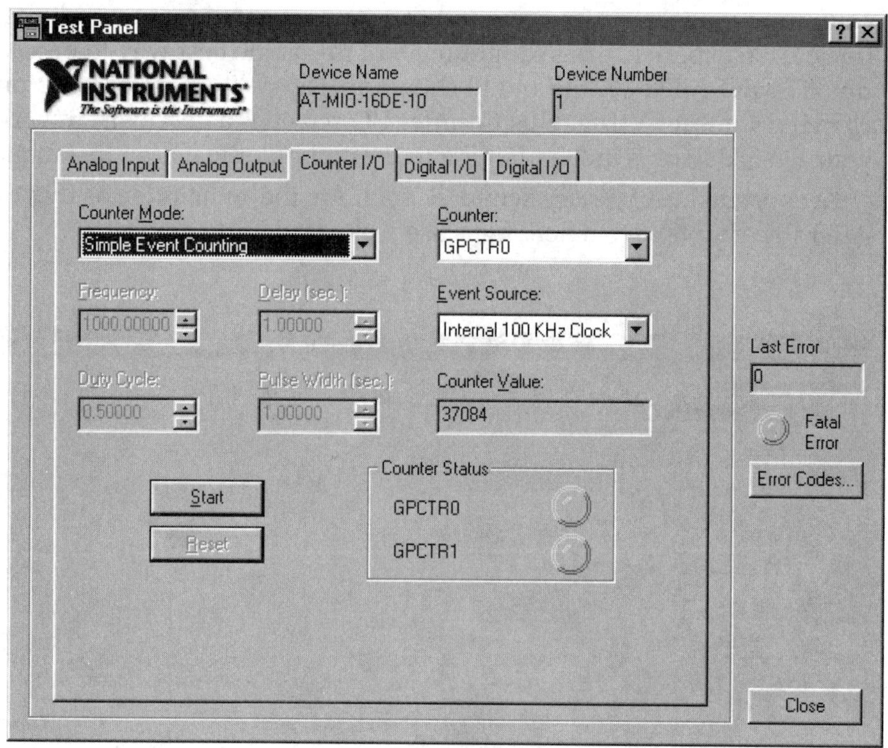

Figure 3–18
A counter test panel.

3.1.5 At Last, A Real LabVIEW/DAQ Example!

We will do this as quickly as possible, where our only goal is to illustrate a LabVIEW/DAQ connection. This section assumes that you have analog input. If not, you'll need to find a different VI in step 4 of Section 3.1.1; select one according to the abilities of your device, referring to *www.LCtechnology.com/noai.htm* for suggestions. Let's go:

1. Make sure you have configured your device with MAX, as described previously in this chapter.
2. Open LabVIEW and create a new VI. If LabVIEW is already open, use the **File»New VI** menu item to create a new VI.

3. Go to your new VI's block diagram if not already there (it's the white window). Use the **Window»Show Diagram** menu item to get there.

4. Pop up on the block diagram (using your right mouse button) and select `AI Sample Channels.vi` as shown in Figure 3–19 (it is hard to see here with all the clutter, but find it under the operating system's cursor, which looks like this).

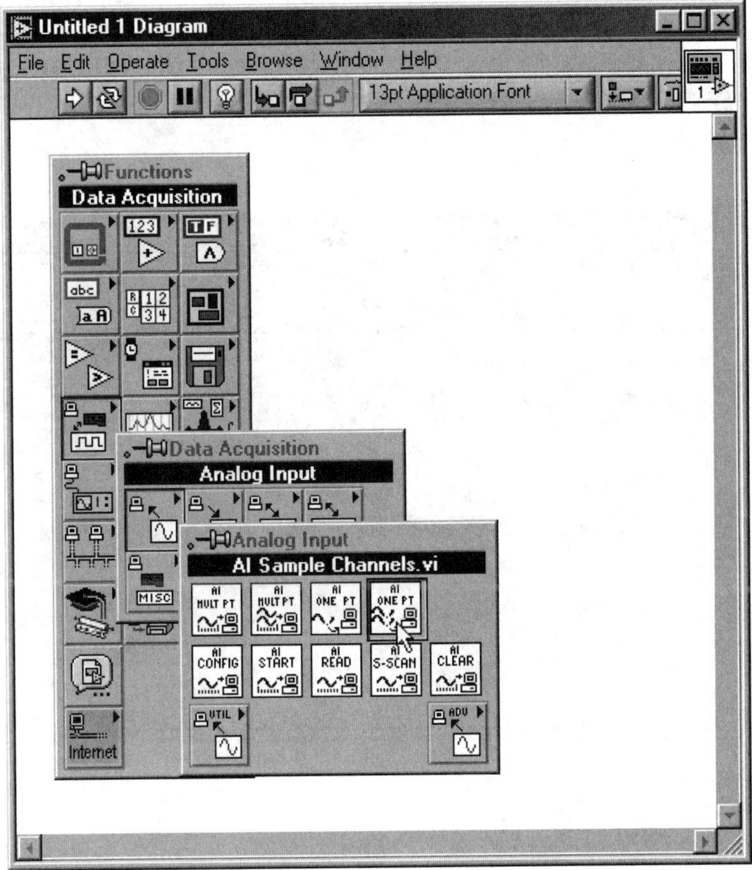

Figure 3–19
Choose `AI Sample Channels.vi`.

5. Once you have this selected, left-click anywhere in the large white area of the block diagram, and you will see the image shown in Figure 3–20.

Figure 3–20
`AI Sample Channels.vi` on the block diagram.

6. Now, double-click that icon on your block diagram, and you will see a window like the one in Figure 3–21 (be careful not to save it!).

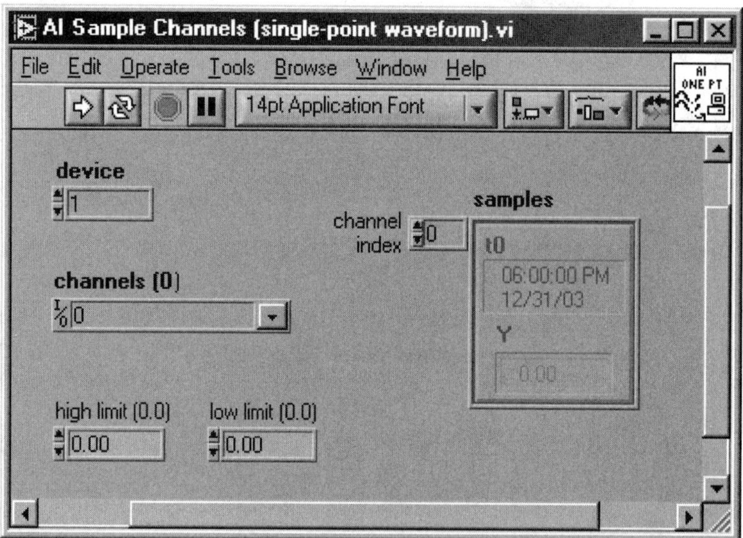

Figure 3–21
The front panel of `AI Sample Channels.vi`.

7. Make sure that device matches the number you set in MAX, as shown in Figure 3–5. If you left-click the **Run** button ⇨, you should see some data from channel 0 showing up in the box with the dimmed Y above it. This data represents the voltage on chan-

nel 0, unless the channels are set up differently so that it represents temperature or something else.

If you see this data, then congratulations, you have your first LabVIEW/DAQ connection! To verify the data is real, and if you have everything wired as described to this point in the chapter, you can change your analog output (or digital output) from MAX to watch the real data appear here. This particular VI allows you to collect data from multiple channels—you can use "0,1" (or "1,0", depending on your DAQ device) and see those channels in index 0 and 1 of the `samples` array.

If instead, you see box pop up with an incomprehensible error message, then any number of things could be preventing the DAQ.

3.2 LabVIEW DAQ Functions: An Overview

Figure 3–22 shows what the current main DAQ palette, the **Functions»Data Acquisition** palette, looks like.

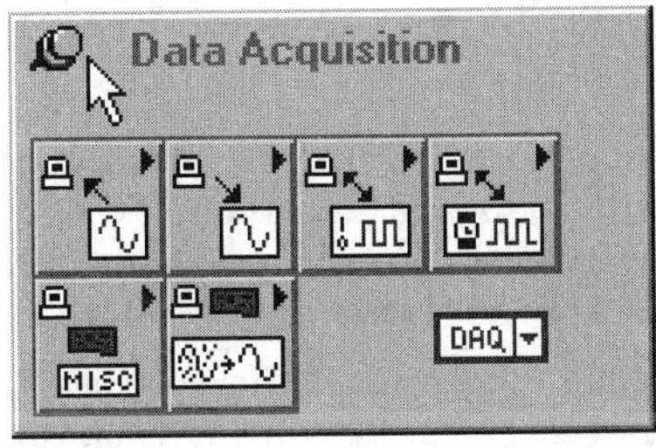

Figure 3–22
The Data Acquisition palette (**Functions»Data Acquisition**).

Within this palette, let's take a look at each of the subpalettes, listed in Table 3.1.

Since this *is* a DAQ book, we should go down at least one more level into some of the more popular items on this palette (analog and digital I/O, plus counter). For the most part, I am blatantly copying text directly from the VIs' documentation (I use quotes, to give the NI folks credit), but I often add my own commentary to clarify the descriptions. Let's look at the DAQ subpalettes in order, starting with the **Functions»Data Acquisition»Analog Input** palette, shown in Figure 3–23.

Table 3.1 *Main DAQ Palette*

Subpalette	Icon	Description
Analog Input		A collection of VIs that can acquire analog input data from DAQ devices.
Analog Output		A collection of VIs that can generate analog output data to DAQ devices.
Digital I/O		A collection of VIs that can acquire digital input or generate digital output data from DAQ devices.
Counter		A collection of VIs that can operate the counters on DAQ devices.
Calibration and Configuration	MISC	A collection of VIs that can set or read a very wide variety of parameters on DAQ devices. These VIs can also perform much of the functionality of MAX.
Signal Conditioning		A collection of VIs that perform the math to convert signals from some common transducers, such as thermocouples, RTDs, thermistors, and strain gauges. These VIs perform various other types of math, including SCXI-specific math.
DAQ Channel Name Constant	DAQ	This is not really a palette, but a constant that specifies channel names. You may refer to your channels by names that you specify (instead of numbers) by using MAX to do so.

If this is your first time reading this book, there will be an overwhelmingly large number of VIs in this section. Even though I consider everything here rather basic, I certainly would *not* if it were my first time reading about it. I suggest you read only the descriptions of those VIs in this section whose names are in **bold** font if this is your first reading, and use this section as a reference later when you need it. Bold font indicates a more commonly used VI.

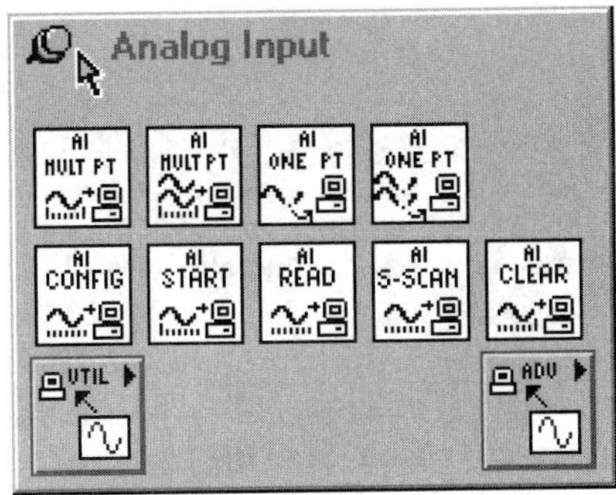

Figure 3–23
The Analog Input palette **(Functions»Data Acquisition»Analog Input)**.

Many VIs can output an array of numeric data or *Waveform* data, where a Waveform is a DAQ-oriented data type described in Section 1.2.3. In this book, I will use a capitalized version of Waveform to distinguish it as the LabVIEW data type, rather than a generic sequence of analog data points. The **Analog Input** palette is described in Table 3.2.

Table 3.2 Analog Input Palette Description

Item	Icon	Description
`AI Acquire Waveform.vi`	AI MULT PT	"Acquires a specified number of samples at a specified sample rate from a single input channel and returns the acquired data." You get a 1D floating-point array of numeric data, or one Waveform from a single channel.[1,2]
`AI Acquire Waveforms.vi`	AI MULT PT	"Acquires data from the specified channels and samples the channels at the specified scan rate." You get a 2D floating-point array of numeric data, or a 1D array of Waveform data from multiple channels.[1,2]
`AI Sample Channel.vi`	AI ONE PT	"Measures the signal attached to the specified channel and returns the measured data." You get a single floating-point data value, or a single-point Waveform from which you can extract such data.[1,3]
`AI Sample Channels.vi`	AI ONE PT	"Performs a single reading from each of the specified channels." You get a 1D array of numeric data from multiple channels, or a a 1D array of single-point Waveforms from which you can extract such data.[1,3]

Table 3.2 Analog Input Palette Description (continued)

Item	Icon	Description
`AI Config.vi`	AI CONFIG	"Configures an analog input operation for a specified set of channels. This VI configures the hardware and allocates a buffer for a buffered analog input operation."[4]
`AI Start.vi`	AI START	"Starts a buffered analog input operation. This VI sets the scan rate, the number of scans to acquire, and the trigger conditions. The VI then starts an acquisition."[4]
`AI Read.vi`	AI READ	"Reads data from a buffered data acquisition."[4]
`AI Single Scan.vi`	AI S-SCAN	"Returns one scan of data directly from the [board's] analog input channels for a non-buffered acquisition or a single scan from the acquisition buffer for a buffered acquisition."
`AI Clear.vi`	AI CLEAR	"Clears the analog input task associated with **taskID in**."[4]
Analog Input Utilities	UTIL	A subpalette containing a collection of lower level analog input DAQ VIs—many VIs above use these as subVIs.
Advanced Analog Input	ADV	Same as above.

1 These first four VIs are the highest level analog input VIs, built upon a number of other relatively low-level VIs.

2 When you acquire a waveform, software times the beginning of the acquisition (usually starts within milliseconds, but there's no guarantee), but hardware times the consecutive points thereafter, so you can be confident that they're equally spaced in time.

3 A "single-point" Waveform is another view of the Waveform, in which only a single data point and time value are shown.

4 These VIs are used when you want to perform continuous timed acquisition (very useful). Continuous acquisition will be discussed later in this chapter.

Now, take a look at the **Functions»Data Acquisition»Analog Output** palette (which is strikingly similar to its neighbor, the **Analog Input** palette) shown in Figure 3–24 and described in Table 3.3.

Now, take a look at the **Functions»Data Acquisition»Digital I/O** palette (shown in Figure 3–25 and described in Table 3.4), which is somewhat similar to the aforementioned analog palettes. The VIs after the first four are lower level VIs, some of which are used by the first four. Others are used in special digital I/O operations, such as buffered I/O or digital *handshaking* situations. Handshaking allows hardware-controlled high-speed digital data transfer into or out of certain DAQ devices.

The next palette to see is the **Functions»Data Acquisition»Counter** palette, which handles counters and is substantially different than the previously-shown palettes. It's shown in Figure 3–26 and described in Table 3.5.

We will not go into detail with the remaining two subpalettes in the **Functions»Data Acquisition** palette, as they're not as often used as the previously described palettes. These two subpalettes, shown in Figure 3–27 are **Calibration and Configuration** and **Signal Conditioning.** If you're using thermocouples, thermistors, RTDs, or strain gauges, you will probably want to look at the top four VIs in the **Functions»Data Acquisition»Signal Conditioning** palette.

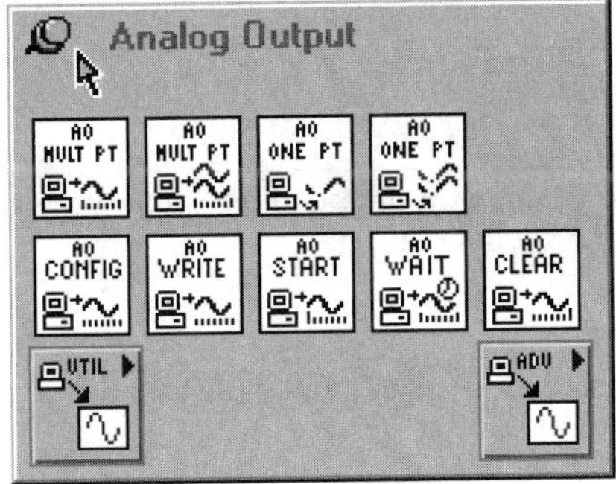

Figure 3–24
The Analog Output palette (**Functions»Data Acquisition»Analog Output**).

Table 3.3 Analog Output Palette Description

Item	Icon	Description
AO Generate Waveform.vi	AO MULT PT	"Generates a timed, simple-buffered waveform for the given output channel at the specified update rate." You send a 1D floating-point array of waveform data or one Waveform for a single channel.[1, 2]
AO Generate Waveforms.vi	AO MULT PT	"Generates timed, simple-buffered waveforms for the given output channels at the specified update rate." You send a 2D floating-point array of waveform data or a 1D array of Waveform data for multiple channels.[1, 2]
AO Update Channel.vi	AO ONE PT	"Writes a specified value to an analog output channel." You send a single, floating-point data value or a single-point Waveform.[1]
AO Update Channels.vi	AO ONE PT	"Writes values to each of the specified analog output channels." You send a 1D floating-point array of waveform data or a 1D array of single-point Waveforms for multiple channels.[1]
AO Config.vi	AO CONFIG	"Configures the channel list and output limits, and allocates a buffer for analog output operation."[3]
AO Write.vi	AO WRITE	"Writes data into the buffer for a buffered analog output operation."[3]
AO Start.vi	AO START	"Starts a buffered analog output operation. This VI sets the update rate and then starts the generation."[3]
AO Wait.vi	AO WAIT	"Waits until the waveform generation of the task completes before returning."

Table 3.3 Analog Output Palette Description (continued)

Item	Icon	Description
AO Clear.vi	AO CLEAR	"Clears the analog output task associated with **taskID in**."[3]
Analog Output Utilities	UTIL	A collection of lower level analog output DAQ VIs—many VIs above use these as subVIs.
Advanced Analog Output	ADV	Same as above.

1 These first four VIs are the highest level analog output VIs, built upon a number of other relatively low-level VIs.

2 These VIs do not return (continue data flow) until output generation is complete.

3 These VIs are used when you want to perform continuous timed generation (very useful when you have the right hardware). Continuous generation will be discussed later in this chapter.

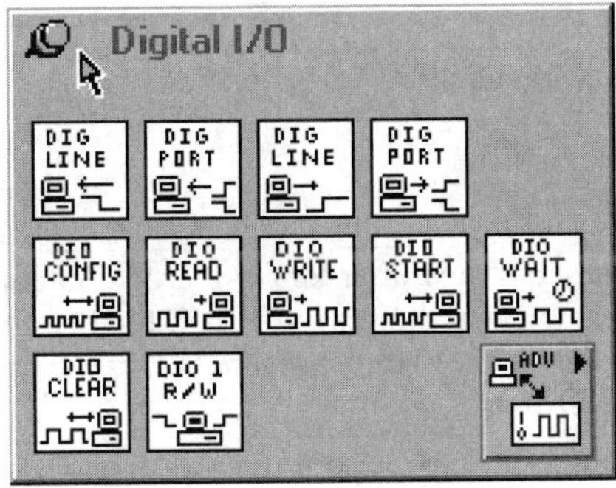

Figure 3–25
The Digital I/O palette (**Functions»Data Acquisition»Digital I/O**).

Table 3.4 Digital I/O Palette Description

Item	Icon	Description
`Read From Digital Line.vi`	DIG LINE	"Reads the logical state of a digital line on a digital channel that you configure."[2]
`Read From Digital Port.vi`	DIG PORT	"Reads a digital channel that you configure."[2]
`Write To Digital Line.vi`	DIG LINE	"Sets the output logic state of a digital line to high or low on a digital channel that you specify."[2]
`Write To Digital Port.vi`	DIG PORT	"Outputs a [binary] pattern to a digital channel that you specify."[2]
`DIO Config.vi`	DIO CONFIG	"This VI configures the hardware and allocates a buffer for a buffered digital input or output operation for a specified group of ports."[1]
`DIO Read.vi`	DIO READ	"This VI reads data from a buffered digital data acquisition."
`DIO Write.vi`	DIO WRITE	"This VI writes data into the buffer for a buffered digital output operation. The data written into the buffer will then be generated (transferred from the buffer to the digital board) at the update rate specified in DIO Start."[1]
`DIO Start.vi`	DIO START	"This VI starts a buffered digital input or output operation. This VI sets the number of scans/updates and the scan/update rate then starts an acquisition."[1]

Table 3.4 Digital I/O Palette Description (continued)

Item	Icon	Description
`DIO Wait.vi`		"This VI waits until the task's digital operation is complete before returning."[1]
`DIO Clear.vi`		"This VI clears the digital I/O task associated with **taskID in**."[1]
`DIO Single Read/Write.vi`		"This VI reads or writes a digital pattern from or to the digital channels specified in the digital channel list."[1, 2]
Advanced Digital I/O		A collection of lower level digital I/O DAQ VIs—many VIs above use these as subVIs.

1 See the real descriptions for these VIs (through the Help window); they're lengthy.

2 "Digital channel" and "port" mean the same thing.

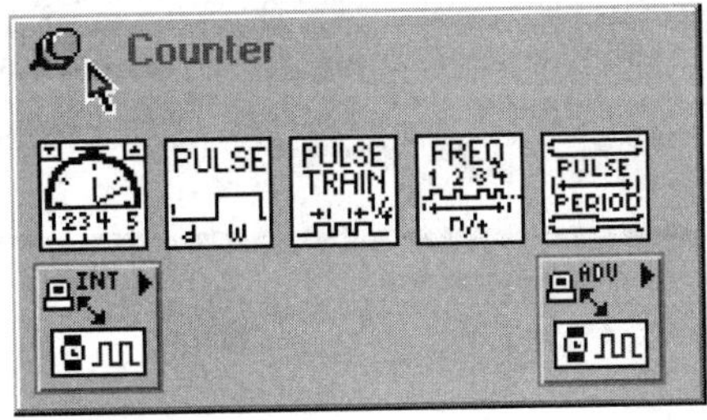

Figure 3–26
The Counter palette (**Functions»Data Acquisition»Counter**).

Table 3.5 Counter Palette Description

Item	Icon	Description
`Count Events or Time.vi`		"Configures one or two counters to count external events or elapsed time. An external event is a high or low signal transition on the specified SOURCE pin of the counter."
`Generated Delayed Pulse.vi`	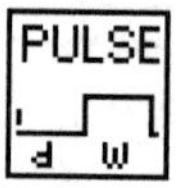	"Configures and starts a counter to generate a single pulse with the specified delay and pulse width on the counter's OUT pin."
`Generate Pulse Train.vi`		"Configures the specified counter to generate a continuous pulse-train on the counter's OUT pin, or to generate a finite-length pulse-train using the specified counter and an adjacent counter."
`Measure Frequency.vi`		Measures the frequency of a TTL signal on the specified counter's SOURCE pin by counting positive edges of the signal during a specified period of time."
`Measure Pulse Width or Period.vi`		"Measures the pulse-width (length of time a signal is high or low) or period (length of time between adjacent rising or falling edges) of a TTL signal connected to the counter's GATE pin."
Intermediate Counter		A collection of lower level counter DAQ VIs.
Advanced Counter		A collection of even lower level counter DAQ VIs—many VIs above use these as subVIs.

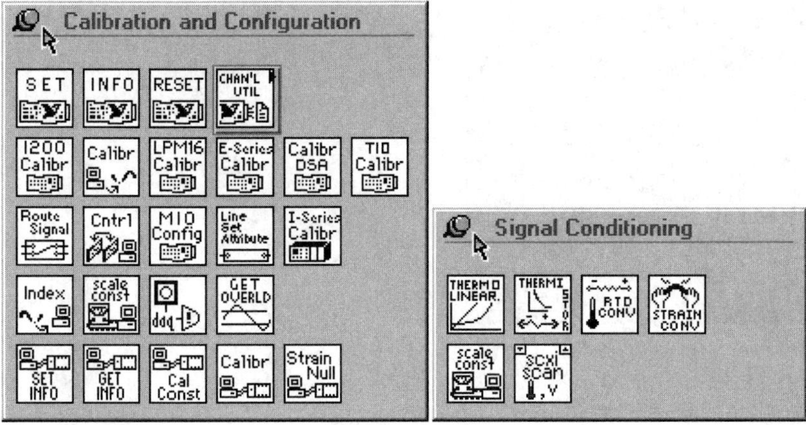

Figure 3–27
Two DAQ subpalettes.

3.3 Using the Basic LabVIEW DAQ Functions to Display Data

In this section, you will see how to use the basic DAQ VIs in these categories:

1. Analog Input
2. Analog Output
3. Digital I/O
4. Counter

These four major DAQ categories correspond directly to the top row of the main DAQ palette, as shown in Figure 3–22.

3.3.1 Analog Input

This analog input section will be fairly long—there are tons of basic concepts to be learned with analog input. We will only be discussing the four most basic analog input DAQ functions in this section, also known as the *Easy I/O* analog input functions. Take another look at the **Analog Input** palette in Figure 3–28.

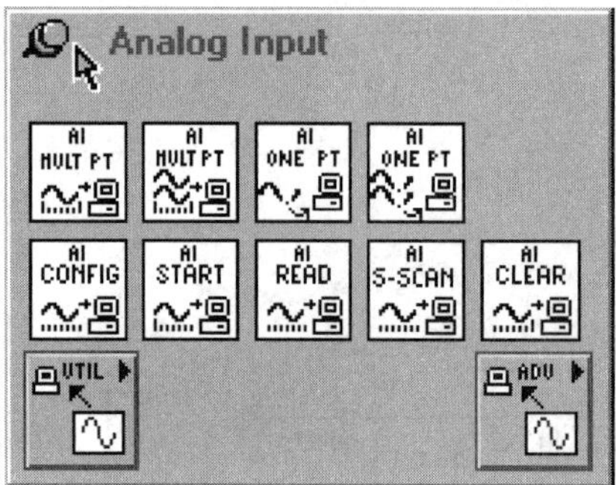

Figure 3–28
The Analog Input palette, once again.

The Easy I/O VIs are used for acquiring one signal or multiple signals through a DAQ device, as illustrated in Figure 3–29.

The black dots represent analog voltages being sampled by the computer on the right, via some sort of DAQ device.

Let's take a quick look at the most basic parts of the `AI Acquire Waveforms.vi`. Once you understand its inputs and outputs, you will be able to apply these concepts to the other three Easy I/O VIs. Figure 3–30 shows the icon/connector part of the current Help window for `AI Acquire Waveforms.vi`.

Table 3.6 provides a description of each of these terminals.

Table 3.6 *Detailed Descriptions of the Inputs of* `AI Acquire Waveform.vi`

Terminal Name	Direction	Data Type	Description
device	input	I16	The device number as assigned by MAX.
channels	input	channel or string	Numbers or signal names from MAX specifying which analog input channels will be used. Commas and colons are allowed. For example, you could enter `3,` or `0,3,` or `0:3` (meaning 0 through 3), etc. Some devices require these channels to be in descending order when the colon or comma is used.
number of samples/ch	input	I32	Number of samples per channel. Since we're acquiring waveforms here, this number is usually much greater than one.

Table 3.6 Detailed Descriptions of the Inputs of `AI Acquire Waveform.vi` (continued)

Terminal Name	Direction	Data Type	Description
scan rate (1000 scans/ sec)	input	SGL	How many times per second will you sample the data? Only your DAQ device or the speed of your computer limits this number.
high limit (0.0)	input	SGL	If left unwired, this will default to the highest value inherent to your DAQ device, as set by MAX. If you know you only need to read a small voltage range, you can use this and the **low limit** input to effectively increase your resolution, as described in Chapter 2.
low limit (0.0)	input	SGL	See the **high limit** input's description.
waveforms	output	[WFM] OR [[SGL]]	Whether this is a 1D array of Waveforms or a 2D array of single-precision numerics, you are effectively getting a 2D array of data either way. One dimension corresponds to the channels, while the other dimension corresponds to the samples per channel.

To begin with, let me show you the quickest way in LabVIEW to acquire a single point of analog input data from channel 0 of a DAQ device. First, make sure your DAQ device is set up through MAX and wired as described in the first section of this chapter.

Create a new VI, then on its block diagram drop `AI Sample Channel.vi` from the **Functions»Data Acquisition»Analog Input** palette. The only object on your block diagram should now be this:  .

Double-click this subVI (with the Positioning tool or Operating tool), and you'll see the front panel shown in Figure 3–31.

In this VI, the **sample** output has a data type of Waveform (WFM), but only the relevant components of the Waveform are shown. The only useful information it really produces is just one voltage number **Y** (first element of the **Y** array) and a time (**t0**). The **sample** output here is polymorphic, so if we wire it to an object that must take a numeric input, its data type will be automatically converted, and the numeric data **Y** will be the voltage passed to the numeric object.

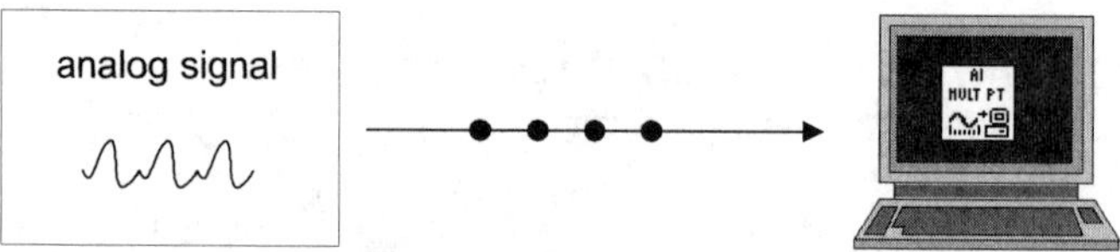

AI Acquire Waveform.vi– one signal, multiple samples

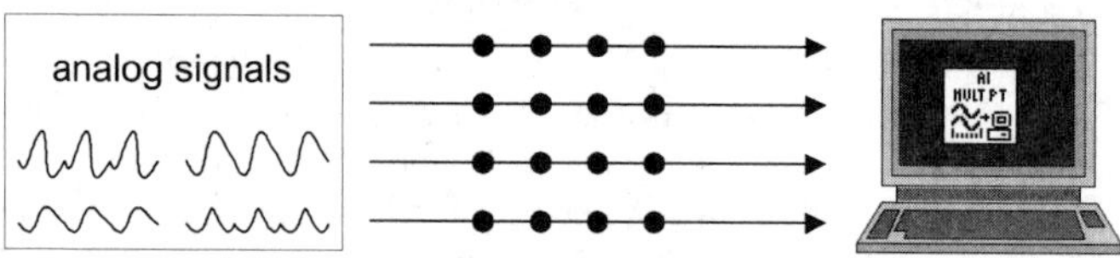

AI Acquire Waveforms.vi– multiple signals, multiple samples

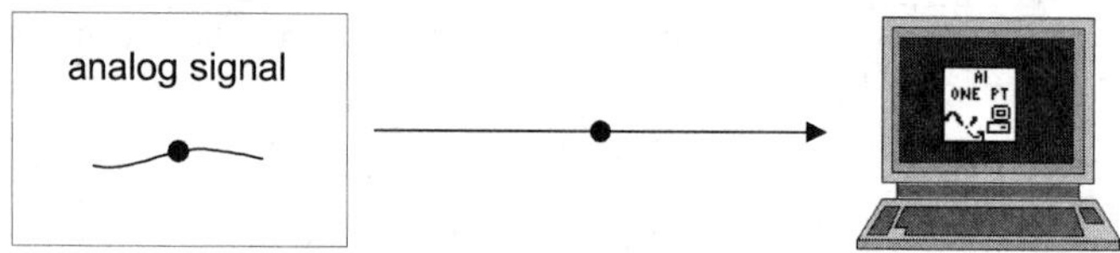

AI Sample Channel.vi– one signal, one sample

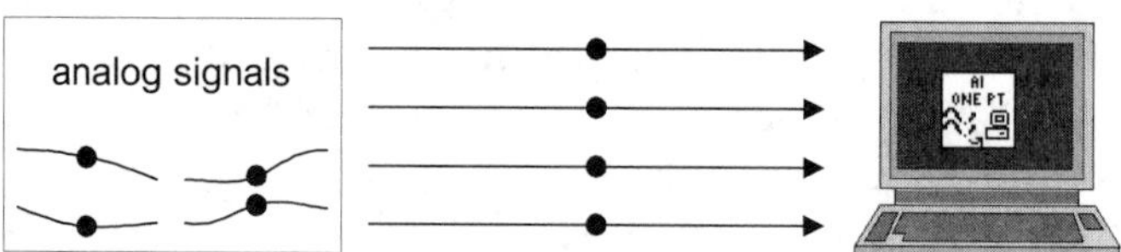

AI Sample Channels.vi– multiple signals, one sample

Figure 3–29
Four basic modes of analog input.

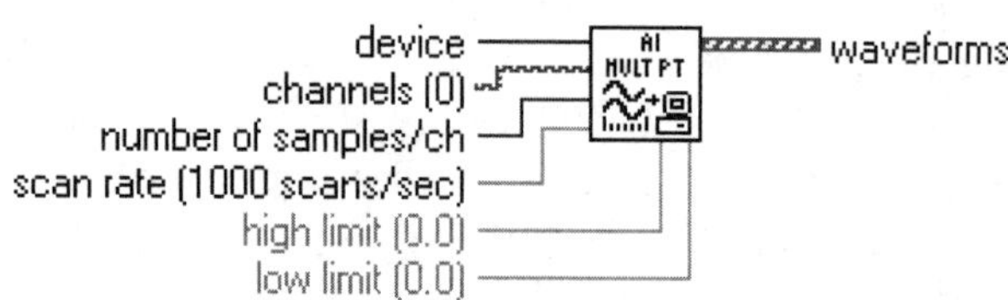

AI Acquire Waveforms.vi

Figure 3–30
The Help window for `AI Acquire Waveforms.vi`.

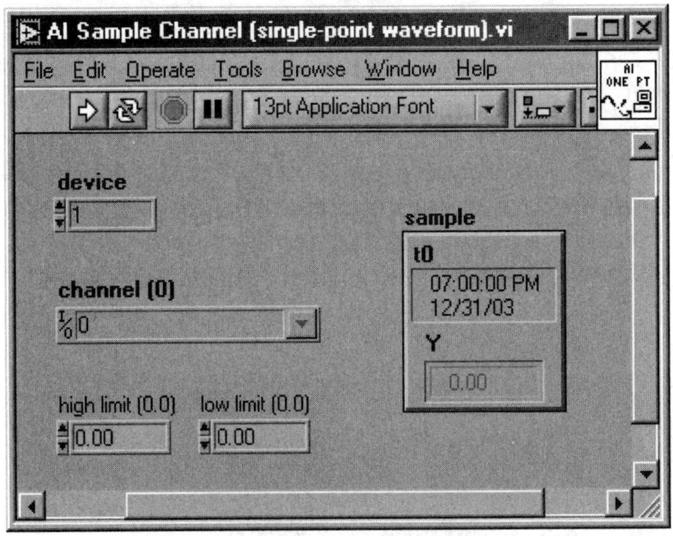

Figure 3–31
The "single-point waveform" flavor of this VI is shown.

Make sure your hardware is wired as described in the first section of this chapter, so that analog output channel 0 is wired to analog input channel 0. Now go to the MAX test panels, as described previously, and set the voltage on analog output channel 0 to `1.00` volts. When you run the above VI, you should see a `1.000` (or something close) show up in the **sample** output.

We have run `AI Sample Channel.vi` as a VI, but when writing programs, you will want to use it as a subVI. Suppose we wanted to acquire eight data points at 10 Hz (10 times per second). In a perfect world, our data would be sampled evenly, as illustrated by the big black dots in Figure 3–32, directly on the 10 Hz time divisions.

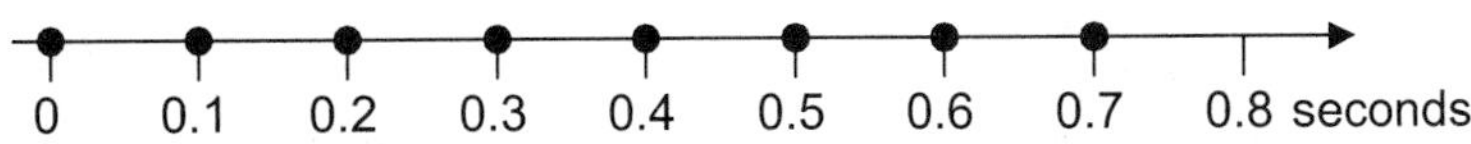

Figure 3–32
Eight data points are sampled at 10 Hz.

Build the VI shown in Figure 3–33, using these tips:

1. Set your point style to a small dot by popping up on the graph's Plot Legend.

2. Make sure autoscaling is on for the graph's x- and y-axes.
3. Use the "scaled" version of the DAQ subVI by popping up on it and using the **Select Type** menu item to get the scaled version. This simplifies the block diagram.
4. Don't bother setting the numbers on the axes on the graph—they will set themselves (and the plot you see will appear) after you run the VI, provided you did step 2.
5. Keep this VI open, as you'll use its graph again soon.

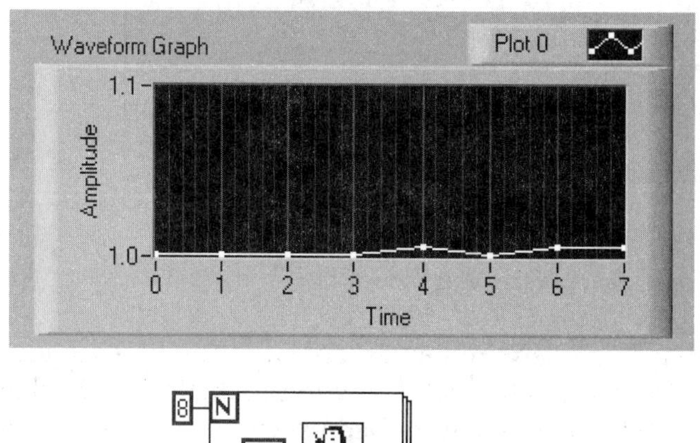

Figure 3–33
Eight data points are sampled at 10 Hz, as in Figure 3–32.

If we had a real-time LabVIEW setup, we could get *very* close to this exact timing. But suppose our acquisition rate was not so perfect, as illustrated in Figure 3–34.

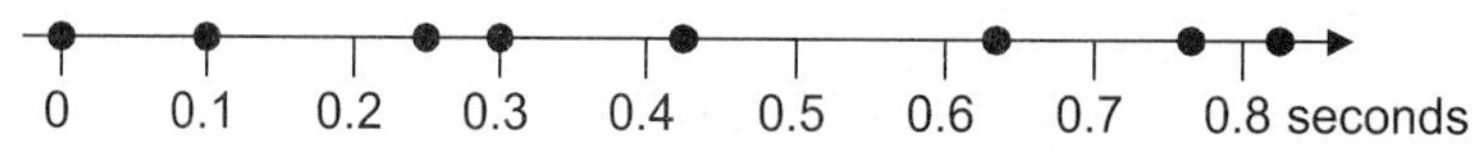

Figure 3–34
Eight data points are sampled at roughly 10 Hz, but the timing is erratic.

On a non-real-time operating system, such as Windows and most others, we could easily be fooled into thinking that we had acquired a perfectly timed waveform if we had built the VI above. But in reality, we will often have underlying timing discrepancies at least as bad as what's shown by those dots. Why? The computer's processor might ignore our DAQ application for several milliseconds at a time, and often several hundred milliseconds at a time, as illustrated by those erratically spaced dots. I've intentionally created such delays of several seconds, and this could happen inadvertently with any LabVIEW/DAQ application relying on the operating system for its timing, especially if other applications are running that involve disk or network activity.

Remember the section on DMA and FIFO in Chapter 2? The same issues are involved here. We showed a picture in Chapter 2 of a sine wave in which the computer went to sleep and ignored a portion of the sine wave (Figure 2–22). The result was a faulty sine wave (Figure 2–23). Since DMA and FIFO are here to save us from this situation, we will modify the block diagram to use them (provided our hardware has these capabilities) for highly accurate timing.

Build the VI shown in Figure 3–35, using these tips:

1. Use the same graph as before—it already has its point style and autoscaling set.
2. Use the Waveform version of `AI Acquire Waveform.vi` (use its **Select Type** pop-up menu item if needed).
3. Since we're using device 1 and channel 0, we don't need to wire those inputs of the AI subVI.
4. The easiest way to wire the inputs to the two lower left terminals is to pop up on each terminal of `AI Acquire Waveform.vi` and select **Create»Constant**—type in the number as soon as the constant appears, before clicking elsewhere. You will also need to change the labels and move everything around, as shown in Figure 3–35.

We have used `AI Acquire Waveform.vi`, which will automatically take advantage of any DMA and/or FIFO resources of the DAQ device (if available). Thus, if we have at least eight samples of either DMA or FIFO, our samples will be very accurately timed. Notice how the graph is polymorphic, as it can accept an array of numbers or the Waveform data type (not to mention many other types).

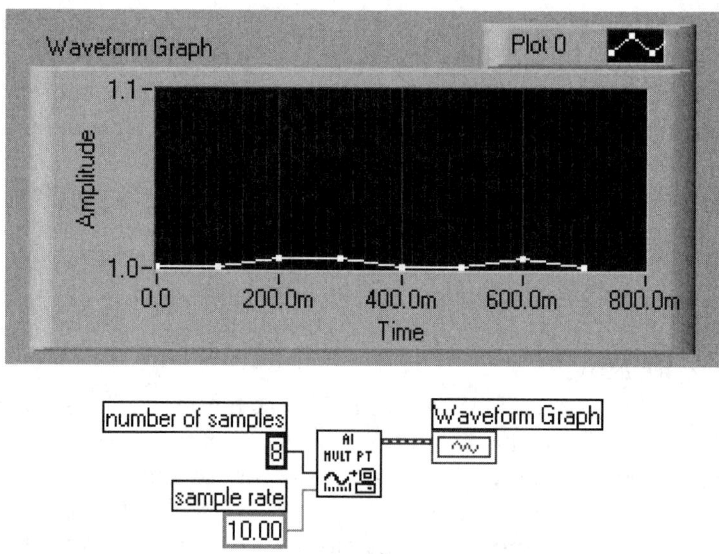

Figure 3–35
Eight data points are sampled at 10 Hz, as in Figure 3–32.

Our `AI Acquire Waveform.vi` gives us a waveform whose `t0` component is a *time* value, as returned by the Get Date/Time In Seconds function, whose Help window, shown in Figure 3–36, describes it well.
As you can see, a time format is automatically used on the x-axis when a Waveform is wired to the graph. If you wanted to see raw numbers, you could change this property of the x-axis.

Get Date/Time In Seconds

Returns the number of seconds that have expired since 12:00 am, Friday, January 1, 1904 Universal Time.

Figure 3–36
The Get Date/Time In Seconds Help window.

At this point, you might think the single-point acquisition via `AI Sample Channel.vi` is practically useless, and you would instead want to use `AI`

`Acquire Waveform.vi` nearly all the time. But despite its real-time performance, it's only useful for short runs because it doesn't allow user interaction until it finishes. If you have a real-time operating system, or if your scan rate is very slow, you may find the `AI Sample Channel.vi` (or `AI Sample Channels.vi`) quite useful. If you're sampling once every minute, you probably do not care about operating system delays that are typically a few hundred milliseconds. If you take care to restrict the disk and networking activities of all programs running, including LabVIEW, these delays will almost never be over one second.

Let's now do a more realistic example using `AI Sample Channel.vi`. Typically, this VI is used in a While Loop, and executed once per second, at most. Build the VI shown in Figure 3–37, using the tips below.

Front panel tips:

1. Use a Waveform Chart, not a Waveform Graph.
2. Turn on your chart's autoscaling for both axes, and set the point syle as shown.
3. On the **STOP** button, make sure you have its mechanical action set to "latching" so that it pops up after it's been pushed. The easiest way to do this is to select the STOP button from the **Controls»Boolean** palette (shown below on the front panel), which already has latching mechanical action.

Block diagram tips:

1. The object labeled `chart` with **History** text is a Property Node of the chart. To create this, you must pop up on the chart, either on the front panel or block diagram, and select **Create»Property Node**. Then you must pop up on the Property Node and select **History Data**. So that data is written to this Property Node rather than read from it, you should pop up on it and select **Change All To Write**. To create the empty array of floating-point numbers to which it's wired, pop up on this Property Node and select **Create»Constant**. This is the easiest way, as you will automatically get the correct data type. What we're doing here is clearing the chart's history prior to each run of the VI. The wire from this Property Node to the While Loop carries no useful data—it ensures that the chart is cleared before the first iteration of the While Loop. Without this wire, we would have no guarantee when the chart would be cleared!

2. Use the scaled version of `AI Sample Channel.vi` (use its **Select Type** pop-up menu item if needed).
3. The logic with the two Get Date/Time In Seconds functions tells the VI to collect a new data point at a rate of roughly once every second. If we changed the 1.00 to 2.00, the data collection rate would change to approximately once every two seconds.
4. The `device` and `channel` constants are most easily created by popping up on the corresponding terminals of the DAQ subVI, and selecting **Create»Constant**.

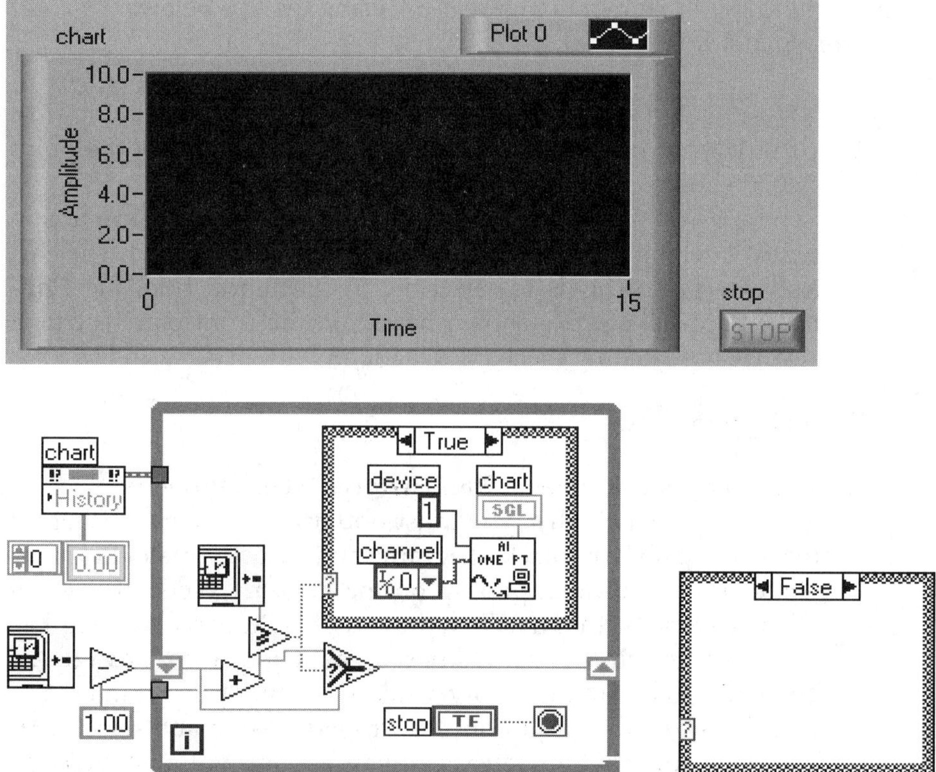

Figure 3–37
This VI will collect data at roughly 1 Hz from channel 0, but its timing will be rendered inaccurate by the operating system.

Save this VI as `AI Single Point and Chart.vi`.

We now have the software to read the data. Next, let's write some data. If you have an analog output on your DAQ device, connect channel 0 of its analog output to channel 0 of its analog input (you should have done so earlier in this chapter). Create a new VI, then drop the `AO Update Channel.vi` on the block diagram from the **Functions»Data Acquisition»Analog Output** palette. Open this analog output VI, which looks like Figure 3–38.

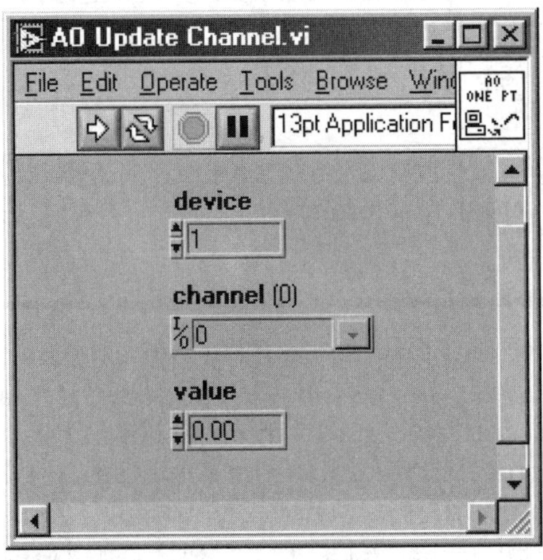

Figure 3–38
LabVIEW's simplest analog output VI.

Now click the usually forbidden **Run Continuously** button of `AO Update Channel.vi`. At this point, you can dynamically change the value to whatever you want, hopefully this voltage will appear on your analog input! If you have a voltmeter and know how to use it, you can monitor the voltage changes yourself, if you like. If you use the up/down arrows rather than typing text, you won't need to hit the Enter key for **value** to actually change. Be careful NOT to save this VI should you accidentally make any changes, as this is one of the built-in DAQ VIs for LabVIEW (see Appendix C).

Now, let's look at that voltage through LabVIEW. Arrange your screen so the front panels of continuously running `AO Update Channel.vi` and your previously created `AI Single Point and Chart.vi` are both fully visible, with both VIs simultaneously running. Now, when you change the `value` of the analog output VI, it should show up in the chart of your input

VI. Be sure to stop both VIs when you're finished, as we'll be running many more VIs in this chapter; it slows down the computer to have many VIs running at once.

Table 3.7 *Data Types for Two of the Basic Analog Input VIs*

Analog Input Easy I/O VI	Icon	Data Type
`AI Acquire Waveform.vi`	AI MULT PT	1D array (or Waveform)
`AI Sample Channels.vi`	AI ONE PT	1D array (or 1D array of single-point Waveforms)

See how both of the VIs above can produce a 1D array? These arrays are interpreted differently! The 1D array in the top VI has elements that correspond to data from one signal at multiple points in time. The 1D array in the bottom VI has elements that correspond to data from multiple signals at one point in time. This concept is important whenever you build your VI. For example, when using `AI Sample Channels (scaled).vi` AI ONE PT with a Waveform Chart, you would hope to see a chart with multiple plots (one plot per channel). However, you'll need to convert the array to a cluster (compare Figure 3–39 and Figure 3–40); otherwise, the chart will display the different data points as different values in time on one plot (oops), rather than single values on different plots (good).

Here, we have used the iteration terminal of the For Loop to create a 1D array of numerics, rather than real DAQ data.

Here's how you could modify a previous VI to accommodate two or more channels of data, instead of one. First, save your `AI Single Point and Chart.vi` as `AI Single Points and Chart.vi`. By doing this, you ensure that your original VI will not be modified by these upcoming changes.

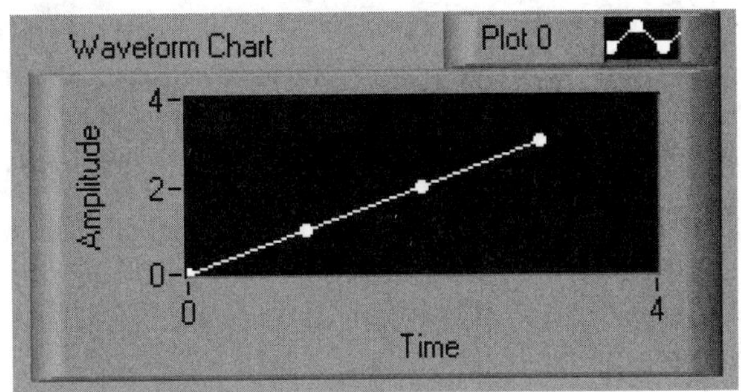

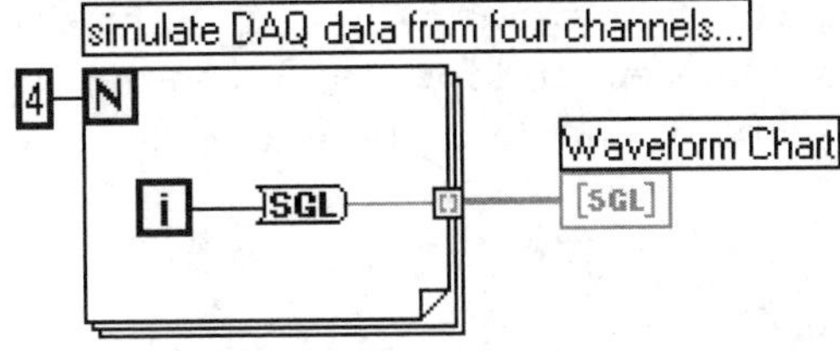

Figure 3–39

The output of the For Loop simulates the output of `AI Sample Channels (scaled).vi`, but is misinterpreted by the chart.

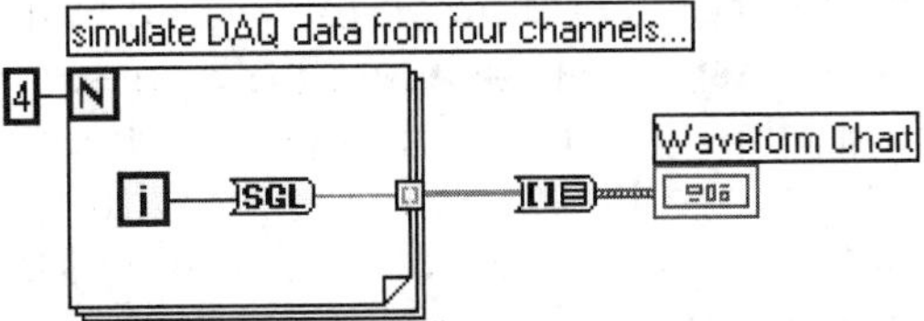

Figure 3–40

This is similar to Figure 3–39, but the chart interprets the data correctly this time.

First, go inside the **True** case of the Case Structure and change things around so that we collect two channels of data rather than one. You will need to replace the DAQ VI with another one, and make sure it's set to the *scaled* version. Make these changes as shown (and *not* shown!), using these instructions:

Front Panel Changes: The chart must now display two different channels. Change the plot's legend, by moving it, growing it, and modifying the second plot's point style so that is looks like Figure 3–41.
Block Diagram Changes:

1. Change your `channel` constant from `0` to `0,1` (some DAQ devices require `1,0`). For this step, and for step 3, you may need to move things around a bit with the Positioning tool .

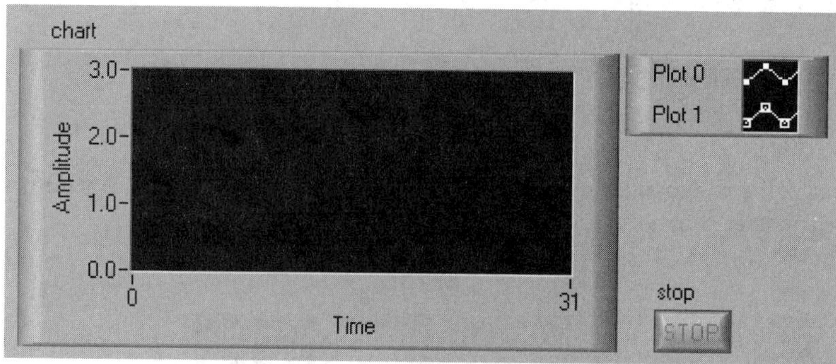

Figure 3–41
Two plots are customized by the chart's legend.

2. Replace the Easy I/O DAQ VI `AI Sample Channel.vi` with `AI Sample Channels.vi`. This allows the collection of multiple channels rather than just one. Pop up on the VI and make sure that it is using the scaled version via its **Select Type** pop-up menu item, so that we're producing an array of numerics rather than an array of waveforms. Although the chart is happy with this, the chart will now misinterpret each array of two numbers as two data samples from one signal at different points in time, rather than data samples from two different signals at the same time. To let the chart know that it needs to interpret the two data points as two different signals, we must send it a cluster of points rather than an array of two points.

3. Insert the Array To Cluster function on the wire to the chart, as shown in Figure 3–42, then pop up on it to make sure that it is creating a cluster of two elements, rather than its default of nine (who picked nine, anyway?). The wire to the chart remains unbroken, since a chart is happy with a cluster *or* an array of numerics. But without this insertion, the chart would then misinterpret each pair of data points as if it were coming from a single channel.

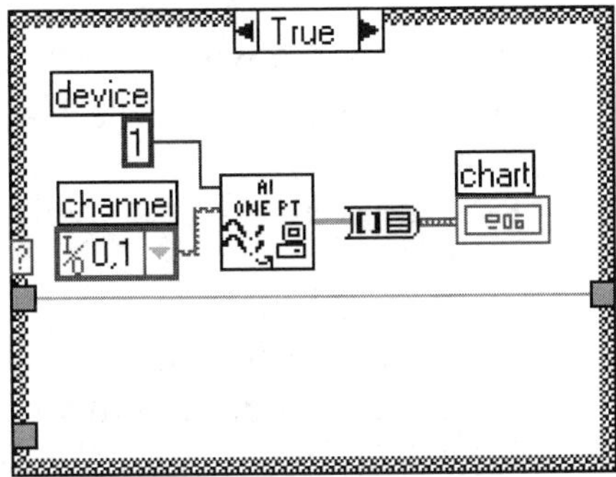

Figure 3–42
This piece of LabVIEW code will collect data from channels 0 and 1.

Note that if you wanted to use channel names like **voltage 0** and **voltage 1** instead of the channel numbers, you could have just typed *voltage 0,voltage 1* into the channel's constant. NI-DAQ software is smart enough to recognize "voltage 0:1." This is very convenient for many consecutively numbered channel names. Alternately, you could use the DAQ Channel Name constant, as described in Table 3.1, or its close relative, the DAQ Channel Name control, available on the front panel, shown in Figure 3–43.

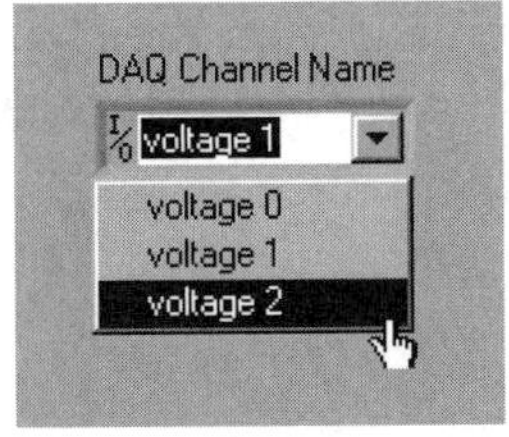

Figure 3–43
The DAQ Channel Name control has a secret affair with MAX, and allows you to select DAQ channels by name.

I made these changes, and even though I've done this sort of thing a zillion times, I actually expected to be ready to run. But no, there's a broken wire over on the left side of my VI connected to the chart's Property Node.

The chart, being polymorphic, now has a fundamental data type of "a cluster of two floating-point numbers," rather than a lone floating-point number. Therefore, its history is now an array of "a cluster of two floating-point numbers." We could tediously create this sort of constant if we wanted, but here's the fastest way to solve this problem:

1. Delete the existing array constant and its broken wire.
2. You can pop up on the chart's Property Node and select **Create»Constant**. Moving stuff around a bit, you should be able to create a ready-to-run VI whose block diagram looks something like that in Figure 3–44.

Now, let's generate some analog data to be charted with `AI Single Points and Chart.vi`. Earlier, we controlled a single channel of analog data by continuously running `AO Update Channel.vi`. We could do something similar with `AO Update Channels.vi`, but its front panel is not designed for easy control of multiple channels—just one array element is showing. Build the VI shown in Figure 3–45, and save it as `Update Two Analog Signals.vi`, using these tips:

1. Those are two Horizontal Pointer Slides on the front panel, found in the **Controls»Numeric** palette.
2. The analog output subVI is `AO Update Channels.vi`.

Arrange your screen so the front panels of `Update Two Analog Signals.vi` and your recently created `AI Single Points and Chart.vi` are both fully visible, with both VIs simultaneously running. By changing the values of your numeric slides, you should see the values on the charting VI change as well. This could be verified with a voltmeter, as before, as these should be the voltages on your two analog wires.

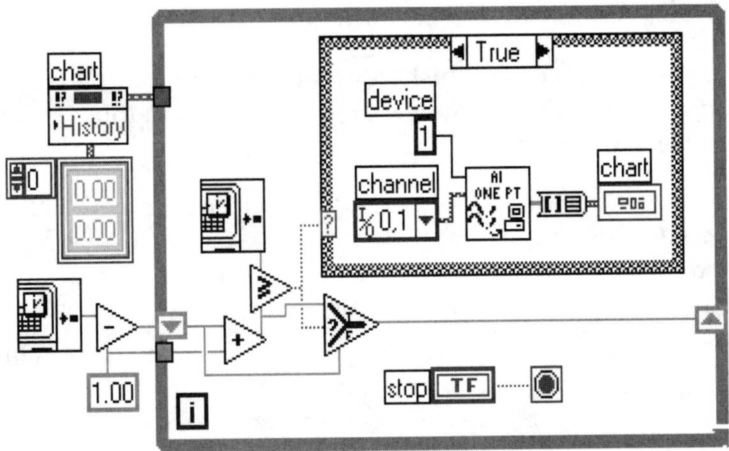

Figure 3–44
This VI will collect data at roughly 1 Hz from channels 0 and 1, but its timing will be rendered inaccurate by the operating system.

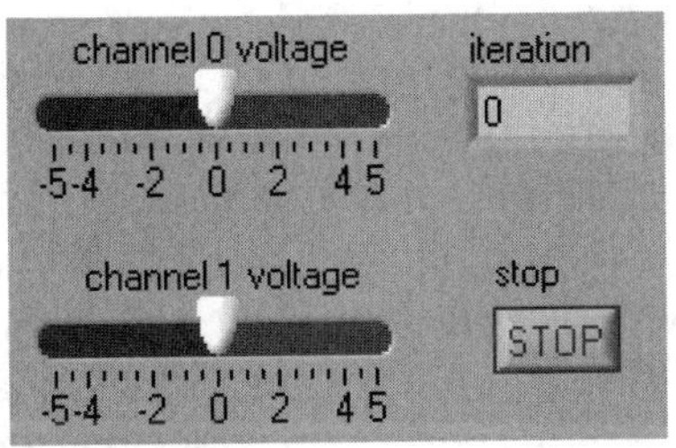

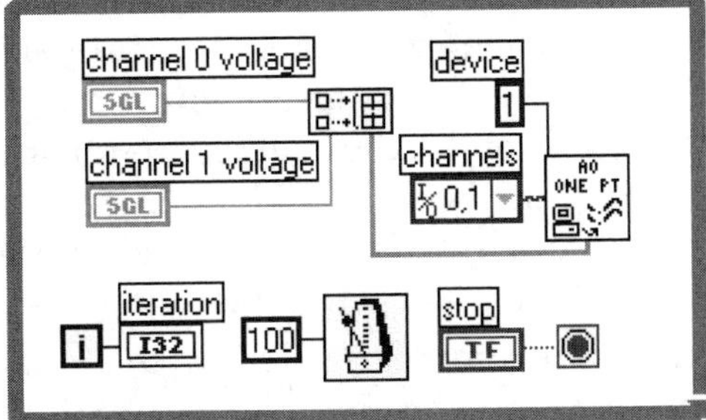

Figure 3–45
This simple VI allows you to continuously control two analog output channels.

Continuous Analog Acquisition (Double Buffering)

Is the section too long for you? Just wait, there's more, but it's really useful! Suppose you want to collect 24 hours of analog data from four channels sampled at 1 Hz. Here are three approaches:

1. Use `AI Acquire Waveforms.vi`, and you immediately get your data! This is the simplest, quickest approach offering you perfectly timed data, but you cannot view your data nor interact with the acquisition during the entire 24 hours.
2. Use `AI Sample Channels.vi` in a While Loop. You will be able to easily view your data during acquisition, but your samples may not be evenly spaced at one second apart.
3. Use Continuous Analog Acquisition techniques described shortly. You get the best of both worlds—fairly simple programming, perfectly timed data, and you can view your data during acquisition.

In order to make continuous acquisition work in LabVIEW, the DAQ device and LabVIEW must share a *hidden* buffer (no block diagram wire), managed by NI-DAQ, that should contain at least a few seconds' worth of samples. This buffer's size is usually set by the `buffer size` input of `AI Config.vi`, as illustrated in the block diagram of Figure 3–48 (`AI Config.vi` and the others used here were summarized in Table 3.2). You will always read from this buffer, *never write to it*, and you will almost always put a subset of this buffer's data into another, second buffer allocated by LabVIEW. This second buffer is usually a wire having a 2D array data type coming from `AI Read.vi`, also illustrated on the next block diagram. This technique is sometimes called *double buffering* or *circular buffering*. The DAQ device writes its data to the buffer in (almost) real time, while LabVIEW reads this DAQ data whenever it gets a chance. Remember that LabVIEW is at the mercy of the operating system and the microprocessor, which might be doing its own thing for a few seconds. Even so, the acquisition must continue uninterrupted. Figure 3–46 is an illustration of the hidden buffer's inherent behavior, which will give you an idea of why it's called circular buffering.

Whenever you want to continuously acquire analog input data at a regular rate, you can start with LabVIEW's examples folder in the `examples\daq\anlogin` area. When hunting around in the `examples\daq\anlogin\` folder, keep an eye out for the key words *Continuous, Cont,* and *buffered* when you're looking for continuous functionality. Here are my two favorite continuous analog input VIs, from which I begin *all* of my continuous analog input projects, along with their paths on my particular computer:

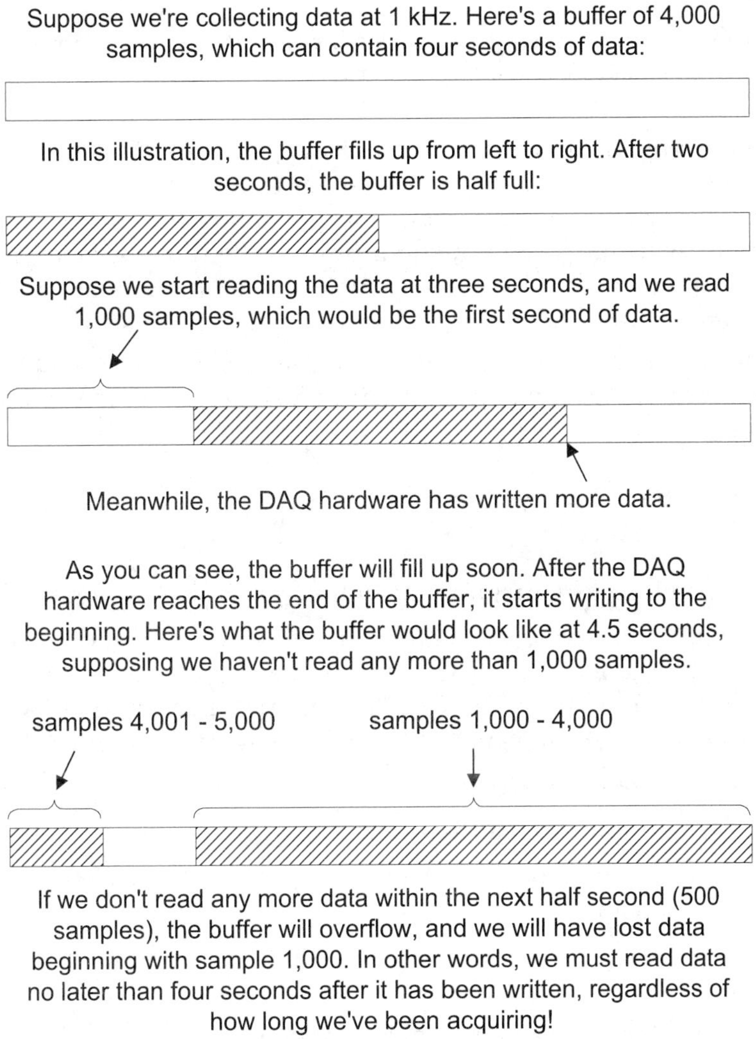

Suppose we're collecting data at 1 kHz. Here's a buffer of 4,000 samples, which can contain four seconds of data:

In this illustration, the buffer fills up from left to right. After two seconds, the buffer is half full:

Suppose we start reading the data at three seconds, and we read 1,000 samples, which would be the first second of data.

Meanwhile, the DAQ hardware has written more data.

As you can see, the buffer will fill up soon. After the DAQ hardware reaches the end of the buffer, it starts writing to the beginning. Here's what the buffer would look like at 4.5 seconds, supposing we haven't read any more than 1,000 samples.

samples 4,001 - 5,000 samples 1,000 - 4,000

If we don't read any more data within the next half second (500 samples), the buffer will overflow, and we will have lost data beginning with sample 1,000. In other words, we must read data no later than four seconds after it has been written, regardless of how long we've been acquiring!

Figure 3–46
The hidden DAQ buffer used with continuous acquisition is exposed. This buffer is typically not seen on any LabVIEW block diagram wire, but it is very important for evenly timed continuous acquisition.

```
C:\Program Files\National Instruments\LV6\examples\daq\anlo-
gin\anlogin.llb\Cont Acq&Chart (buffered).vi
```

and

```
C:\Program Files\National Instruments\LV6\examples\daq\anlo-
gin\anlogin.llb\Cont Acq&Graph (buffered).vi
```

Concerning the two VIs above, look at their block diagrams and notice the *Graph* version acquires the same number of data points per iteration of the main loop, while the *Chart* version acquires however many data points are currently available. I recommend acquiring the same number of data points per iteration of any loop, as it makes data manipulation much easier. Whether you send the data to a chart or a graph is irrelevant.

Since we're wired up already, we may as well go ahead and use these VIs. Open `Cont Acq&Chart (buffered).vi`, and make the following changes:

1. Save this VI as `My Cont Acq&Chart (buffered).vi` with an icon like this [my acq].
   Open the block diagram, open this subVI [MY DATA PROC], and save it as `My Data Processing 2.vi` in the same directory where you saved the main VI. If you do not make these saves, you run the risk of fouling up LabVIEW's examples, per Appendix C. While you're at it, change the icon of your new subVI to this: [my data proc 2].
2. Change **scan rate** to `10`.
3. Change **scans to read at a time** to `2`.
4. Change **channels** to `0,1` or `1,0`—whichever makes your DAQ device happy.
5. Change the point style on the first two channels as shown in Figure 3–47.
6. Leave the block diagram alone for now.
7. Select the **Operate»Make Current Values Default** menu item, and save the VI again, to spare you some typing if you want to use this VI again. The chart's plots won't be saved with this VI, but if you had a graph instead of a chart, its plots would have been saved had you made its value default.

8. You won't see any plots until you run the VI while changing your analog output values, as we'll do shortly.
9. Set autoscaling on your y-axis, but not on the x-axis.

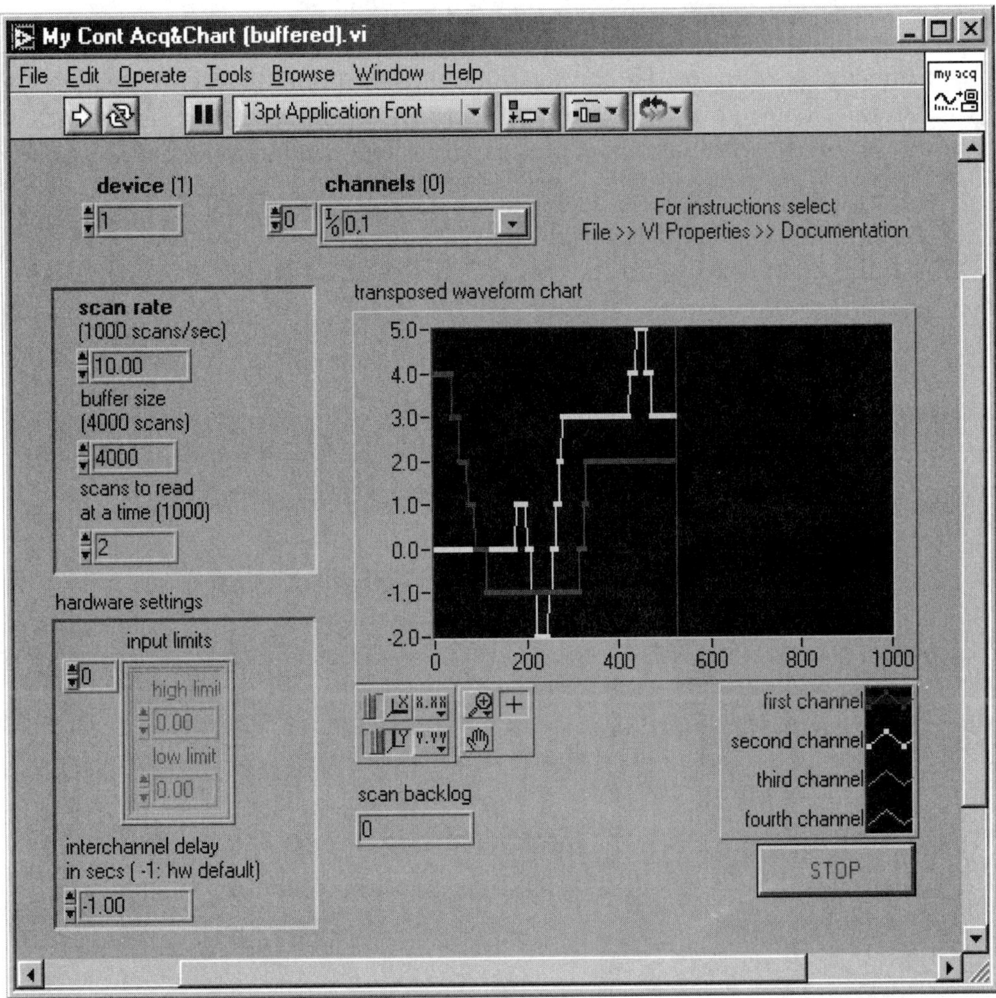

Figure 3–47
This VI can be found in LabVIEW's example folder.

Arrange the front panels of this VI and your recently created `Update Two Analog Signals.vi` so that they are both visible. Simultaneously run `Update Two Analog Signals.vi` and `My Cont Acq&Chart (buffered).vi`. Change your first two analog output channels to different values in the +/- 5V range, and you should see the plots changing in your charting VI as shown in Figure 3–47. Stop both VIs when you're finished, so they don't slow the performance of VIs we'll create later.

With these continuous acquisition VIs, it's a good idea to make sure that the number of scans to read at a time is about four or five times smaller than your scan rate. Thus, your main loop will iterate four or five times per second. If it loops more slowly than this, your user interface will be noticeably sluggish.

Let's take a quick look at the block diagram of `My Cont Acq&Chart (buffered).vi`, which should be identical to the example version of this VI at this point, except for its icon and its subVI, as shown in Figure 3–48.

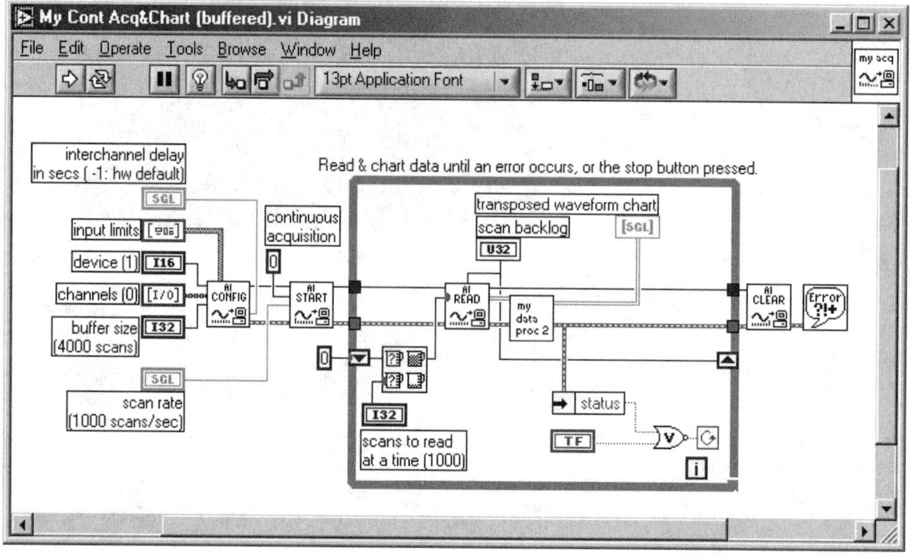

Figure 3–48
This is the block diagram of the front panel seen in Figure 3–47.

These analog input subVIs labeled "AI-whatever" were introduced in Table 3.2; they are more advanced than the Easy I/O VIs, but they will be used any time you want to do continuous acquisition. The key component to using these VIs for *continuous* acquisition is the constant *zero* being passed to `AI Start.vi`. Remember the "hidden" buffer that is filled up in a circular fashion a few pages earlier? Its size is determined by the **buffer size** constant here, where the number refers to scans, not samples or bytes.

Your program can get rather complicated when you want to do any moderately complex data processing on data from continuous acquisition because your data is acquired piecemeal. You must build the logic to analyze separate pieces of incoming LabVIEW data. Since this chapter has the word "basic" in its title, we will defer this logic to Chapter 6, which has no pretense of being basic.

3.3.2 Analog Output

In Figure 3–49, compare the **Analog Output** palette to our recently discussed **Analog Input** palette.

Figure 3–49
Analog Output palette and Analog Input palette.

The top row of the **Analog Output** palette is analogous to the top row of the **Analog Input** palette in terms of number of signals and waveforms versus single-point acquisition. The only significant difference is that we have data going out of the computer rather than coming in. The underlying data types are quite similar, especially when you use the scaled versions of your analog input VIs.

I know this is a software chapter, but take care not to attempt to drive too much current whenever you do analog output (or digital output).

Okay, back to software. In Figure 3–50, we have illustrated the top row of the analog output VIs, also known as Easy I/O VIs.

The black dots represent analog voltages being generated (or *updated*) by the computer on the left, via some sort of DAQ device.

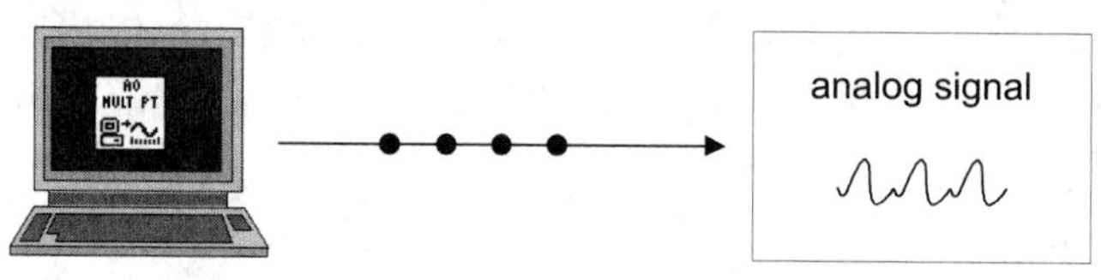

AO Generate Waveform.vi– one signal, multiple updates

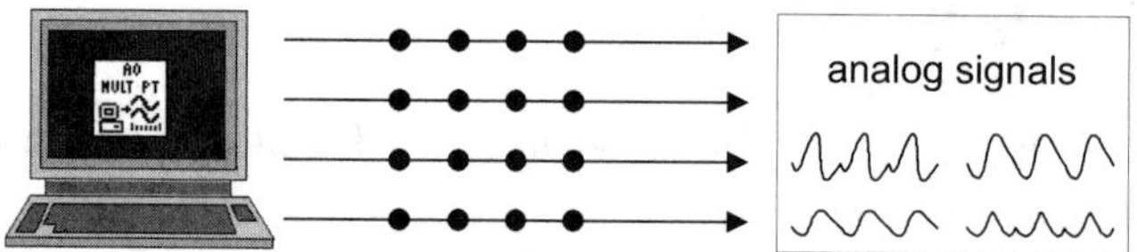

AO Generate Waveforms.vi– multiple signals, multiple updates

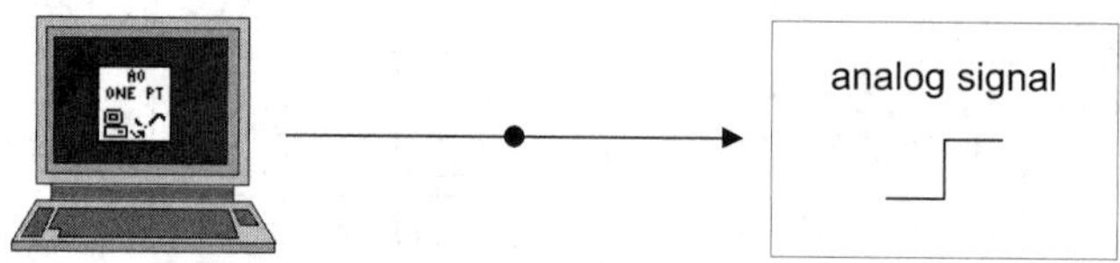

AO Update Channel.vi– one signal, one update

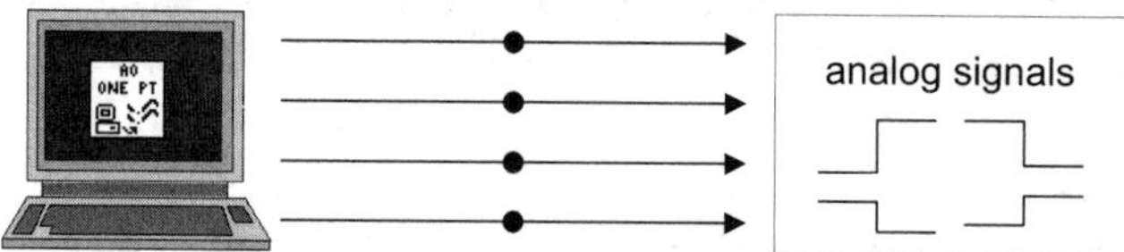

AO Update Channels.vi– multiple signals, one update

Figure 3–50
Four basic modes of analog output.

Compare Figure 3–50 to Figure 3–29, and notice the similarities. The big difference is whether the analog data is going into or coming out of the computer.

As we have used some basic analog output VIs in the last section, I will not drag you through a series of analog output VIs, as they are analogous to the analog input VIs. However, here are the main differences:

1. The terminology is switched around a bit, even though the fundamental idea is that data goes out of your computer rather than into it. When switching the wording from analog input to analog output, the word *sample* becomes *update* (for single points of data), and the word *acquire* becomes *generate* (for waveforms of data).
2. Many DAQ devices do not support DMA for analog output. If this is the case, your high-speed output may experience timing irregularities.

If you want to generate analog output waveforms at a regular update rate, first make sure you have DMA on the DAQ device's analog output channel. Next, look through LabVIEW's `examples\daq\anlogout` directory for VIs with the word *Continuous* or *Cont* to find relevant examples.

3.3.3 Digital I/O

A digital signal is one that can take on only two values. At the DAQ device, these values are usually 0 volts (off) or 5 volts (on). This digital voltage scheme is sometimes called *TTL* (transistor-transistor logic). The physical aspects of digital I/O are detailed in Chapter 2, so let's devote the rest of this section to LabVIEW with digital I/O.

LabVIEW refers to a single digital signal as a *line*, and a group of digital signals as a *pattern*. DAQ devices have "ports" to which you can write patterns. Depending on the DAQ device, a port is currently a group of 4, 8, 16, or 32 digital lines. You will generally be working with one of the following three digital signals: (1) one line at a time; (2) a pattern (a number of lines as a group); or (3) a digital waveform (a regularly timed series of patterns). For these three digital signal types, the naturally corresponding LabVIEW data types are shown in Figure 3–51.

Probably the most straightforward way to manipulate bits in a pattern (an integer in LabVIEW) is to use Boolean logic functions, which can also operate on integer data types to read or write individual bits. Figure 3–52 illustrates such logic.

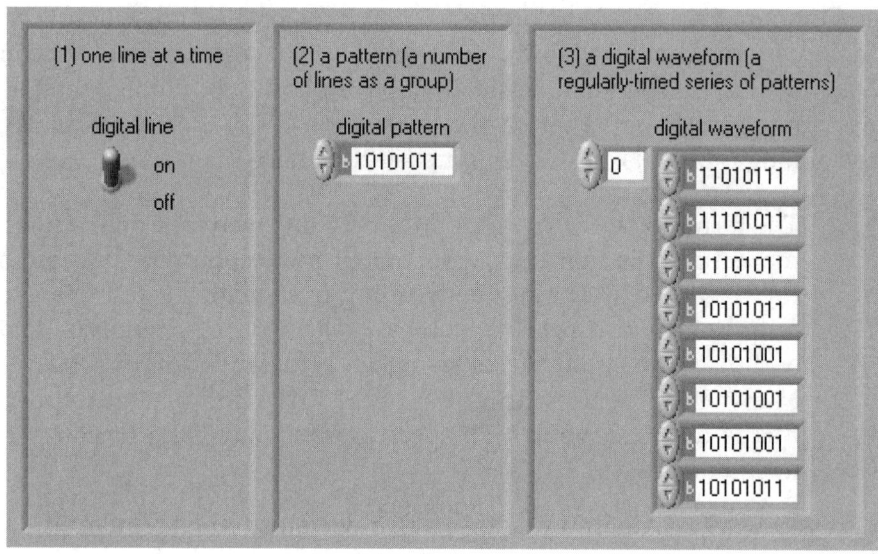

Figure 3–51
Three common data types for LabVIEW's digital data are shown; (1) Boolean, (2) integer, and (3) array of integers.

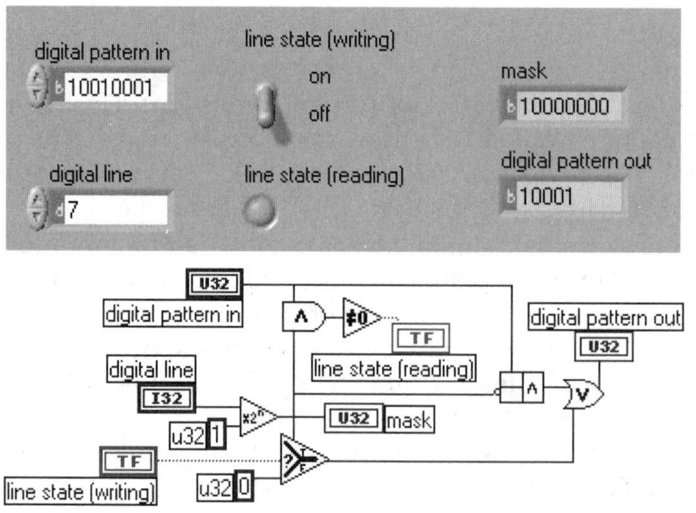

Figure 3–52
One way to manipulate bits inside a digital pattern. "Mask" is an intermediate value, but the objective is to read and/or write a particular bit ("digital line") from "digital pattern in".

First, look at the example VI in Figure 3–53 showing off LabVIEW 6i's new Digital Waveform Graph, which is designed to immediately display digital data as presented by the DAQ functions.

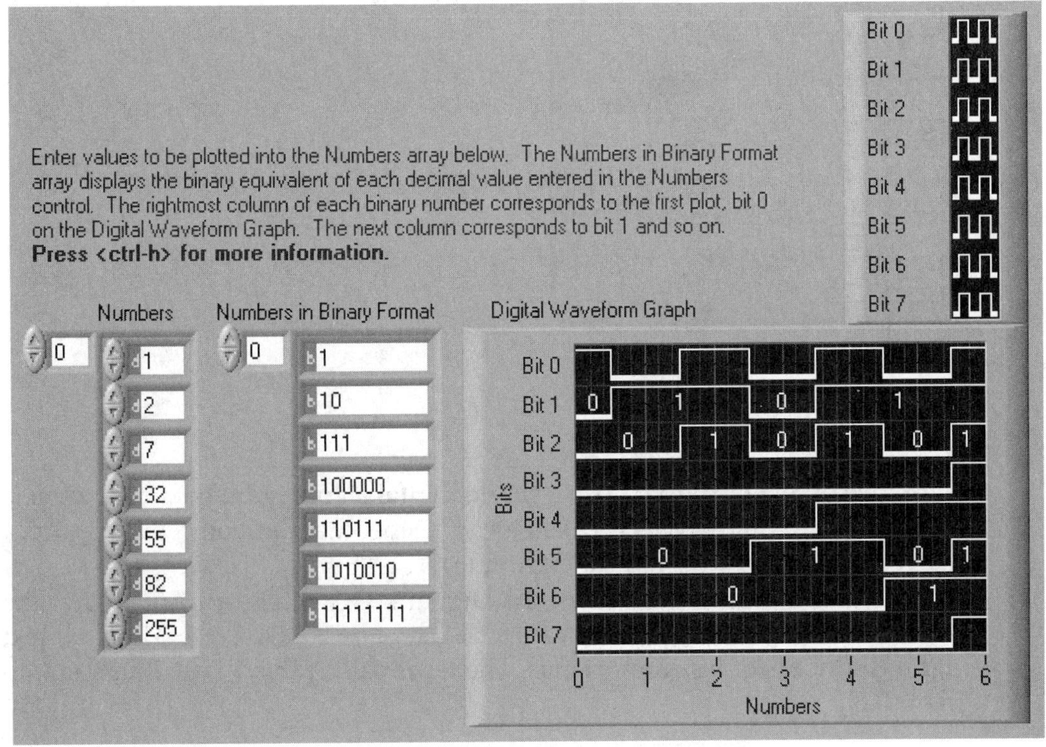

Figure 3–53
An example VI shows off the Digital Waveform Graph.

This VI can be found in LabVIEW's examples folder. Here's the particular path to this VI on my machine:

```
C:\Program Files\National Instruments\LabVIEW 6\examples\gen-
eral\graphs\DigitalWaveformGraph.llb\Simple Digital Waveform
Graph.vi
```

Next, take another look at the **Digital I/O** palette in Figure 3–54.

Like the two analog palettes, the top four VIs are considered Easy I/O. The categorization is somewhat different than that of the analog VIs, however. For these VIs, one pair reads data and the other pair writes data. Within each of these pairs, one handles a single digital signal and the other handles multiple digital signals.

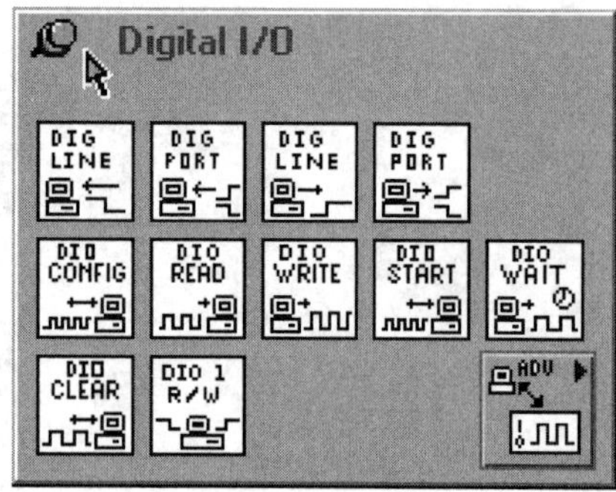

Figure 3–54
The Digital I/O palette, once again.

In the context of digital I/O, a single digital signal is called a *line*, while a grouping of digital signals is called a *port*. Depending on your DAQ device, a port usually consists of four or eight lines.

Let's take a quick overview of the Digital I/O VIs. Figure 3–55 shows the icon in the Help window of `Read From Digital Line.vi`. The gray text indicates inputs of less importance. The terminals of this VI are illustrated in Table 3.8.

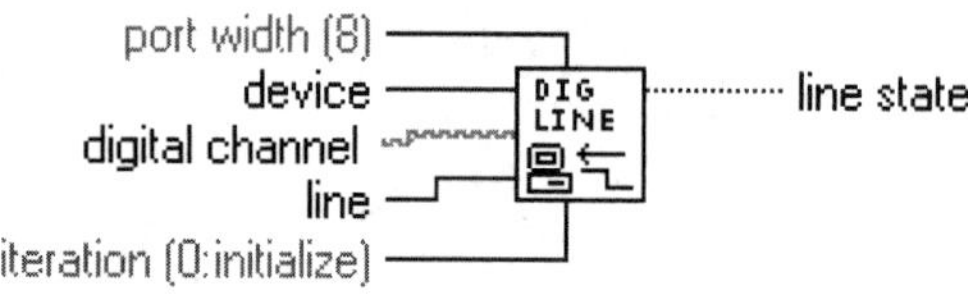

Figure 3–55
The icon of `Read From Digital Line.vi`

Compare this to the icon on the Help window of `Read From Digital Port.vi`, shown in Figure 3–56. The inputs are the same, except there is no input for **line**, since we're now reading all the lines of the specified digital port (called **digital channel** here). The **pattern** output is an unsigned 32-bit integer, in which each bit represents the state of one of the port's lines. Bit 0 represents line 0, bit 1 represents line 1, and so on. As most ports contain less than 32 lines, there are usually some high bits left unused.

Table 3.8 Detailed Descriptions of the Inputs of `Read From Digital Line.vi`

Terminal Name	Direction	Data Type	Description
port width (8)	input	I16	Number of lines in the port; this varies per DAQ device, but is usually 4 or 8.
device	input	I16	The device number you should have already assigned with MAX.
digital channel	input	string	Which port are you using? "Digital channel" means "port." This will be a number, like 0, 1, 5, etc. This number can vary per DAQ device.
line	input	I16	Which line of the port specified by **digital channel** are you using? For an eight line port, this number will be in the range 0 to 7.
iteration (0:initialize)	input	I32	If 0, the VI will initialize the specified digital I/O port to input. This input is designed to be connected to the iteration terminal of a While Loop or a For Loop.
line state	output	Boolean	True if the line is low, False if the line is high.

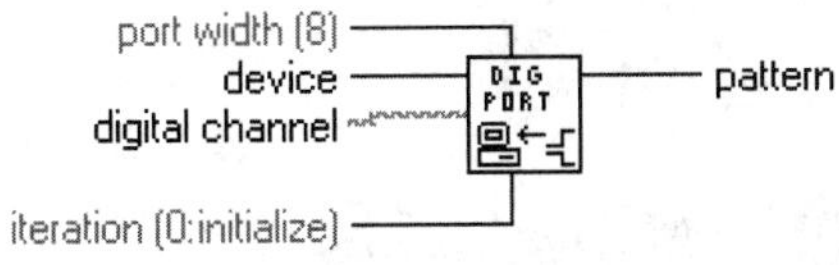

Figure 3–56
The icon of `Read From Digital Port.vi.`

Let's go ahead and use these VIs, just to watch them work. This is basically a repeat of what we saw in the first section of this chapter, only we'll be using VIs instead of the test panel in MAX.

Don't run the following VIs until instructed! Open a new VI and drop the `Write to Digital Line.vi` and `Read from Digital Port.vi.` Open these two sub VIs and arrange them so that you can see them both, as in Figure 3–57.

I have shrunk these VIs a bit, to save space as usual. Now, save each VI as `My (whatever their names are).vi` so you won't be changing LabVIEW's VIs! On `Write to Digital Line.vi`, change **digital channel** to your second digital port, so you won't be frying any hardware. MAX can tell you what your digital ports are—if they are 2, 3, and 4, then 3 is your second digital port. Make this the default value for the **digital channel** control, and save the VI again so that if you open it again, the second digital port will be there waiting for you.

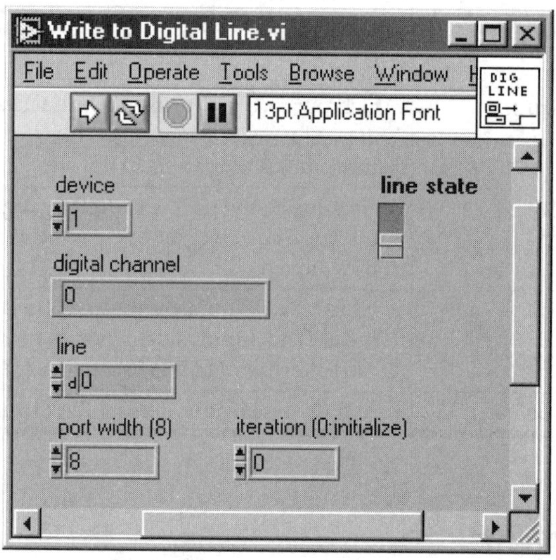

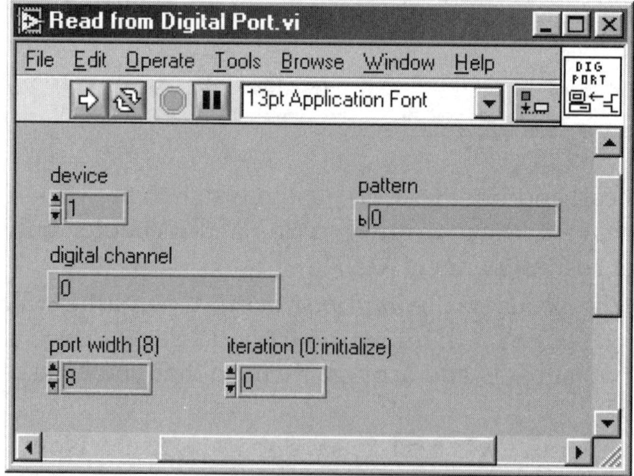

Figure 3–57
Two high-level digital I/O VI's front panels.

When you are using multiple digital ports, you should initialize your ports first, assume nothing about port state, then begin reading and writing. Why? For some crazy reason, initializing any port can set unwanted values to other digital ports (it certainly does on my Lab-PC+)!

First, realize that some DAQ hardware will invert its logic, so when you read about a digital line value of 0 here, you might see a 1, and vice versa. With that in mind, follow these steps:

1. On `My Read from Digital Port.vi`, change **digital channel** to your first digital port, and change **port width** if your DAQ device has a different port width. Make sure **iteration** is 0, then run this VI, knowing that it has initialized your first port to be an input port. Ignore **pattern** for now, as it is undefined upon initialization.
2. On `My Write to Digital Line.vi`, change **port width** if your DAQ device has a different port width. Make sure **iteration** is 0, then run this VI, knowing that it has initialized at least line 0 of your second port to be an output port.
3. Set **iteration** on *both* VIs to anything but 0. Leave them that way in the following steps; they have been initialized from steps 1 and 2, and we will use them as they are from now on.
4. Set the **line state** of `My Write to Digital Line.vi` to False, then run this VI.
5. Run `My Read from Digital Port.vi`. If you wired your connectors as described earlier in this chapter, the **pattern** indicator should indicate that bits 0, 1, and 2 of your first port are 0. Bit 0 of the first port should be hard-wired to 0, and bits 1 and 2 were set to 0 from step 4. You need to understand binary numbers in order to see that this is true (see Appendix A).
6. Set the **line state** of `My Write to Digital Line.vi` to True, then run this VI.
7. Running `My Read from Digital Port.vi` now should show you, through the **pattern** indicator, that bits 1 and 2 of your first port have changed to 1, while bit 0 is still 0. We've wired the two ports to obtain this exact result.

If you want to work with the individual lines of a digital port, there's a Number To Boolean Array function ([#[...]]) that can make it easier for you to access the individual bits from the `Read from Digital Port.vi`. With this function, you can easily display **pattern** as a Boolean array with multiple

Booleans showing on your front panel. Conversely, there's a Boolean Array To Number function that can piece together Booleans into an integer; this is useful for passing data to `Write To Digital Port.vi`.

3.3.4 Counter

Take another look at the **Counter** palette, shown in Figure 3–58.

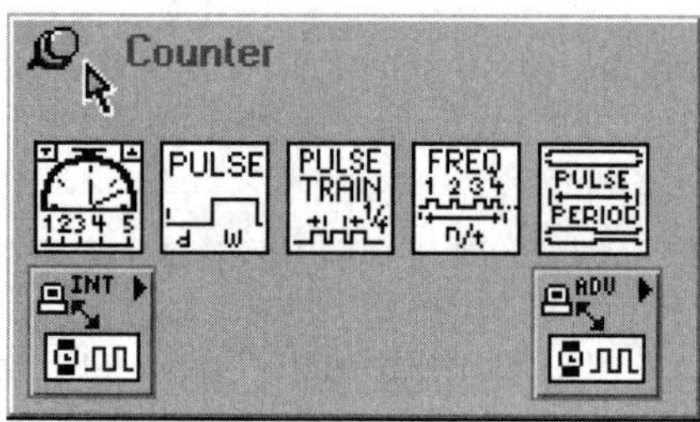

Figure 3–58
The Counter palette.

Counters work strictly in the digital domain. No analog signals are involved—at least not from our point of view. Probably the most common uses for a counter are counting digital events and generating pulses or pulse trains of digital data.

Older DAQ devices may use an older counter chip that is not supported by the standard counter VIs in the **Counter** palette. My Lab-PC+ counter chip is an 8253, whereas the only counter chips supported by the counter VIs in the palette of Figure 3–58 are the Am9513 and the newer NI-specific DAQ-STC, which is shown in Figure 3–59.

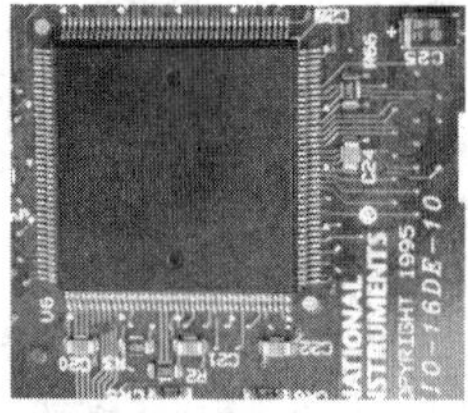

Figure 3–59
DAQ-STC, a counter chip custom-made by NI.

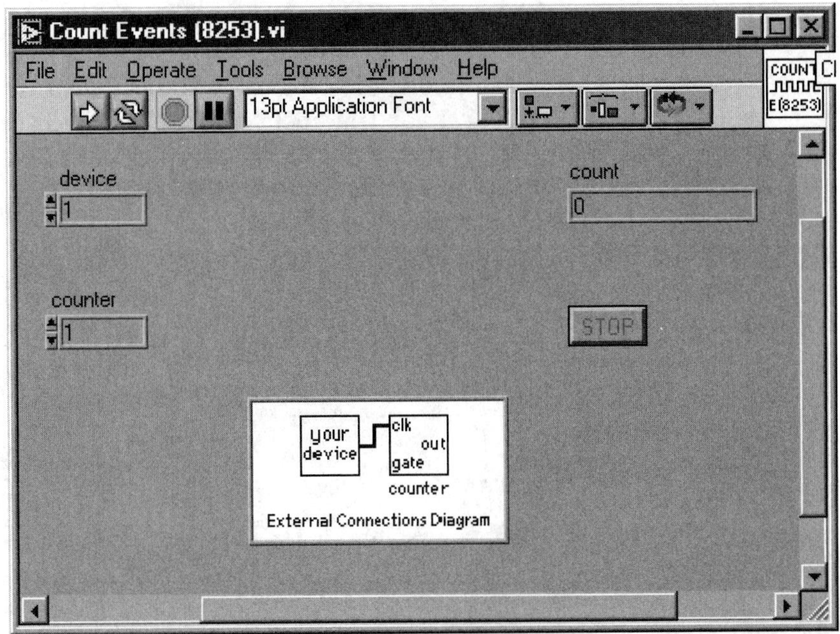

Figure 3–60
This example VI only works with DAQ devices having an 8253-compatible counter chip.

If you have an 8253 (or 8254 or 82-whatever) counter chip, find `Count Events (8253).vi` in LabVIEW's examples directory (see Figure 3–60).

If you have a newer device with the Am9513 or the DAQ-STC chip, simply use the first VI in your **Counter** palette. As usual in this chapter, we will run the VI directly, so drop it on the block diagram of a temporary new VI, then open it (see Figure 3–61).

Since you should have your digital output line connected to one of your counter's CLK lines, you should be able to open up `My Write to Digital Line.vi` and cause that counter's clock to count pulses that you manually create. Here's how to do this:

1. Set the **counter** input of whichever of the two previous VIs you're using to match the counter you've wired. If you're using `Count Events or Time.vi`, set your **source edge** and **counter size** inputs as shown in Figure 3–61. If you're using `Count Events (8253).vi`, no adjustments are needed.

2. If you've been working straight through this chapter, you should have your DAQ device set up to write to output line 0 via `My Write to Digital Line.vi`. This line should have been initialized to output, as described in Section 3.3.3. This output line should also have been connected to the **counter** input you've wired.

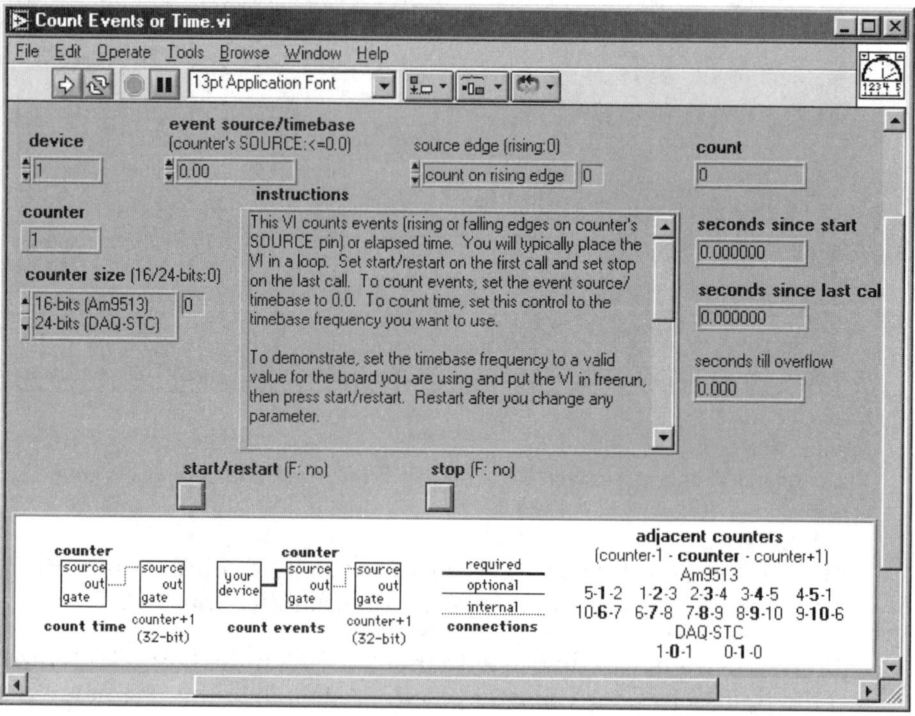

Figure 3–61
This example VI works with DAQ devices having counter chips newer than the 8253.

3. Make sure the **iteration** input of `My Write to Digital Line.vi` is anything but 0; otherwise, we'll be initializing the line. Run the appropriate counting VI, one of the two shown above (either Figure 3–60 or Figure 3–61 depending on your DAQ device), then run `My Write to Digital Line.vi` a few times while changing **line state** from True to False to True to False, and so on, every time you run it. The **count** output of your counting VI will increment for every rising (or falling) pulse of your digital signal.

If you have done this much, you have demonstrated the most commonly used function of your counter—counting. To avoid lengthening this chapter any further, I will not demonstrate how to generate a pulse or a pulse train, nor will I demonstrate the many other uses of the counter, although it's tempting.

You might wonder, why not just use LabVIEW to continuously acquire digital data, and do the counting in software? In a perfect world with infinitely fast hardware and software, we would not need any counters. But here in the real world, most DAQ devices cannot continuously acquire digital I/O, and even if they could, many digital I/O events happen more quickly than can be handled by even the fastest of computers and/or DAQ devices.

As of this book, all of NI's counters are either 16 bits or 24 bits. If this isn't enough for you, you can *chain* two or more counters together, so you can have as many bits in one virtual counter as your DAQ device has counter bits! Your LabVIEW math could get tricky with more than 32 bits, but it's possible. See Chapter 2, Section 2.2.4 for counter-chaining details.

3.4 LabVIEW File I/O with DAQ

You'll often want to save your DAQ data to files so you'll have a record of your testing. Or at least you'll have the record until your hard drive crashes! In this section, we will use some of our simple VIs created earlier in the chapter. We will cover the most basic file I/O techniques in this section. More complex file I/O techniques are addressed in the more complex Chapter 6.

First, open your previously created `AI Single Points and Chart.vi`, and focus on the part of the block diagram shown in Figure 3–62.

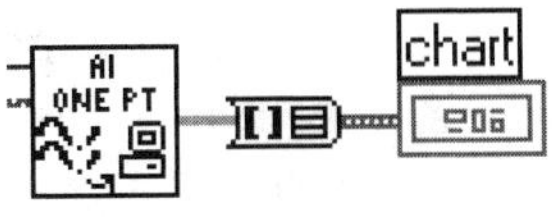

Figure 3–62
Part of the block diagram of our previously-created `AI Single Points and Chart.vi`.

We're going to attach a little "file I/O" action to the array of numbers coming from that DAQ VI. We will write data to disk in both binary and ASCII format (these formats are described in the preface). Keep this block diagram open; we'll get back to it.

Build the VI in Figure 3–63 by using the Array To Spreadsheet String function (in the **Functions»String** palette) and popping up on its terminals to create the constants, the indicator, and the control as shown. Be sure to make your array 1D.

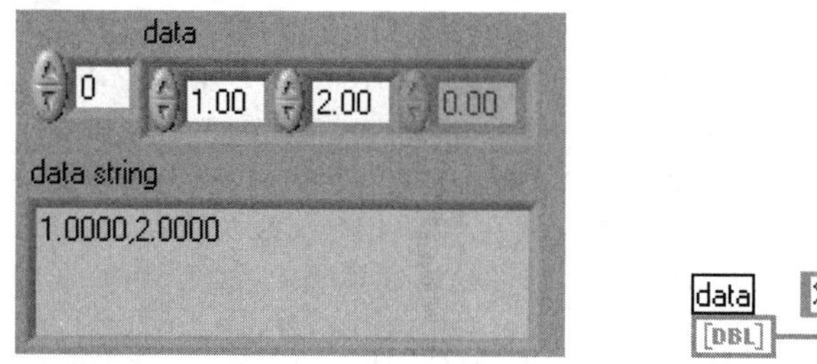

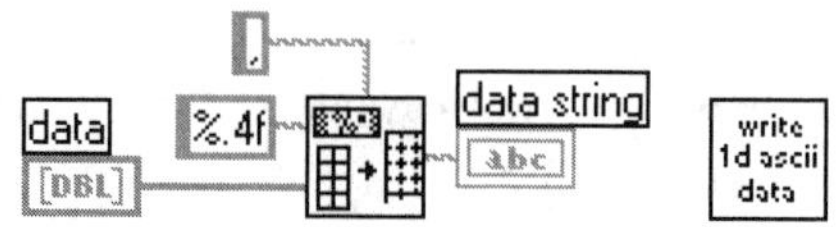

Figure 3–63
We create this simple VI from scratch, named `Write 1D Ascii Data.vi`.

Save this VI as `Write 1D Ascii Data.vi`.

We're using the Array To Spreadsheet String function to convert our array to a string, which is compatible with most file I/O functions. Run the VI as shown, and you should see `1.0000,2.0000` in the `data string` indicator. The Array To Spreadsheet String function, wired as shown, gives us a string of numbers with four digits of precision, hence the four zeros. A comma separates these numbers, and there's an *end of line* character at the end. You can see this end of line character if you pop up on the `data string` indicator and select the backslash option—you will see `1.0000,2.0000\r\n`. On a PC, this end of line character is really two bytes, but if you are developing on a non-PC platform, you might see only the `\r` or the `\n`. If you're not clear on this, reread Section 1.13.3 in Chapter 1, particularly the fourth paragraph.

Now let's make `Write 1D Ascii Data.vi` earn its name. Drop a Path control on the front panel, add `Write Characters To File.vi` (from the **Functions»File I/O** pallette) to the block diagram, and make the changes shown in Figure 3–64 (your path will be different—point it to an existing folder, but a nonexistent file). Go ahead and try this VI to write some simulated DAQ data, by following these steps:

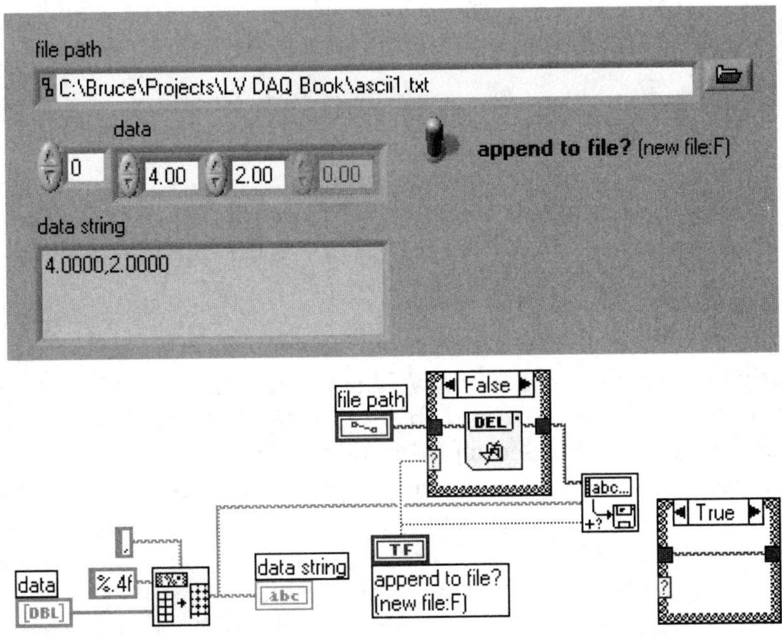

Figure 3–64
We create a VI using the high-level file I/O controls.

1. The Delete function can be found in the **Functions»File I/O»Advanced File Functions** palette
2. Specify the path of a file called `ascii.txt` that you will soon create; the folder should exist already, but not `ascii.txt`.
3. Specify two or three numbers in your "data" control.
4. Make sure `append to file?` is set to False, so we can create a new file. The key controls on my screen now look like Figure 3–65.

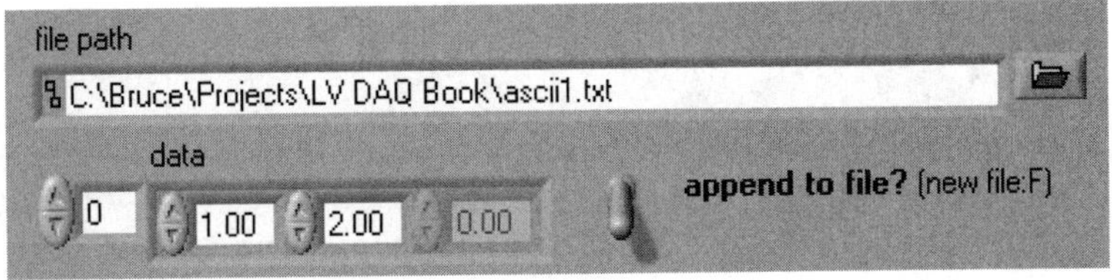

Figure 3–65
Front panel items for the VI we're creating.

(Unless your name is Bruce, your `file path` will likely be different.)

5. Run the VI.
6. Set `append to file?` to True, and change the values of your data, while keeping the array size (two or three) from step 2.
7. Run the VI a second time.
8. Now, open the text file with your favorite text editor. The ability to read the file with *any* text editor is the big advantage of using ASCII, rather than binary data.

This VI will write data to an ASCII file specified by `file path`. We will shortly be putting it in a loop. We'll set `append to file?` to False on the first iteration of our loop, and this will erase any file that's there and begin writing a new file. Subsequent iterations of the loop will have `append to file?` set to True, and this, as you might have guessed, will append data to our file rather than create a new file.

If you're experienced with LabVIEW, or if you have an amazing memory, you might be wondering why I didn't use the `Write To Spreadsheet File.vi` mentioned in Chapter 1. If we had used it, we would have a simple block diagram. But in the interest of saving time and reducing the page count of this book, we will be using a *slightly* modified version of the VI we just created in Chapter 6 to show you how to use binary data with file I/O.

I wish the `Write Characters To File.vi` had an input to suppress dialog boxes, because if we don't use the Delete function as shown in Figure 3–64, we will be asked if we want to replace the file if it already exists.

Let's now use `Write 1D Ascii Data.vi` as a subVI in a real DAQ VI. Wire its connector pane as in Figure 3–66 (such wiring is described in Chapter 1, Section 1.10), noting that the `data string` indicator is unwired.

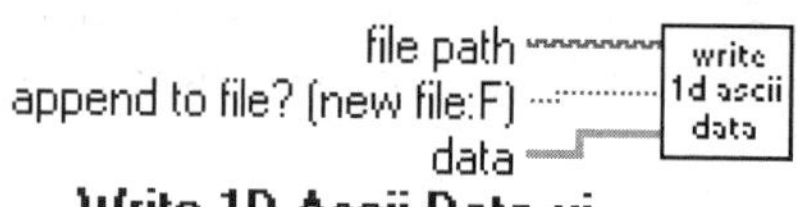

Figure 3–66
The connector pane of the VI we're creating.

I got the image in Figure 3–66 from the Help window of my new subVI—you could add further information as described in Section 1.3.3, if interested.

Incorporate your new VI as a subVI as shown in Figure 3–67 into your `AI Single Points and Chart.vi`, as shown.

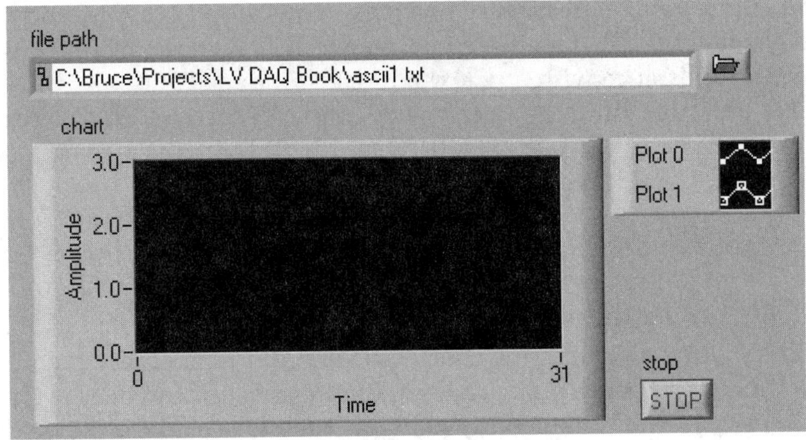

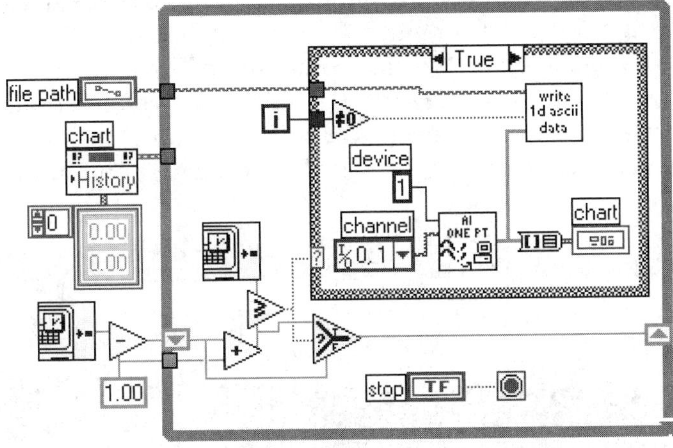

Figure 3–67
We incorporate our new VI as a subVI into a previous VI we've created.

Again, your path will likely be different than the one shown.

If you have your DAQ device wired as instructed in Section 3.1 of this chapter, you can now generate some real DAQ data and save it to disk; then we'll read it later. Open `Update Two Analog Signals.vi` and `AI Single Points and Chart.vi`, then arrange their front panels so that you can fully see them both. Make sure `file path` is valid. Run both VIs for about 10 to 20 seconds, while changing the voltage controls on `Update Two Analog Signals.vi`. You should see the chart on `AI Single Points` and `Chart.vi` reflect these changes. Now, stop both VIs, and you should have a new `ascii.txt` file containing real DAQ data on your disk (you can verify its existence with any text editor).

We will now build an extremely simple VI to read and display the data from `ascii.txt` on a LabVIEW graph. Build this new VI, using these instructions and hints:

1. You can drag the `file path` control from `AI Single Points and Chart.vi` to the front panel of a new VI, so you don't need to reenter the path.
2. You can also drag the chart from one VI to the other, so you won't need to set the point styles and autoscaling. However, we want a graph here (not a chart) so we are sure to see all the data—so pop up on the chart and use the replace function to make it a graph.
3. That's the Spreadsheet String To Array function (in the **Functions»File** pallette) on the block diagram in Figure 3–68. By default, it expects to produce a 2D array of DBL data, which is exactly what we have, so we don't need to explicitly wire it as such. The %f string generically converts any style of floating-point number.

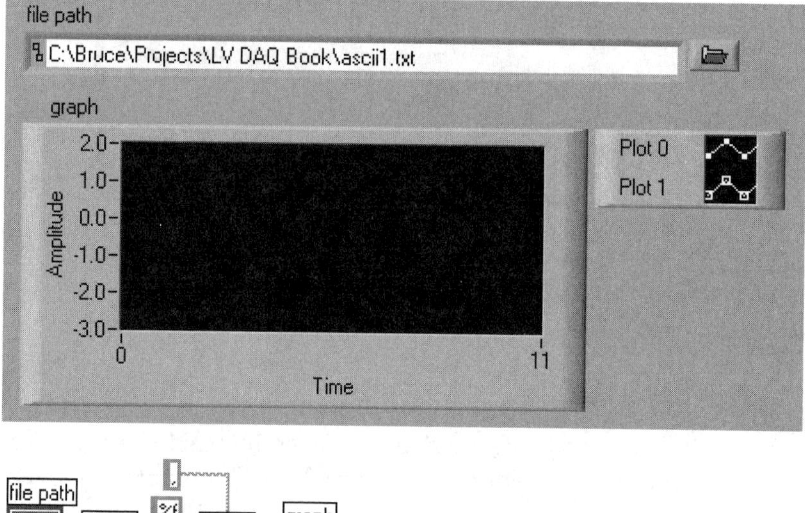

Figure 3–68
This VI reads a certain type of spreadsheet file (CSV, numeric) and graphs it.

Run this VI, and you will likely see some really weird-looking data like this on the graph (see Figure 3–69).

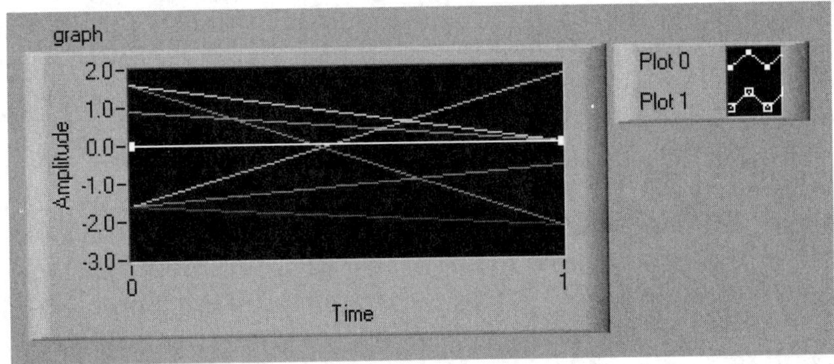

Figure 3–69
The VI shown in Figure 3–68 initially shows this data.

This is many plots of two samples of data, whereas we want to see two plots of many samples of data. To make this happen, simply pop up on the graph and select the **Transpose Array** menu item, which swaps the rows and columns of a 2D array. Your data should look much better now, something remotely like Figure 3–70, although your patterns will be different depending on how you moved your slide controls.

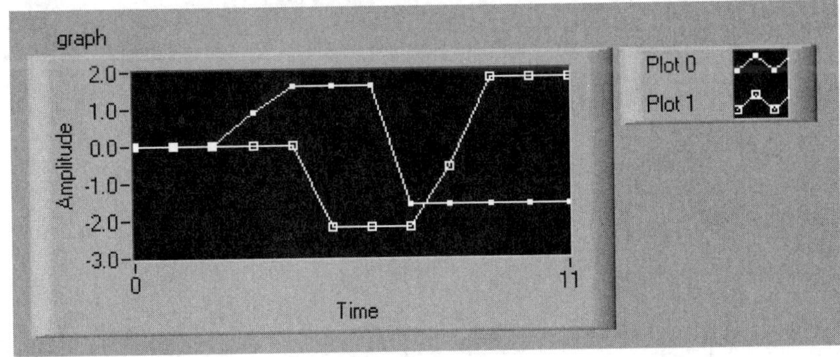

Figure 3–70
The data shown in Figure 3–69 looks like this when the graph's **Transpose Array** menu item is invoked.

Save this VI as `Read and Display CSV Data.vi`. It would have made sense to end our file name with `.csv`, as well as `.txt`, as many spreadsheets recognize both extensions.

If you want to save data in binary format instead of ASCII, Chapter 6 addresses this. Your computer processes binary data far more rapidly than ASCII, although you cannot read binary data with a text editor or word processor. Also, binary data produces smaller data files.

If you want to save data from continuous analog input, you will probably want to work with a 2D array rather than a 1D array, as multiple channels are often used. Although this chapter seems long, saving 2D data is a part of many DAQ projects, so let's do this in the last exercise of this chapter.

Start by opening `Write 1D Ascii Data.vi`. Save it now as `Write 2D Ascii Data.vi`, update its icon to reflect its new name, then change the front panel objects to look like Figure 3–71, using these tips:

1. Your `file path` will be different than the one shown, but use an existing folder and a nonexisting file name like `ascii2.txt`.
2. `data` has been changed to a 2D array.
3. Make sure `append to file?` is False.
4. The block diagram doesn't change a bit!

Run this VI, which will create a new file containing the data you see. Change `append to file?` to True, and change the data to that in Figure 3–72.

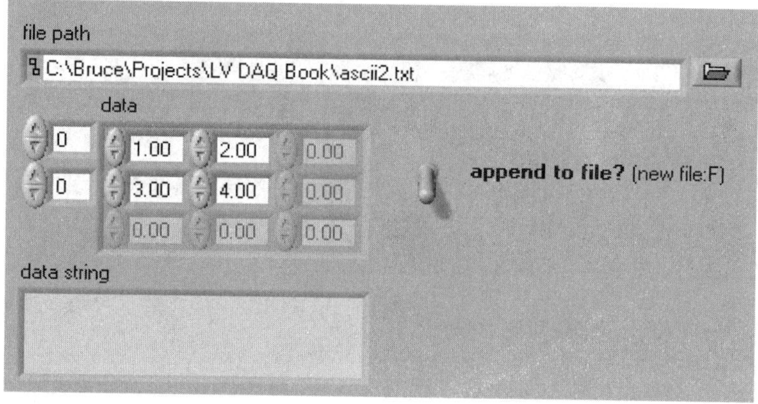

Figure 3–71
Building a front panel for file I/O.

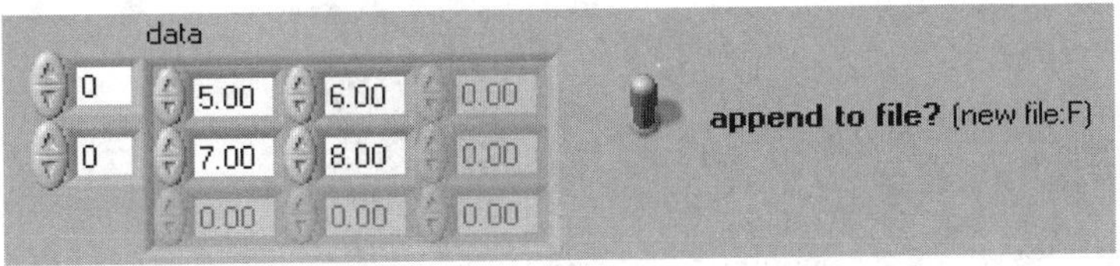

Figure 3–72
Appending to a file rather than overwriting it.

Then, run the VI again. This will save data in the same format as our earlier 1D example, so you could use the same `Read and Display CSV Data.vi` to read the data from your new `ascii2.txt` file, as in Figure 3–73.

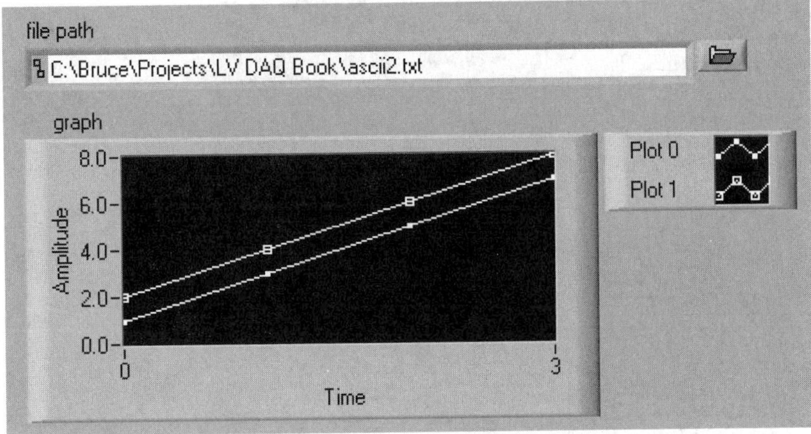

Figure 3–73
The data from `ascii2.txt` is shown on a graph.

We will now incorporate our new `Write 2D Ascii Data.vi` into our previously built continuous acquisition example, `My Cont Acq&Chart (buffered).vi`; open this VI, then open the `My Data Processing 2.vi` subVI in its block diagram.

With all these VIs open, follow these steps:

1. Modify the front panel, block diagram, and icon of `My Data Processing 2.vi` as in Figure 3–74 (`file path` may be empty, as shown, as we'll be using this as a subVI).

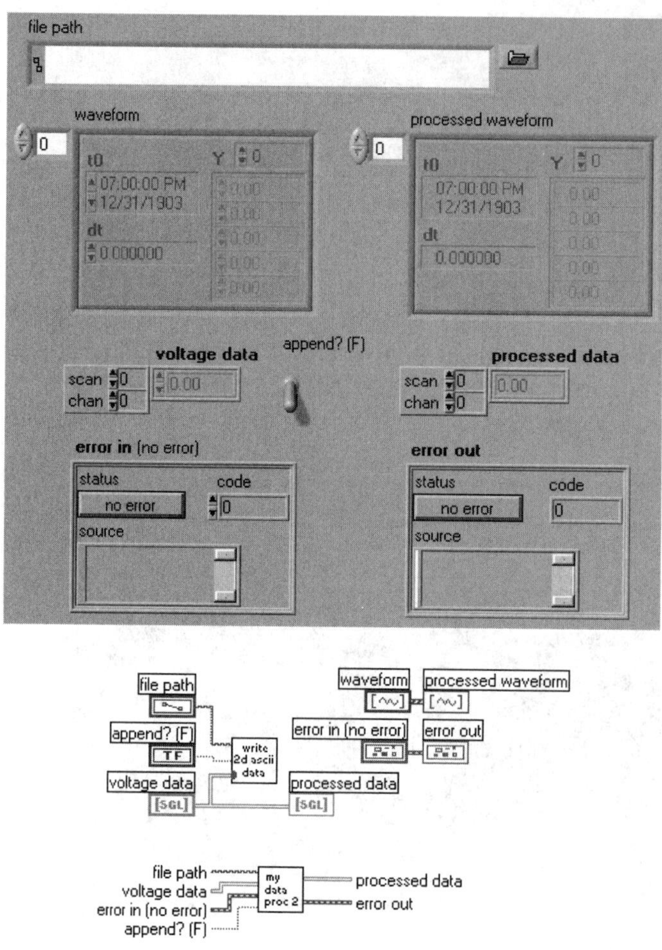

Figure 3–74
An extremely simple file I/O VI.

2. Modify `My Cont Acq&Chart (buffered).vi` as shown in Figure 3–75—add the Path control, the shift register wit its constants, and the history-clearing Property node. Your `file path` will be different, but make sure its folder name exists and its file name is a nonexistent `ascii3.txt`.

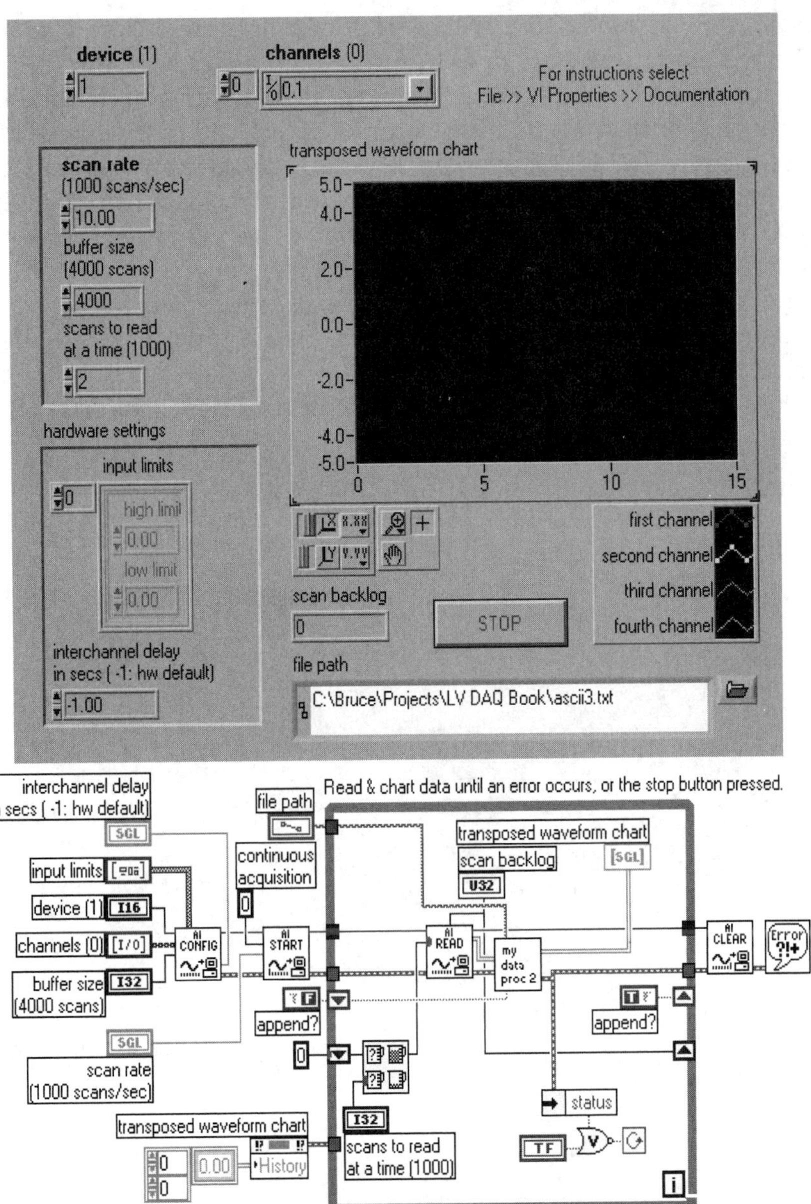

Figure 3–75
A complex DAQ VI with file I/O.

You are now ready to test this VI. Open `Update Two Analog Signals.vi` along with this one. Run `Update Two Analog Signals.vi`, then run this VI. As this VI is running, slowly move each slide control on your analog output VI for a few seconds apiece, alternating between the two slides, running for a total of about 30 or 40 seconds. Stop the analog input VI, then stop the analog output VI. To verify that your data has been saved properly, you can use `Read and Display CSV Data.vi` to show the data as it is stored on disk.

The VI in Figure 3–75 is not doing any error checking. If you were to implement a VI similar to this demo, you would want to add an error cluster output to the file-writing subVI and use it.

Close any VIs you might have open with the **File»Close All** menu item.

Simulation Techniques

4

Simulation is an important tool for DAQ software development that permits you to test your program's functionality without having any DAQ devices connected. This can be convenient for many reasons. Suppose your DAQ system will ultimately be running in a different city, where your testing would require travel. Unless you really like to travel on business, you will want to simulate the system near your home before the system is finished. In this book, running your program without any DAQ hardware (other than a LabVIEW-capable computer) will be termed *soft simulation*. To take it a step further, suppose you have the DAQ device you'll be using, but you want to run your program without connections to the real signals. We will also cover a variety of *hard simulation* techniques, in which you generate real electrical signals that simulate the actual signals that your DAQ device will ultimately be handling.

It is especially nice to have a large monitor when simulating, because you can then temporarily expand the size of your front panel to accommodate the display of your simulated inputs and outputs. When you're satisfied everything is working properly, you can delete the simulated inputs and outputs from the front panel, then shrink it to its original size.

This chapter will first discuss soft simulation, then hard simulation.

4.1 Soft Simulation

Here we discuss the different categories of signals in order, as the simulation techniques in each category vary. First, analog input will be discussed, then the other forms of DAQ signals. Many concepts will be presented in this upcoming analog input section that will apply to all sections, so don't miss it.

4.1.1 Analog Input

Let's further divide analog input simulation into *single-point simulation* and *waveform simulation*. We will begin with the simplest form of analog input: single-point simulation sampling a single channel. We've already built such a simple VI in Chapter 3, but have another look at the DAQ portion of `AI Single Point and Chart.vi`, shown in Figure 4–1.

Possibly the simplest way to simulate this type of data would be to replace the DAQ function with a random number generator, as in Figure 4–2.

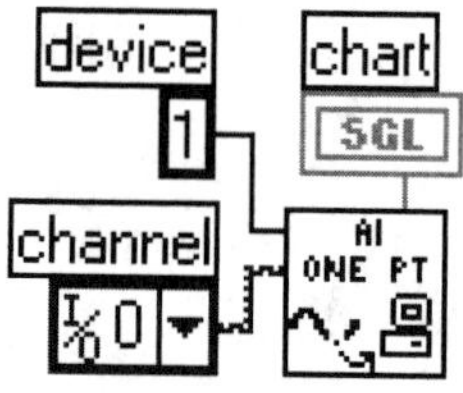

Figure 4–1
The DAQ portion of `AI Single Point and Chart.vi`.

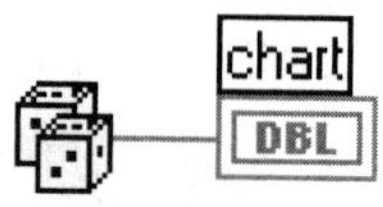

Figure 4–2
The DAQ function has been replaced with a random number generator.

This is our first example of simulating DAQ data. Unfortunately, it's not a very powerful one, because we have no control over the data as the VI is running. Place a Horizontal Pointer Slide control on your front panel from the **Controls»Numeric** palette and label it `Slide`. We'll use it to manually control our simulated data by wiring the DAQ portion of our block diagram as in Figure 4–3.

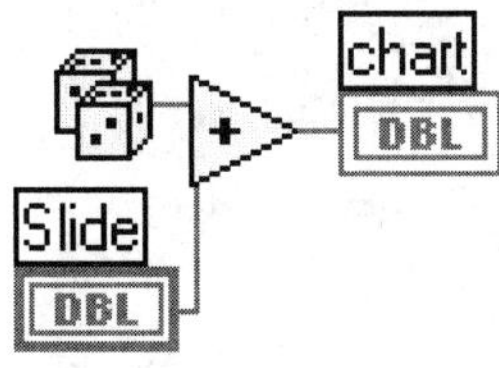

Figure 4–3
The numeric control lets us control our simulated data.

To generalize even further, the above example illustrates that we can use front panel controls to control our simulated DAQ data. If you were acquiring data from multiple channels, or if you were acquiring waveforms, you could use arrays on the front panel to simulate this.

Simulating analog waveforms is usually not very difficult, provided you can define your waveform as a function of time. For example, suppose you want to simulate a 1 Hz sine wave that you're sampling at 20 Hz, and your acquisition loop is running at 5 Hz, so you would have 4 data points per iteration of your LabVIEW loop.

Figure 4–4a shows a VI that will generate just such a 4-point section of a simulated pure sine wave per iteration of the While Loop.

Build this VI, as we will use it again in Chapter 6. The **Functions»Numeric»Conversion»To Single Precision Float** function is used, as this is the data type used by real DAQ functions. Use a Waveform Chart on the front panel, turn autoscaling on for the y-axis, and set the point style as shown. For simplicity and clarity, we are sending an array of numerics rather than the Waveform data type to the chart, and we are not scaling the x-axis to seconds or milliseconds—so the units seen simulate 1/20 of a second.

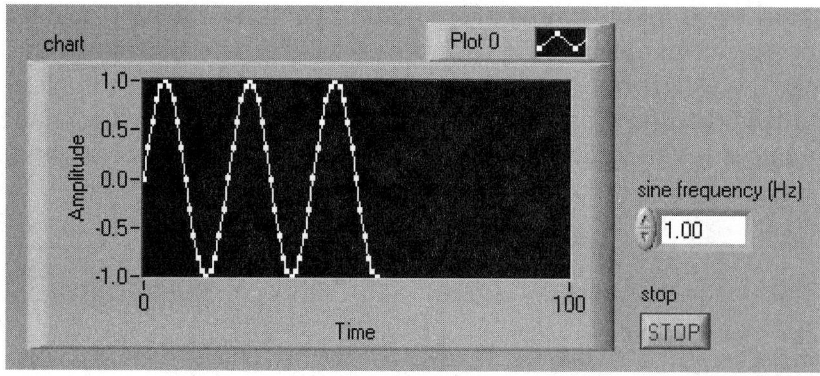

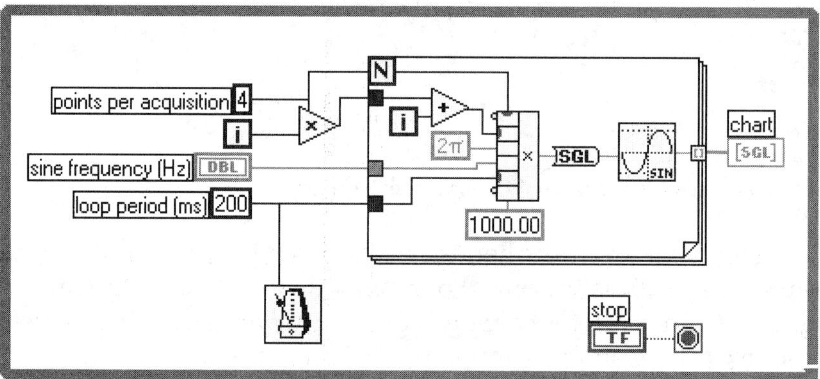

Figure 4–4a
This VI simulates continuous analog acquisition.

If you arrange the block diagram items as shown in Figure 4–4a, you can easily select the For Loop, the Multiply function, and the wires in between to create the subVI seen in Figure 4–4b (use **Edit»Create SubVI** with those items selected). Save this subVI as `Simulate Sine Section.vi`. The subVI should look like Figure 4-4c; if you need to rearrange its terminals through its connector pane, take care to make the right connections! Run the VI to verify that a sine wave is being drawn on the chart four points at a time. Finally, save the calling VI as `Simulate AI Waveform.vi`.

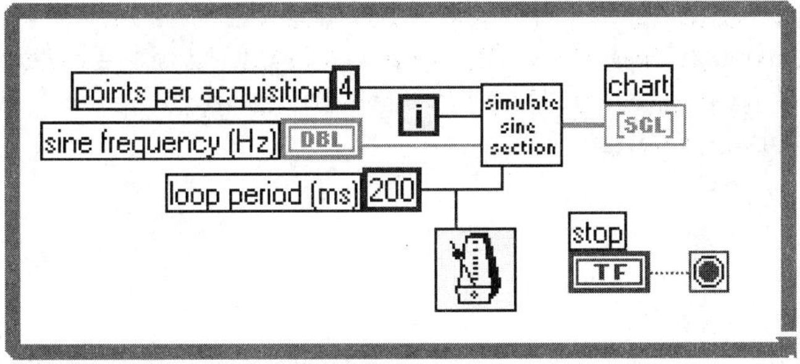

Figure 4–4b
The new block diagram of `Simulate AI Waveform.vi` is shown with its new subVI `Simulate Sine Section.vi.`

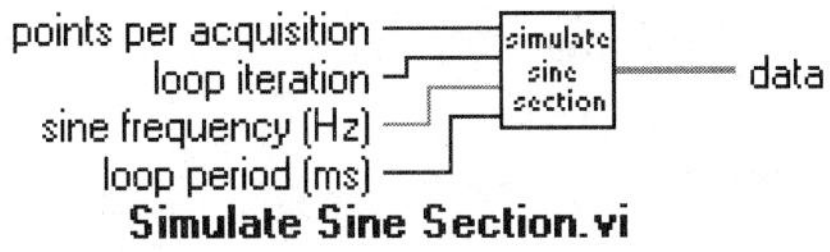

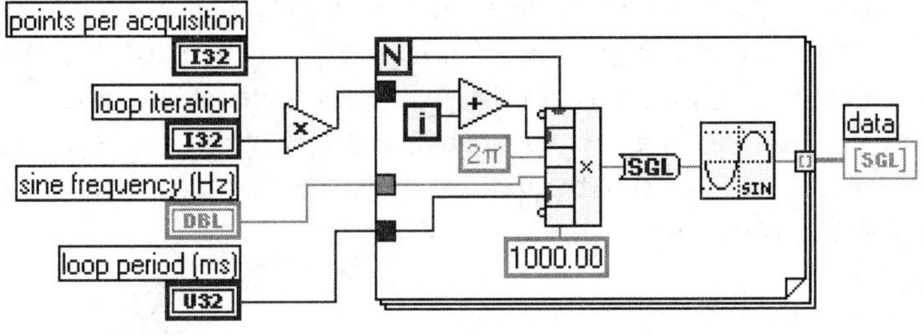

Figure 4–4c
This should be your terminal pattern and block diagram of `Simulate Sine Section.vi.`

4.1.2 Analog Output

Analog output simulation techniques are very similar to analog input simulation techniques. Rather than passing the data to a real DAQ function, pass it to a front panel indicator of your choice so you can monitor this value. Charts are especially useful for this because they show a history of data, unlike other indicators.

For analog output simulation, you may want to dedicate a numeric global variable per analog output line, since like real DAQ hardware, a global variable can be accessed from any VI. Heed my warnings in Chapter 1, Section 1.14.3 concerning globals, though.

4.1.3 Digital I/O

Digital I/O simulation is also very similar to analog I/O, except you generally use Booleans or integers to simulate your data, depending on whether your program operates on digital lines or digital ports, respectively.

For digital output simulation, as with analog output simulation, you may want to dedicate a Boolean or integer global variable per digital output line or port.

4.1.4 Counters

The heart of a counter is an N-bit integer, also called a *register*. For counting purposes, such an N-bit register can most easily be simulated with an unsigned N-bit integer global variable. Suppose you are simulating a 16-bit counter. You could create a VI that doesn't directly correspond to any DAQ VI, but rather simulates M digital pulses physically coming into your counter. Figure 4–5 is an example of such a VI.

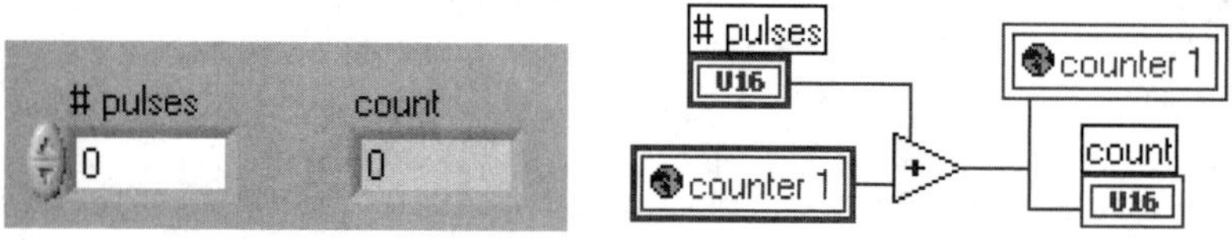

Figure 4–5
An example of a VI that simulates M digital pulses physically coming into the computer.

As it is the nature of unsigned N-bit integer addition to roll over to zero after the count 2^N-1 is reached, the simple structure shown in Figure 4–5 accurately simulates a 16-bit counter.

Suppose you're trying to simulate an N-bit counter that does not conveniently have 8, 16, or 32 bits, like LabVIEW's integers do. Suppose you're trying to simulate a 24-bit counter. In this case, you can programmatically modify the VI in Figure 4–5 with the Remainder & Quotient function, as in Figure 4–6.

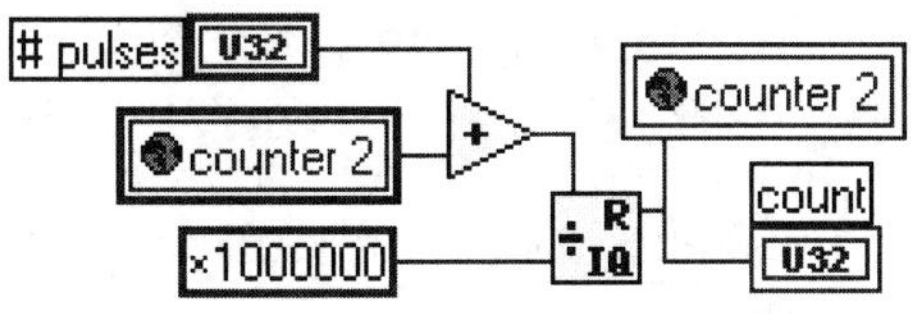

Figure 4–6
A VI modified with the Remainder & Quotient function.

The hexadecimal number `1000000` shown in Figure 4–6 is really 2^{24}. We are using 32-bit integers, since 32 is greater than 24, whereas 16 is not.

If you're not simply counting events, simulating timers is not always as straightforward as everything else mentioned in this section. If the timers are timing events that are slow enough, you can use LabVIEW's timing functions, do a bit of math, and try to approximate your timer's behavior. If the timers you're simulating are too fast to be simulated with LabVIEW's timing functions, you will need to come up with your own math.

4.1.5 Soft Simulation for Complicated VIs

As your DAQ VIs get more and more complex, you may find the DAQ functions buried deeply inside your code, many subVIs down. In this case, it is not always easy to wire directly from the front panel to the block diagram, so you may want to use the two following tricks: (1) Create simulated versions of all of LabVIEW's DAQ VIs that you're using, and (2) pass the data from your front panel simulation controls to these simulated DAQ VIs via global variables.

I will expand these two steps into many, mentioning useful details. Be patient—they'll pay off if you need to do this.

1. **Create simulated DAQ VIs.** Drop **AI Sample Channel.vi** from the **Functions»Data Acquisition»Analog Input** palette onto the block diagram of a new VI. Choose its "scaled value" version with its **Select Type»Scaled Value** pop up menu item. Open it and save a copy with its file name preceded by `Sim`, which would

be `Sim AI Sample Channel (scaled value).vi`, *in the same directory* as the rest of your application's VIs. This method saves such simulated DAQ VIs' connector patterns, which will maintain the wiring geometry should you choose to later replace them with the original DAQ VIs. For each simulated DAQ VI, give a small unique symbol to its icon that will catch your attention at a glance. My personal preference is a small bright green s in the upper left-hand corner (probably looks like dull gray blob in this illustration): AI ONE PT becomes AI ONE PT.

2. **Create global variables for simulation control and for simulated data.** In your new VI from the previous step, create a Boolean global variable for overall DAQ simulation control. Set this global variable to True or False at the *very beginning* of your overall program, and *never* write to it thereafter. While you have this global's front panel open, go ahead and create another global or globals that will represent the actual data you are simulating. Since we are simulating just one number in this example (one channel, single-point acquisition), create a lone floating-point number. Your global's front panel, for our example, should now look something like this: daq sim, sim volts 0.00.

3. **Rewire simulated DAQ VIs.** Go to the block diagram of your newly created simulated DAQ VI, select everything, then delete everything with the <Delete> key so that only the front panel terminals remain. My copy of LabVIEW is showing a bug that forces me to save, close, then re-open my simulated DAQ VI at this point. Next, drop the original real DAQ VI in the middle of this block diagram, and wire the corresponding terminals (see Figure 4–7).

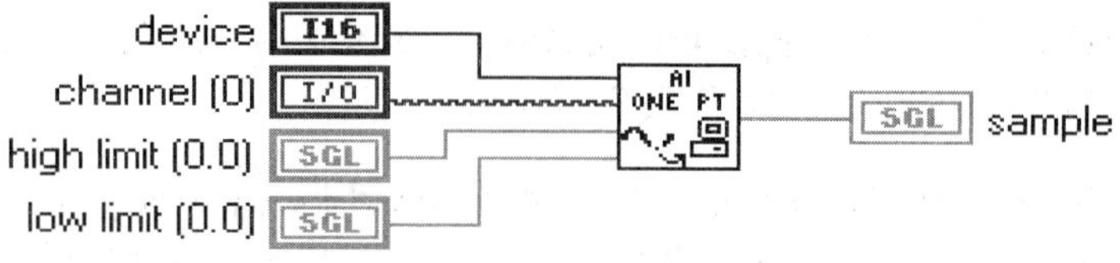

Figure 4–7
A real DAQ VI is used as a subVI in preparation to build a simulated DAQ VI.

As it stands, we've really done nothing so far; our simulated DAQ VI acts just like the real one. So let's make it different, using your global variables from Step 2, as in Figure 4–8.

4. **Use your new simulated VIs by writing to your simulated globals.** The VI shown in Figure 4–9 uses our new simulated DAQ VI; we control our simulated data with the `sim volts` control. Note that without the Sequence Structure around the `daq sim` global, the loop might execute before that global is set.

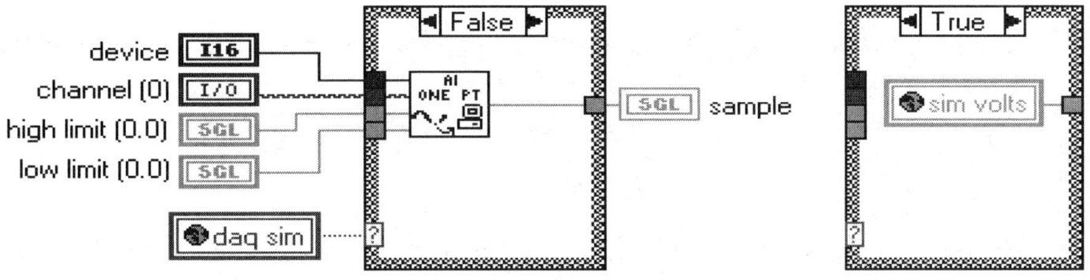

Figure 4–8
Global variables are useful in building simulation VIs.

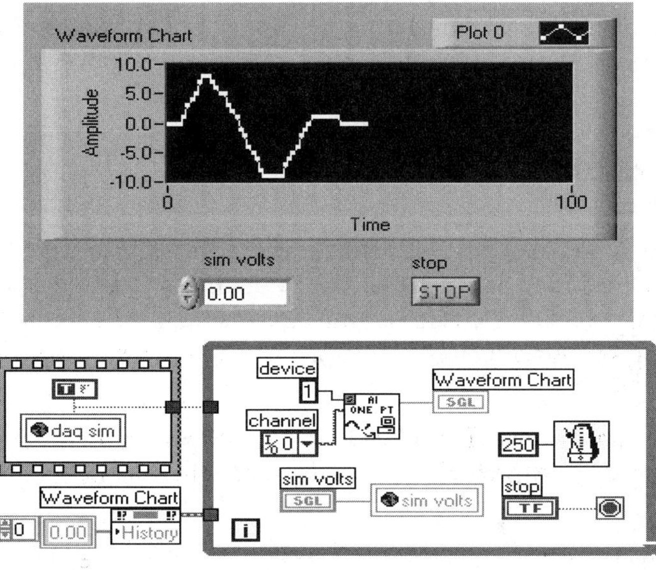

Figure 4–9
Our simulated DAQ VI in action.

To get the chart's plot you see in Figure 4–9, I ran the VI with `sim volts` at `0.00`, then changed the value of the `sim volts` control with its up/down arrow keys during the next 20 seconds or so, then stopped the VI.

4.2 Hard Simulation

WARNING—In this section, always make sure the simulation signals' voltages do not exceed the specs of your DAQ device.

Suppose you have your DAQ hardware, but you don't have the real signals to which this hardware will ultimately be connected. Why not simulate these signals? As a consultant who prefers to work at home, I do this for many DAQ projects to minimize time spent onsite. Not that I don't like my clients. I really do! Keep paying me, clients!

This section will cover real simulation signals from DAQ devices, from lab equipment and from homemade devices.

4.2.1 Simulation Signals From DAQ Devices

Luckily, many of NI's DAQ devices are multifunction devices, meaning they have many different kinds of inputs and outputs. If you have unused output signals on any of your DAQ devices, you can use them to simulate external inputs. An analog output, when set to 0 volts or 5 volts, could simulate a TTL digital signal.

Keep in mind that many DAQ devices cannot produce accurate waveforms on their analog outputs due to the lack of DMA associated with that analog output.

Counters can be used to generate all sorts of digital signals, including square waves of varying duty cycles. LabVIEW has examples of this, some of them right on the **Functions** palette (such as `Generate Pulse Train.vi`), so I won't bother repeating them in this book.

4.2.2 Simulation Signals From Other Lab Equipment

Companies like Agilent, Tektronix, and WaveTek make function generators whose primary purpose is generating analog voltage waveforms. As with any analog signal, they can also be used to simulate digital signals. Even the most basic function generators can generate sine waves, square waves, and triangle waves. They will also allow you to adjust the frequency, DC offset, and amplitude of these waves.

Some of the more advanced function generators allow you to simulate arbitrary waveforms that you define, should your project require nonstandard waveforms.

If you are measuring signals that are slow enough, a simple variable power supply can simulate your analog signals. Be especially careful here not to exceed the voltage limits of your DAQ device, as most variable power supplies easily exceed the basic DAQ devices' voltage specs. A voltage divider made of two resistors is a good safety mechanism in this case, as can be seen in Chapter 2, Figure 2–19.

"Hamfests" and Internet auctions can be good sources for used lab equipment of this type. No guarantees, usually, but these sources can pan out.

4.2.3 Simulation Signals From Low-Cost, Homemade Devices

Some of the information in this section might be dated, but even so, it should give you some ideas.

I've put together a list of tools I personally use quite often in DAQ work:

1. a voltmeter that can read Ohms as well (Fluke makes reliable models)
2. small, flat-blade screwdriver
3. 22-gauge solid wire
4. wire cutter
5. wire stripper
6. needle-nose pliers
7. several test leads with tiny hook clips on either end
8. solderless breadboard
9. IC extractor (*only* if you're using ICs, like the one shown in the breadboard in Figure 4–10)

Figure 4–10 shows my own versions of these tools for assembling test circuits.

The solderless breadboard allows you to quickly assemble test circuits. I've put in a little DIP (Dual In-line Pin) chip just to show you how these are used, but you can easily add more common discrete components, such as resistors, capacitors, and so on. Additional helpful tools would be a DC power supply (if your DAQ device can't provide enough current for your test circuit), a Philips screwdriver, screwdrivers with larger handles for more torque, soldering equipment, a wire stripper for very tiny wires, and a second wire stripper for very large wires. Where do you get these parts? See the Electronics Suppliers section of *www.LCtechnology.com/daq-hardward.htm.*

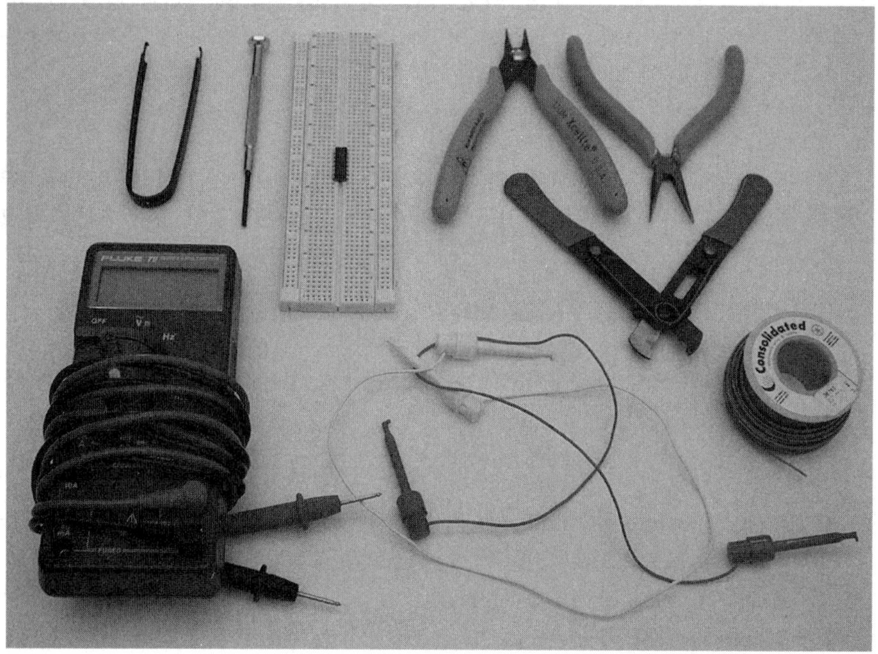

Figure 4–10
My own well-used DAQ tools—the basics.

If you need a sine waveform or other simple waveform in roughly the 20Hz to 20KHz range with a peak-to-peak magnitude of 3 volts or less, and don't want to spend hundreds of dollars on a function generator, a *common sound card* will do this! See Chapter 8, Section 8.2.1 for more details.

If you want to build a homebrew circuit from scratch, Radio Shack has a couple of useful handbooks, *Basic Semiconductor Circuits* and *Op Amp IC Circuits*, both by Forest M. Mims III, published by Radio Shack, which talk about generating sine waves and square waves. Jameco sells an inexpensive function generator kit. This kit is based upon a dedicated function-generating IC, so you may want to look around and see what types of function-generating ICs exist; Digi-Key is currently a good source for these. If you only want to generate digital waveforms, simple square waves of varying duty cycles can be built from parts mentioned earlier in this paragraph, or highly complex patterns can be generated from a microcontroller (from Microchip, Motorola, and others). Microcontrollers, as fun as they are (for me), can be time-consuming to learn and use.

Low-speed digital signals can be read from or written to a computer's parallel port. See Chapter 8, Section 8.2.2 for details.

DAQ Debugging Techniques

5

Is LabVIEW not doing the DAQ things you were expecting? Maybe you have PEBCAK (Problem Exists Between Chair And Keyboard), meaning you've made a mistake (quite easy), or maybe there's another problem with the software or the hardware that's not your fault. We have already reviewed LabVIEW's basic debugging tools in Chapter 1, Section 1.11. Let's now look at some additional techniques for tracking down LabVIEW/DAQ problems in general.

5.1 LabVIEW's Debugging Tools

LabVIEW's probe and toolbar buttons were created specifically for debugging. They are described in Chapter 1, Section 1.11, as they're rather fundamental to LabVIEW. Please review those few paragraphs quickly, then we'll mention some other techniques.

5.2 Tracking Down Bad Data

If you can isolate your bug to a certain condition in terms of LabVIEW's data, then this section contains some ideas you should try first. For example, suppose you have some LabVIEW data on a block diagram that should never be negative, yet it's occasionally negative. If this is the case, you can use the Case Structure to catch that negative data only when it occurs, plus any other useful data from your block diagram that might be causing the problem.

Whether you're using the Case Structure to catch your debugging data only during certain conditions, or you want to continuously watch the data, there are a number of ways you can see this data. Here are some useful techniques:

1. The *probe*, as described in Chapter 1, Section 1.11, is the simplest way to view the data. Its obvious disadvantage is that you can only view one data point at a time.
2. The *state cluster*, as described in the next chapter, is a single cluster that you should often create on your main While Loop that contains all important state information. For debugging purposes, it is easy to add an extra data element to this state cluster for display on the main panel, then remove it later.
3. *Global variables* can be used like the state cluster, but should be used as a last resort, since it's sometimes tough to know exactly when they're updated.
4. *File I/O* can be used when you need to analyze a large amount of data for debugging purposes at a later time. File I/O can slow down the normal operation of your program, as it takes some time to access the disk.
5. *Serial I/O* (the RS-232 port) can be used to stream the data to another computer. Although tougher to implement than file I/O, serial I/O does not slow the program nearly as much. Actually, any such data transmission to another computer will work as well, but serial I/O is currently the easiest to implement.

If your computer should lock up, the first three techniques do not always guarantee that you're seeing the most recent data, particularly if your screen goes blank. The last two techniques are rather crash-proof, particularly the serial port strategy, since file writes are often buffered in memory unless the file is flushed or closed. The second computer will have control of the data as soon as it leaves the serial port of the first computer.

5.3 Solving Crashes

Is your computer crashing for no apparent reason? Maybe it has a virus. See *www.pcmag.com* for *PC Magazine's* opinion of the best virus protection software—they're usually right on target with their reviews. I've found that certain virus protection software substantially slows down your computer.

I've been counting. LabVIEW itself does not crash very often. As I've written this book, my many untested *beta* versions of LabVIEW 6i have crashed 37 times, including the times I've repeated certain tricky procedures to *try* to make them crash, and I've reported all of these bugs to the responsive NI LabVIEW beta team. By contrast, my "shipping" version of Microsoft Word, which I'm using to type this document, has crashed well over 100 times (I've given up counting!). With Word, I'm doing my utmost to avoid crashes, as they often require a reboot. Based on my knowledge of LabVIEW shipping criteria, I expect it will be relatively difficult to make the "shipping" version of LabVIEW 6i crash. My point? LabVIEW crashes are usually a result of something other than LabVIEW itself. A virus, or some other piece of software, hardware, or hardware driver is usually the culprit. The most common cause of LabVIEW crashes is a lack of graphical system resources while editing. If you have a number of other graphics-intensive programs running with LabVIEW, you are at risk.

The worst kind of bug you can have is a crash that happens infrequently, yet ruins your test. First, try running your software on a completely different computer and see if that solves your problem. If not, your crash may indeed be caused by LabVIEW.

5.3.1 Hardware Induced Crashes

If you really must use the computer that is crashing, try replacing various pieces of hardware. Try removing different cards inside your computer, starting with the easiest to install and uninstall, then progressing to the most difficult. Also, try using a mouse with a different bus (PS/2, serial, or USB); I've run into crashing mice more than once. In the unlikely event that more than one component is causing a crash, your crash hunting will be very difficult! If you track down the crash to a particular piece of hardware, you *may* be able to solve your problem by updating the hardware's driver from the hardware's vendor.

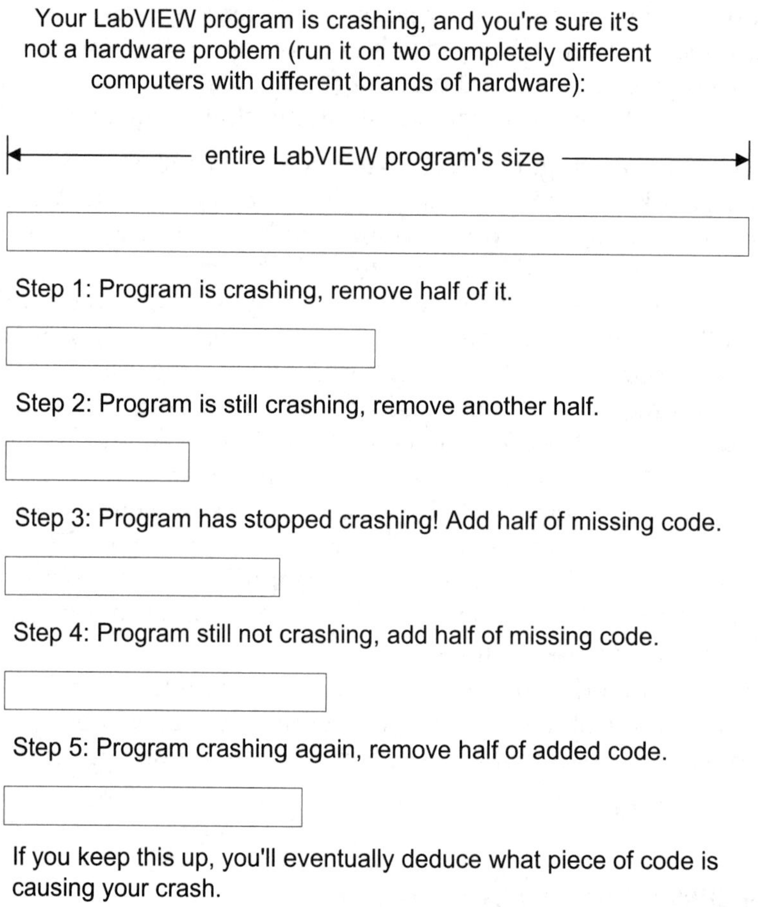

Your LabVIEW program is crashing, and you're sure it's not a hardware problem (run it on two completely different computers with different brands of hardware):
entire LabVIEW program's size
Step 1: Program is crashing, remove half of it.
Step 2: Program is still crashing, remove another half.
Step 3: Program has stopped crashing! Add half of missing code.
Step 4: Program still not crashing, add half of missing code.
Step 5: Program crashing again, remove half of added code.
If you keep this up, you'll eventually deduce what piece of code is causing your crash.

Figure 5–1
Here's my best shot at a graphic illustration of tracking down a crashing bug when no other techniques are working.

5.3.2 Software Induced Crashes

Your crash may be a result of some other piece of software running on your computer. First, try removing programs that are running simultaneously with your LabVIEW program. Some of them may be easy to stop, others not. For Windows, many programs are automatically started in Window's startup folder. To find other sneakier Windows programs, run `msconfig` via Microsoft Windows' **Start»Run** menu. If extremely desperate, save your important data, wipe your hard drive(s) clean, then reinstall the operating system.

If you've tried running your LabVIEW program on two computers with completely different brands of hardware, the crash is likely coming from some component of LabVIEW or NI-DAQ. First, back up your LabVIEW program in a different folder! Next, try removing all DAQ calls completely from your program and simulating your data with random numbers. If the crash still happens, try stripping down your program in chunks roughly half the size of the entire program until you isolate the crash (see Figure 5–1). Then, rebuild smaller pieces of the program until the crash starts again, at which point you start stripping it down again in even smaller pieces. This technique is applicable to any programming environment, not just to LabVIEW. Figure 5–1 is a crude diagram (don't know how to make an elegant one) of how your crash hunting should proceed.

5.4 Last Resort

If you still cannot track down your problem, try getting outside help through a newsgroup or by calling NI. See the preface for more details. NI has some very sharp people working technical support. Though some of them may be new, they will eventually answer most of your questions.

Before moving on to the next chapter, close any VIs you might have open with the **File»Close All** menu item.

Real-World DAQ Programming Techniques

6

In this section, we will cover LabVIEW programming techniques that are especially helpful for DAQ programming. The following areas will be addressed:

1. Programming Structure for DAQ
2. Analog Waveform Analysis
3. File I/O for DAQ
4. Averaging
5. Speed and Efficiency
6. Display Techniques
7. Alarms
8. Power Losses

6.1 Programming Structure for DAQ

This section is applicable to many types of LabVIEW programming, but is specifically tailored for DAQ.

6.1.1 Handling Large Projects

Suppose you're building a really large LabVIEW application, and you run into the usual problem of having many items cluttering your block diagrams with too many wires strewn about. Sound familiar? If not, you have not yet tried to build a really large LabVIEW application. Here's a quick summary of tips for managing such large LabVIEW projects (also useful for small ones!):

1. **MOST IMPORTANT TIP:** For your main While Loop, as most LabVIEW programs have, construct the loop with a single shift register containing a *typedef* custom control we'll call the *state cluster* (described soon) containing all your data that may be used in subVIs.
2. Cluster multiple controls or indicators on your front panel whenever it helps; this reduces the number of terminals on the block diagram.
3. Avoid complex wiring on your block diagrams, and for this purpose, avoid crossing wires wherever possible.
4. Avoid overlapping objects on your block diagrams.
5. Restrict your block diagrams to 800 × 600 pixels if there's any chance they will ever be viewed on other monitors. See Appendix E, item 7 for more details.
6. Add comments to your block diagrams describing the "big picture"—a high-level description of what's happening. Use these sparingly. I recommend coloring them all the same color; yellow or another light color is ideal for printing. I will not use such comments in this book, because I'm trying to keep my block diagrams simple.

Now, let's demonstrate these tips (do not follow along in LabVIEW, until we create the typedef custom control). Figure 6–1 shows an example of not following tips 2 and 3.

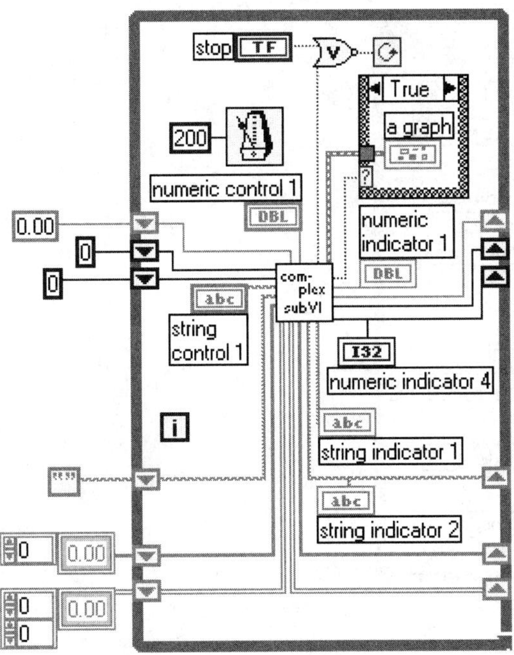

Figure 6–1
A working but sloppy VI—too many shift registers.

Ouch! This hurts my eyes just looking at it. I can simplify this greatly by combining my shift registers into one, as suggested by tip 1; see Figure 6–2.

Ahh—this is much easier on the eyes. Before I describe the typedef business of tip 1, let's apply tip 3. This involves creating clusters of controls and clusters of indicators on the front panel, so as to simplify the block diagram, as shown in Figure 6–3.

As you can see, I've created one front panel cluster for most of the controls, and one for most of the indicators. For cosmetic or logical reasons, you may not want to combine all of your front panel objects as such, but bundle them into clusters whenever you can. I could have also included the `stop` button in the `controls` cluster, then dragged the Not Or function inside the subVI, but I usually prefer to leave it visible for high-level clarity. I have also not included my graph with the indicators, for two reasons:

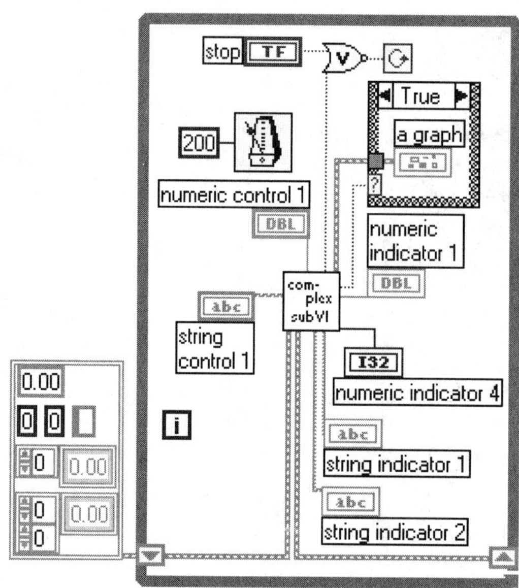

Figure 6–2
Greatly improved from Figure 6–1—only one shift register.

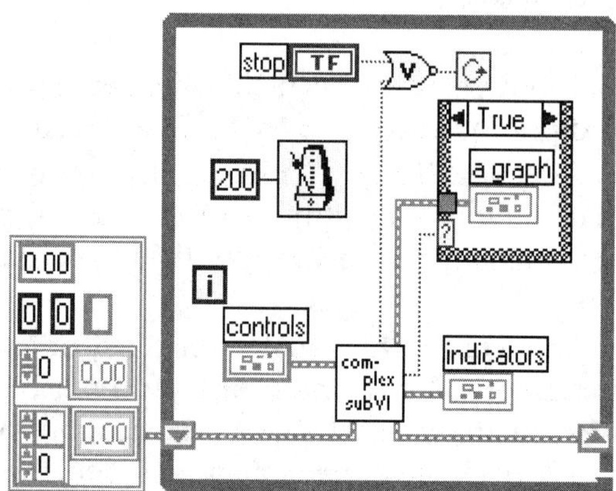

Figure 6–3
Improved even further from Figure 6–2—clustered controls and indicators.

1. Most people like the looks of a graph by itself, not in a cluster.
2. I now have control over the update rate of the graph, as it's in the Case Structure (I may want to update it only at certain times, as a large graph can take a long time to draw).

Now let's cover the typedef issue. The mechanics of creating a typedef are rather involved, so don't be swamped by all of these tedious steps—once you know how to create a typedef, it's not so hard.

So what is a typedef, and why should you want one? Suppose you spend a fair amount of time modifying a control (or indicator) to look and behave just the way you want, and you want to save it because you think you'll use it again. A typedef allows you to change just one saved instance of this control, and the others will automatically change themselves to match the saved one. For example, do you see where the wire from the shift register enters the subVI in the block diagram of Figure 6–3? Figure 6–4 shows what the front panel of that subVI might look like, and the wire connected to the calling VI's shift register is connected to the `cluster in` and `cluster out` terminals.

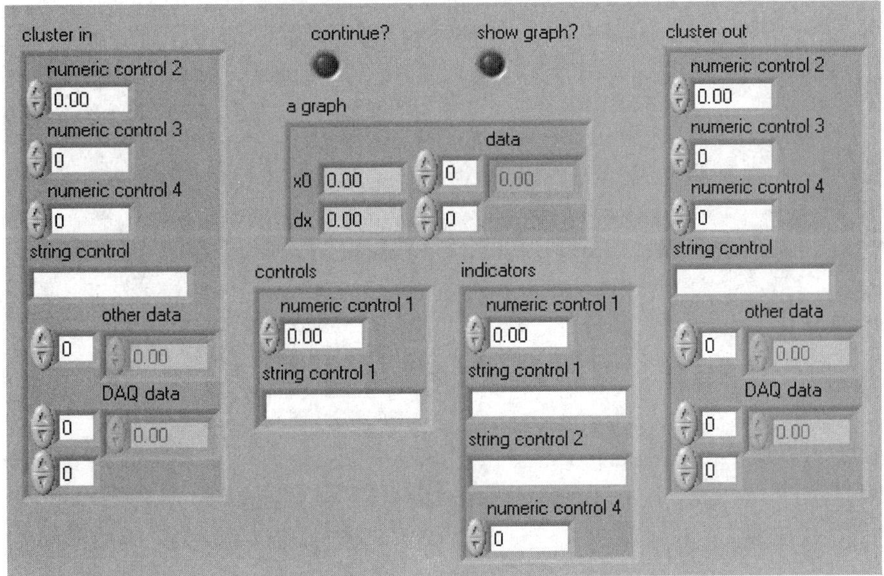

Figure 6–4
Front panel objects illustrating the need for the *typedef.*

The exact same cluster type is used in the constant of the block diagram of Figure 6–3. That's three separate places, just in this simple example, where the same typedef can be used. Your complex LabVIEW projets will often benefit by using a typedef as such to synchronize the data type of a cluster when it is being used by many subVIs.

To create a typedef, we must first create a custom control, which is a front panel object that we customize to our application, then save to disk, usually with a `.ctl` file extension. A typedef is a particular type of custom control—we will create one shortly. A *strict typedef* is like a typedef, but it updates cosmetic properties like color and size, as well as data type, of the control wherever it's being used.

Start following along in LabVIEW now.

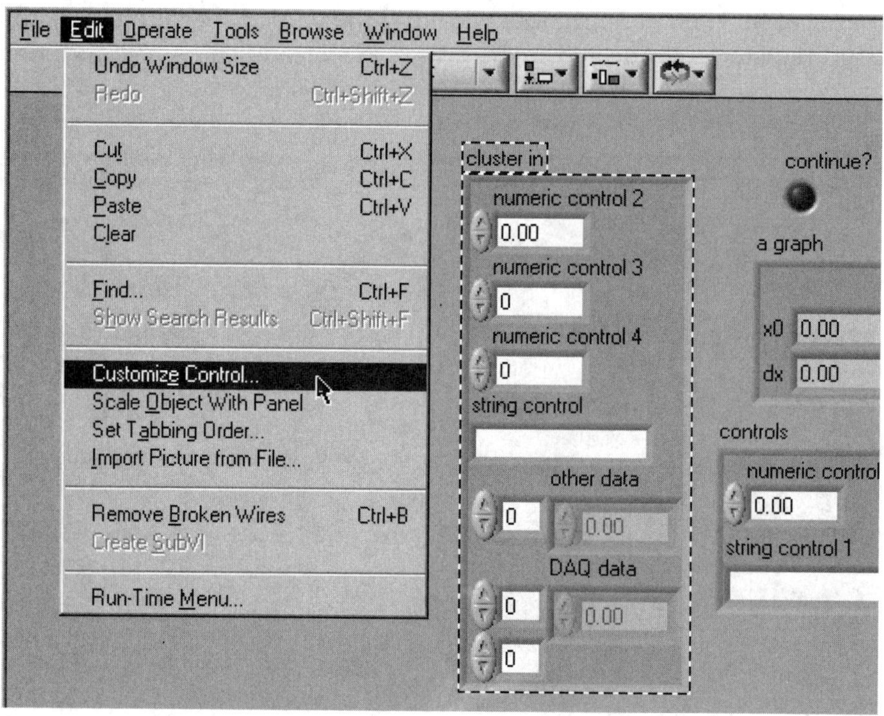

Figure 6–5
Preparing to customize a front panel control.

1. Create a new VI with a cluster control containing the items shown in Figure 6–6, or whatever you want in the cluster. In general, you should *label all of your cluster's elements*, so you can manipulate them by name later on. Copy the control and change the copy to an indicator, then label the two cluster objects `cluster in` and `cluster out`, as shown in Figure 6–4.

2. To change this cluster control into a typedef, *select the cluster control* from your front panel, *then* select the **Edit»Customize Control...** menu item while the cluster control is selected; see Figure 6–5.

   Once you select this menu item, up pops the *control editor*, shown in Figure 6–6, featuring your control (you can do some serious cosmetic surgery here, but that's not the point of this book).

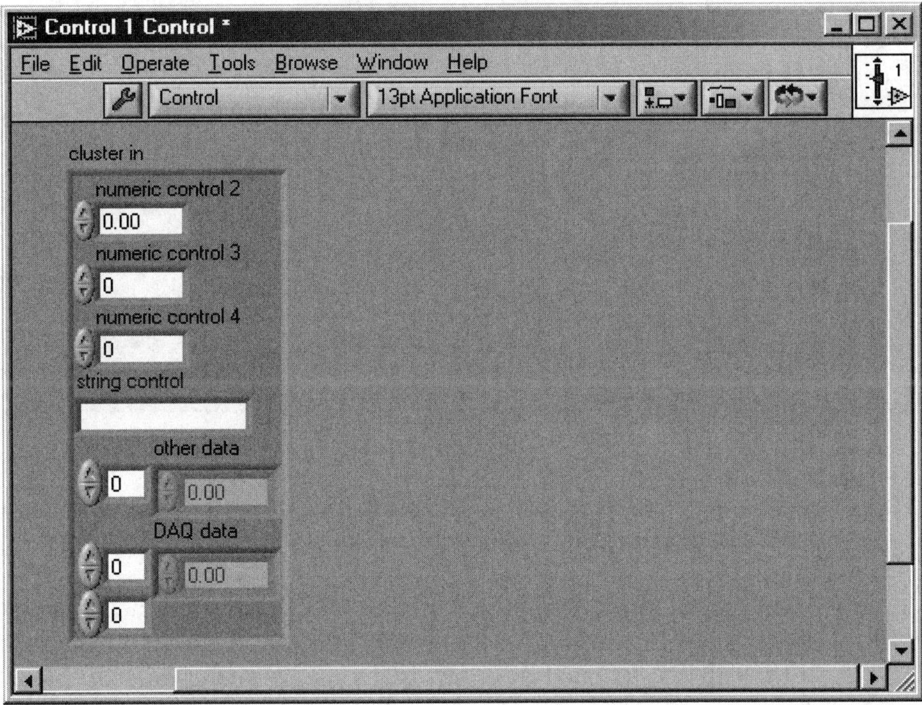

Figure 6–6
A front panel control, ready to be customized.

You can now save this control as a *custom control,* much like you save a VI. Our custom control will soon be saved as a file with `.ctl` extension, not the usual `.vi` extension.

3. Pop up on the **Type Def. Status** ring from the tool bar (which should say "Control" at first), then select the **Strict Type Def.** item, as shown in Figure 6–7.

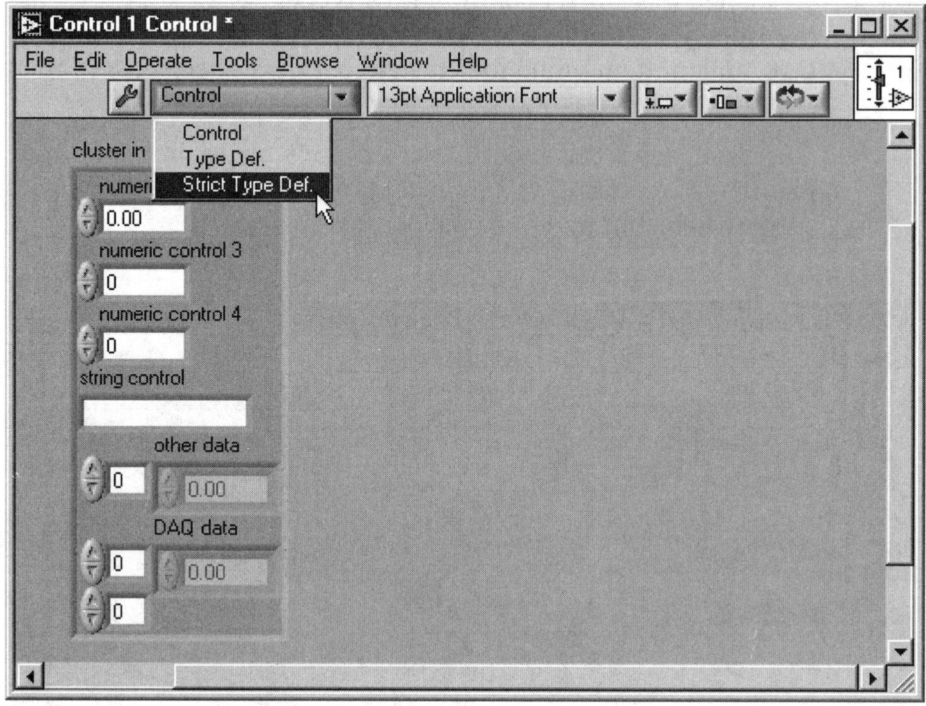

Figure 6–7
Selecting this option, "Strict Type Def.", will force the control to not only take on the same data type, but the same graphical look wherever it's used on any front panel.

4. Save this custom control as `Cluster.ctl`. I recommend giving it an icon, but this is not quite as important for custom controls as it is for subVIs, as custom controls' icons are never seen in block diagrams. Their icons *are* seen in LabVIEW's hierarchy window, however.
5. Close this control editor—when asked to replace the original control, say yes. In the front panel from which this control came, the

`cluster in` has just been replaced with this strict typedef control, but the `cluster out` indicator has not. So pop up on the `cluster out` control, and replace it with your recently saved `Cluster.ctl`, as in Figure 6–8.

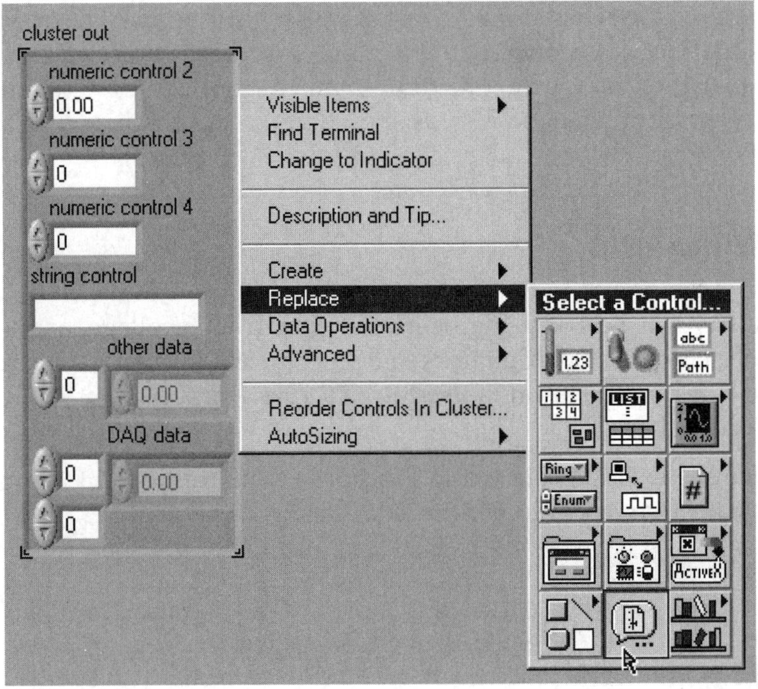

Figure 6–8
A file dialog box will pop up once you select this item, from which you can select your custom control (`Cluster.ctl`), thus replacing `cluster out` with your typedef.

Why is this typedef so special? If you use it in all your subVIs that use this cluster, and you want to change them all at once, you can do so automatically and quickly by (1) opening the typedef custom control, (2) modifying it, (3) saving it, then (4) closing it. You must execute *all four of these steps* before all instances of this typedef will automatically update.

A couple of final tips in this section: When using your cluster typedef in subVIs as a *state cluster,* use the Unbundle By Name function to pick out the elements by name, and use the Bundle By Name function to modify the elements. For this to work, you must label any cluster elements you want to use in your typedef. Also, if you have one of these typedefs on a front panel and you want to quickly edit it, even if you've forgotten its file name, just pop up on the typedef and select **Open Type Def.**

One word of warning—when you increase the height or width of a strict typedef cluster control, and all of its instances are updated, many unseen instances of the typedef may overlap nearby front panel controls. To remedy this situation, it is convenient to place your strict typedef cluster controls such that they may grow vertically, but not horizontally, without overlapping other front panel objects, as shown in the front panel of Figure 6–4.

6.1.2 State Machines

If you have an electrical engineering degree from a college, or if you're some sort of self-educated digital designer, chances are you've heard of a *state machine*. If not, this term is used to describe a process that is always in one state or another.

A simple example, shown in Figure 6–9, is a traffic light, which has three basic states—green, yellow, and red.

Figure 6–9
A traffic light showing a green light (please pretend to see a green light on the bottom in this black and white book).

Figure 6–9 is really a cluster of three Booleans on a LabVIEW front panel that I've created to look something like a traffic light. I've colored the True states of the Booleans green, yellow, and red, from bottom to top. I've colored all the False states black.

Figure 6–10 presents a graphical way to think of this state machine (with just *one* traffic light), supposing we have a really speedy light.

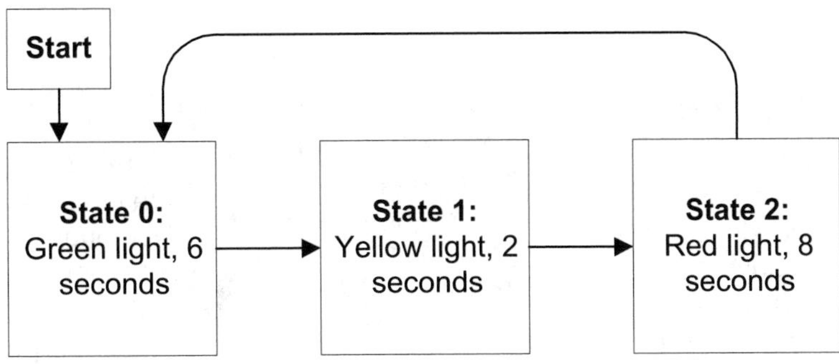

Figure 6–10
Our first state machine.

Real traffic lights had better not be this fast, but we'll build a fast one in LabVIEW so you don't get bored waiting for the light to change!

To implement a state machine in LabVIEW, simply use a Case Structure inside a While Loop, then add an integer in a shift register to cycle through the states. We cannot have a long delay inside any of these LabVIEW states, because we must respond to user events quickly. We should only stay in any given state for a split second. So, let's reconsider a slightly more complex state machine for our single traffic light that doesn't linger in any one state for very long. To do this, you *could* consider the traffic light as a state machine having *six* states: (0) green, (1) changing from green to yellow, (2) yellow, (3) changing from yellow to red, (4) red, and (5) changing from red to green, as illustrated in Figure 6–11.

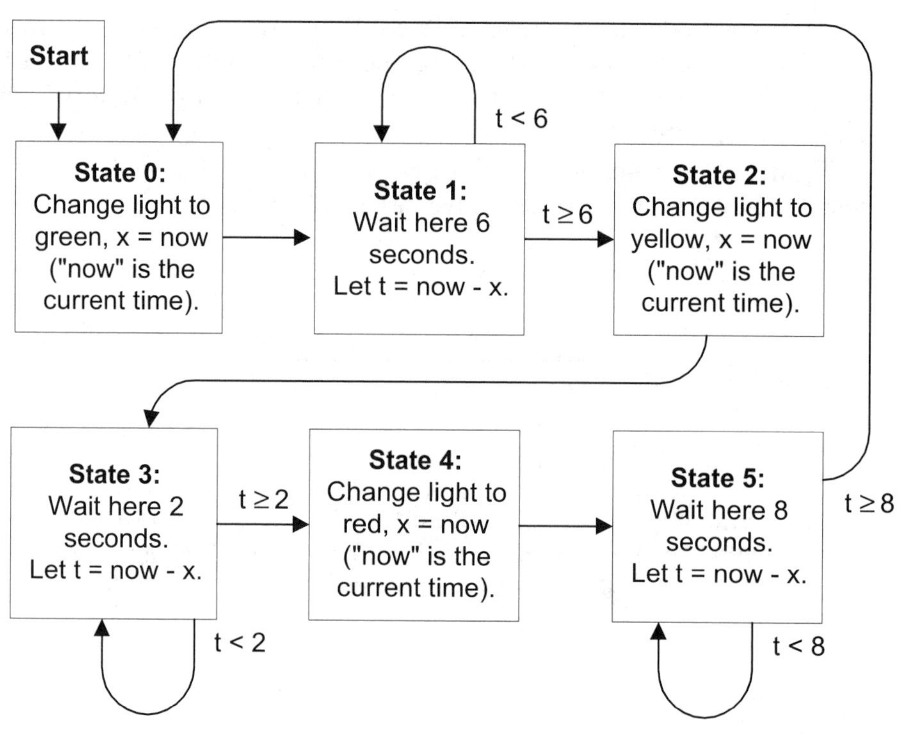

Figure 6–11
A useful rendition of our first state machine, based on Figure 6–10.

LabVIEW has the ability to get the current time in seconds, via its Get Date/Time In Seconds function. For example, I just ran this function, and it returned 3039173669.54, which is the number of seconds since some specific time in 1904, as its Help window indicates. The **x**, **now**, and **t** variables in the diagram use this function to calculate delay times while waiting for a light to change—if this variable business is clear to you, skip the next paragraph.

Start with State 0. The light is changed to green, and the current time is recorded as **x** by the statement **x** = **now**. Let's pretend the current time is 10.0 (it will really be a much larger number, but we'll use small numbers here for simplicity), so the statement **x = now** means that **x** is now 10. The **x**, **now**, and **t** are *variables*, or storage spaces, for numbers. In this state machine, **x** always records the time just before we begin a delay. We then move to State 1; **now**, as always, is the current time. Supposing our main loop occurs every 0.2 seconds, **now** is now 10.2, so **t = now - x** is now 0.2. Effectively, in State 1, **t** tells you how long you've been in State 1. If you have been in State 1 less than 6 seconds, **t < 6** is True, so the **t < 6** arrow indicates that you return to State 1. Once you've been in State 1 for 6 seconds or more, **t ≥ 6** is True, so you proceed to State 2, which will change the light to yellow. The same logic that applies for States 0 and 1 also applies for States 2 and 3, as well as States 4 and 5.

Whenever we're delaying for a light, we're really quickly looping back into the same state, as is usually happening in states 1, 3, and 5 in Figure 6–11.

Let's build this traffic light state machine in LabVIEW. Although we won't be directly using the DAQ VIs, the timing concepts herein are applicable to many DAQ applications. Create the following VI, using these tips:

1. Create a cluster of three Boolean indicators on the front panel, such as the traffic light shown in Figure 6–9. If you don't want to go to the trouble of coloring them, just label any three Booleans as green, yellow, and red, and drop them into a cluster on the front panel. If you do want to create this beautiful cluster control, the Boolean indicators are the Round LED found in the **Controls»Boolean** palette—make sure that their colored states correspond to True, and their black states correspond to False (this can be tricky). In any event, the cluster order is such that green is cluster element 0, yellow is 1, and red is 2. Also drop the standard Stop Button from the **Controls»Boolean** palette, as shown in Figure 6–12.

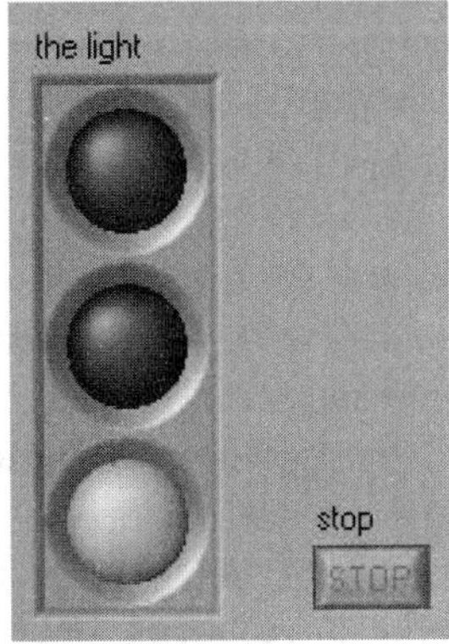

Figure 6–12
The front panel of our VI, with a cluster of three Booleans trying to imitate a traffic light.

2. You can quickly create the block diagram constant `the light` (the cluster of three Booleans inside the `state cluster`) by popping up on the terminal of `the light`, and selecting **Create»Constant**.
3. Start by building the partially complete VI as shown in Figure 6–13, wiring the unseen Case 1 just like Case 0. Make sure `x` has DBL representation, as SGL would not be able to represent time.

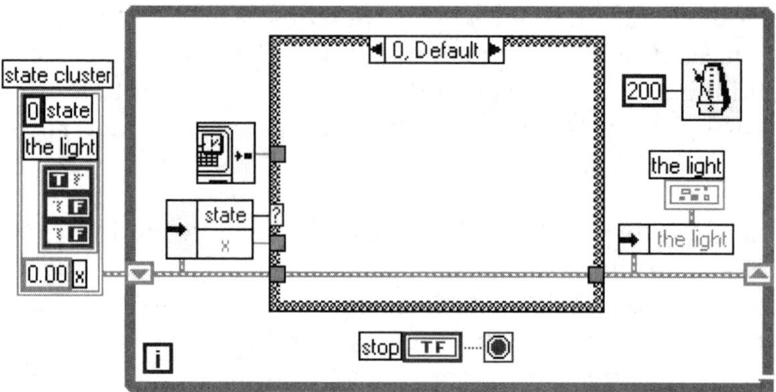

Figure 6–13
We begin to build a "state machine" VI.

Notice that you must label the three cluster elements, `state`, `the light`, and `x`, for their names to appear in the Unbundle By Name functions. I have not shown case 1 in the illustration, but it's wired just like case 0 for now.

4. We will set up the Case Structure cases 0 to 5 to directly correspond to states 0 to 5 in the state machine diagram in Figure 6–11. Create states 0 and 1 in cases 0 and 1 of Case Structure, as in Figure 6–14.
5. Since we need two more cases similar to case 0, and two more similar to case 1, duplicate each of these cases twice—we want a total of three cases similar to case 0 and three cases similar to case 1. To duplicate a case, pop up on the wall of the Case Structure and select **Duplicate Case**.

   The more proper way to build this would be to select everything in case 1 except the two constants, and create a subVI from it before we duplicate it. This would effectively eliminate the redundancy of having the duplicate math in different states. However, we will tolerate this small amount of redundancy in this example for the sake of brevity and clarity.

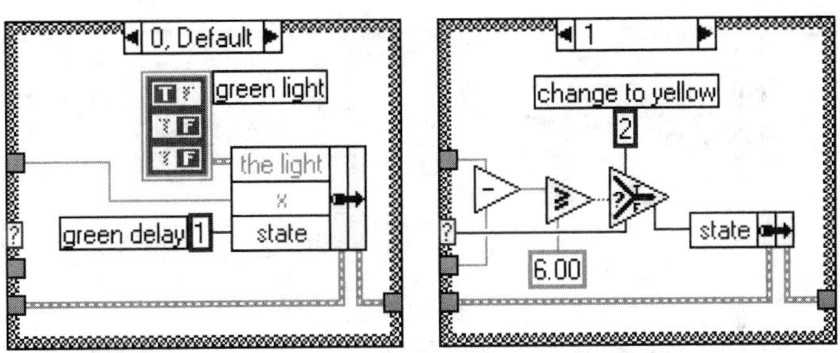

Figure 6–14
Various states of our state machine are being created as cases of a Case Structure.

This paragraph is a little tricky, so follow closely. We should now have three cases whose contents (not case numbers) look exactly like the case 0 , and three like case 1 (refer to Figure 6–14). However, since we've used the **Duplicate Case** function, the numbering will likely be all out of order. Using the little arrow buttons on the Case Structure, and changing the number between those arrows, set the three cases that look like case 0 above to cases 0, 2, and 4. Similarly, set the three cases that look like case 1 above to cases 1, 3, and 5. To make cases appear in order, should you pop up on the Case Structure, you can now sort these cases by selecting the Case Structure's handy **Rearrange Cases...»Sort** item.

6. Modify your cases as shown in Figure 6–15; then you should have a working VI.

   To get super-compact, we could have also combined the cases as such: 0-1, 2-3, 4-5, or 0-2-4, 1-3-5; but either of these formats would have been much more difficult to illustrate and understand.

   Compare Figure 6–11 to Figure 6–15, and you should be able to see an exact, one-to-one correspondence between states and cases.

7. Run the VI, and you should see the traffic light on your front panel spend 6 seconds on green, 2 seconds on yellow, 8 seconds on red, then repeat the cycle until you stop the VI. If you're still having trouble understanding how this works, you can run the VI with execution highlighting while watching the data in the block diagram, paying special attention to the input of the Case Structure's selection terminal.

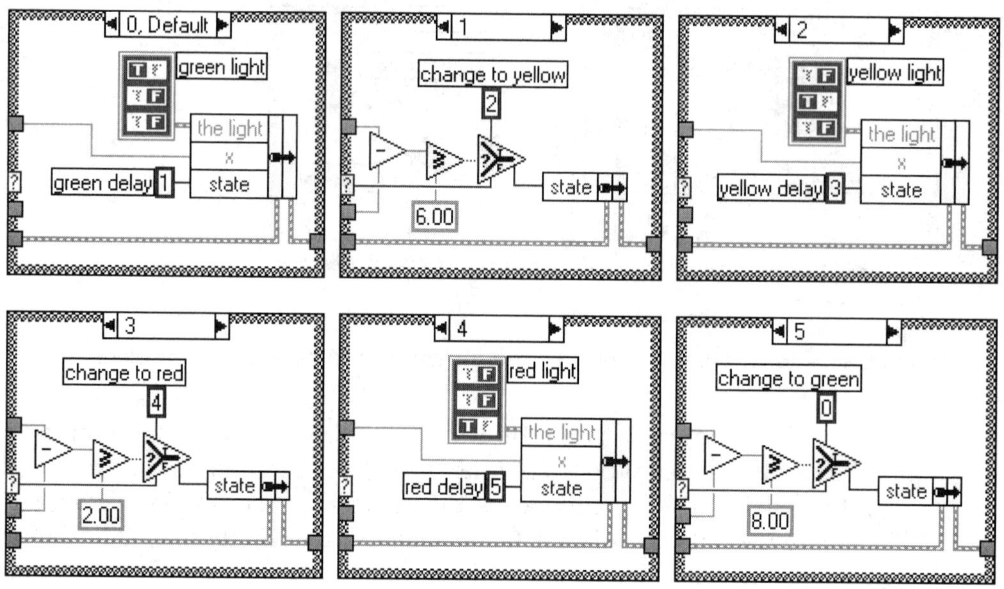

Figure 6–15
All six states of our state machine are implemented as cases of a Case Structure.

For simplicity, I've used integers to represent the states of a state machine. When designing a state machine in LabVIEW, you can use the Enum Constant from the **Functions»Numeric** palette so as to put the *name* of the state at the top of the Case Structure. The underlying data type will still be an integer ranging from 0 through whatever, but the Enum data type associates text with those numbers. When using an Enum Constant to control the state of a state machine in LabVIEW, it is wise to make it a typedef, as described in the previous section. To do this, strange as it may sound, you must start on the front panel and pick the **Controls»Ring & Enum»Enum** control, as shown in Figure 6–16.

After you drop this Enum control on any front panel, type in the following values, using the <Shift-Enter> trick when you want to add a new value:

0: `change to green`

1: `green delay`

2: `change to yellow`

3: `yellow delay`

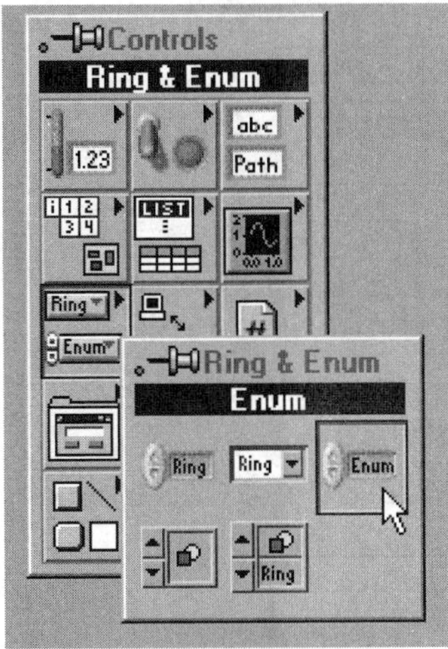

Figure 6–16
Selecting the Enum control on the front panel.

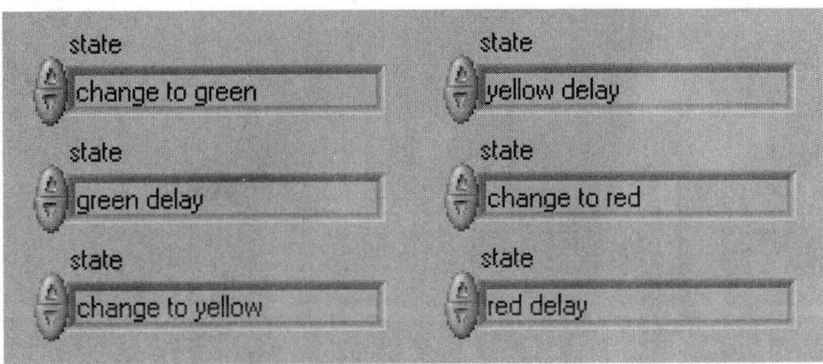

Figure 6–17
Six ring items on the front panel describe the states of our state machine.

4: `change to red`

5: `red delay`

If you have added these six ring items properly, you should be able to see any one of these six values on the front panel object, as shown in Figure 6–17.

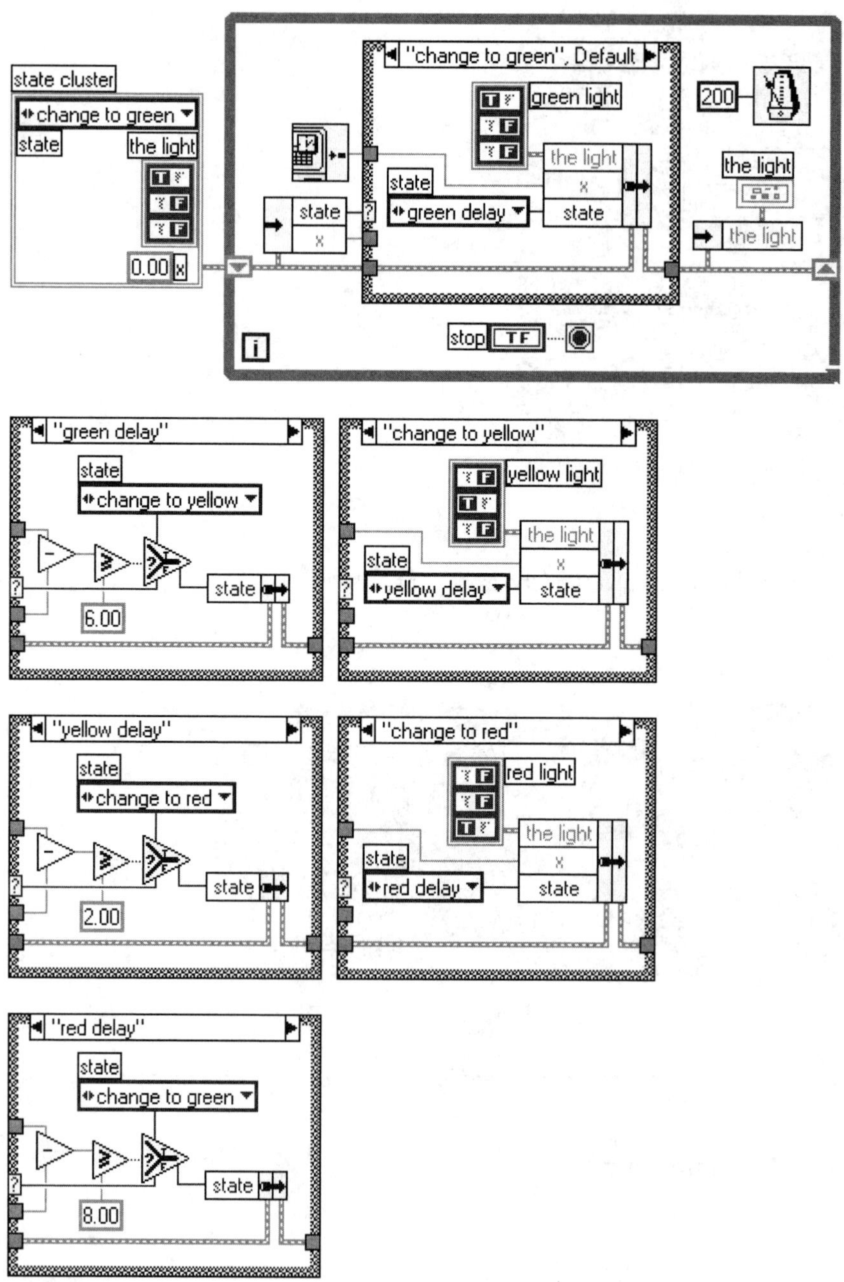

Figure 6–18
This block diagram is easier to understand and maintain than the one without the Enum.

Save this Enum control as a strict typedef `State Index.ctl`, as described with the cluster in the previous section, then delete if from your front panel. At this point, since it's been saved, you can use it on any front panel, or as we're about to see, on any block diagram. A typedef has another advantage in state machines; if you want to add or remove a state later, the typedef will describe each state with text instead of with a number, making it easier to modify your states. Figure 6–18 shows how you should use your new Enum `State Index.ctl` to represent the state (it's in the cluster constant on the left, too). Any custom control can be added to a block diagream as you would a subVI, with the **Functions»Select a VI...** menu item.

Technically, any time you have a Case Structure inside any loop, you have a state machine. Build the VI shown in Figure 6–19, using these tips:

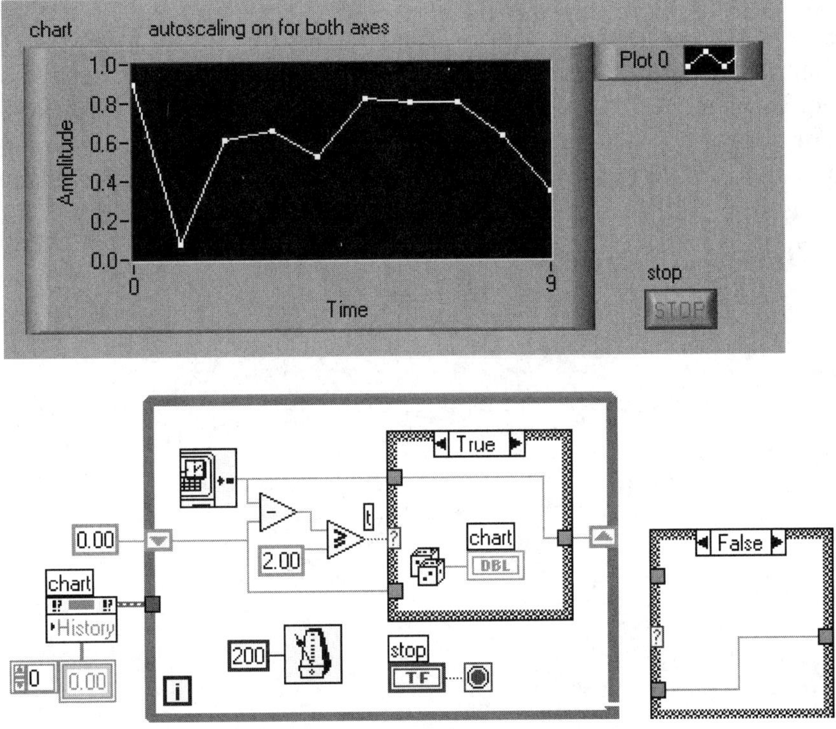

Figure 6–19
A VI simulated simple, slow DAQ is shown.

1. Drop a Waveform Chart on the front panel. Pop up on the chart, turn autoscaling on for both axes, then create the free label as shown in Figure 6–19. Make sure the point style is set as shown. The plot seen will be created later.
2. After you create the chart's Property Node with the History Data property, you can pop up on this node and select **Create»Constant** to create the array of the proper data type.
3. Save this VI as `Simple Slow DAQ.vi`.

Run this VI for about 20 seconds, so a plot like the one in Figure 6–19 is shown. Since these are random numbers, your plot will probably look different.

This block diagram illustrates how to implement very slow DAQ by executing a particular case of a Case Structure only at certain times—in this case, once every two seconds. If you wanted to collect once per minute, or once per hour, or at any rate slower than about once per second, this is a good technique. Notice how this timing differs from that in `AI Single Point and Chart.vi`, shown in Figure 3–37, and consider the difference in behavior when the computer's processor delays for several times the desired sampling rate. In the earlier case, several samples might be taken in quick succession right after the processor's delay, but for long tests, you will get a relatively constant number of samples over a longer period of time. I prefer this style.

Figure 6–20 is a somewhat simplified interpretation of this VI as a state machine, where **t** is the time since the last execution of the True State:

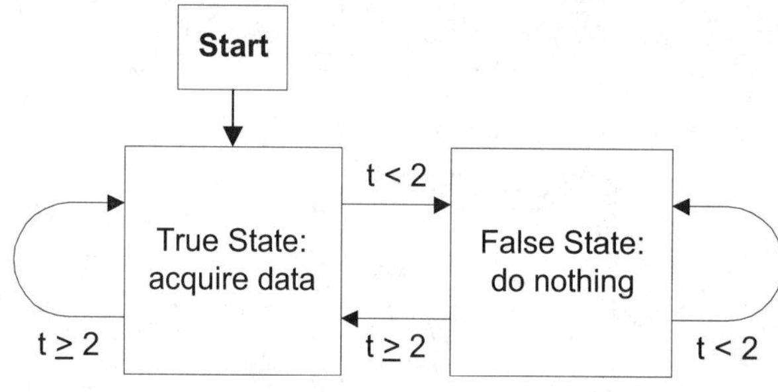

Figure 6–20
A simple state machine is shown.

Many times, you may want to use some DAQ input to control your program. If so, it sometimes helps to consider your VI as a state machine. Suppose you have a refrigerator or some other type of cooling mechanism, with a thermostat that turns the cooler on when the temperature drops below a certain value, then off when it rises above that same value. This is not a good implementation, because when the temperature is right at the critical point, any noise in the temperature-reading mechanism can cause the cooler to turn on and off very rapidly, thus reducing the life of the cooler—and irritating anybody within earshot.

For this reason, most temperature control mechanisms have *hysteresis* built in, which prevents rapid switching, as described in this paragraph. For temperature control, hysteresis defines a temperature range in which the cooler might be on or off *depending on its previous value*. For our imaginary refrigerator, let's define this range as 35° to 40° F, and assume the ambient temperature is 70° F. If you turn the refrigerator on for the first time, the cooler will stay on until the temperature drops below 35°. Then, the cooler turns off until the temperature rises above 40°, at which point it turns on again. In other words, the refrigerator must *remember* whether the cooler was last on or off whenever the temperature is within the range 35° to 40° F. Figure 6–21 shows how it might work.

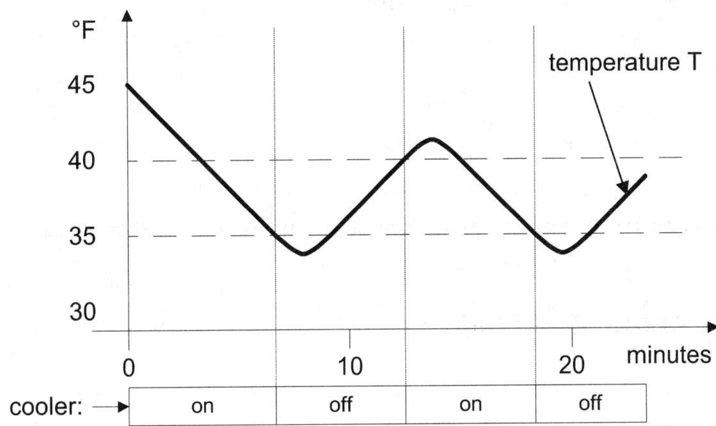

Figure 6–21
Typical temperature control with hysteresis is graphed.

Figure 6–22 shows how a state machine could be designed to provide this type of control.

And finally, Figure 6–23 shows a VI that implements this functionality, where we are simulating the hardware with a front panel control (`thermometer`) and indicator (`cooler`). I've used a local variable in the block diagram so I can write to the `cooler` indicator from two separate cases—if this were a real DAQ control, we would likely have a digital output in the block diagram wherever we are writing to the `cooler` indicator. But the important point here is to notice how we first design a state machine's functionality, then implement the states as cases in a Case Structure, just like we did earlier with the traffic light.

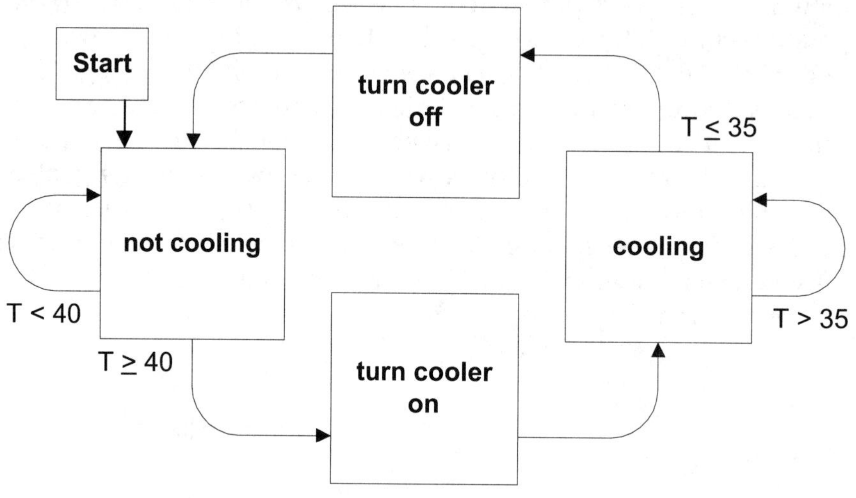

Figure 6–22
A state machine performing temperature control with hysteresis has these states.

6.1.3 Using Examples as Templates

In Chapter 3, we saw how to use an analog input example as a basis for our continuous acquisition VIs (your path may differ from mine):

```
C:\Program Files\National Instruments\LV6\examples\daq\anloin\
anlogin.llb\Cont Acq&Graph (buffered).vi
```

The most important point about using the examples is that you should immediately save them under a different name once you open them (see Appendix C). Failure to heed this warning can result in much time wasted. Figure 6–24 shows a quick overview of LabVIEW's current `examples` folder. Table 6.1 provides information about the folders. See also **Help»Examples...** .

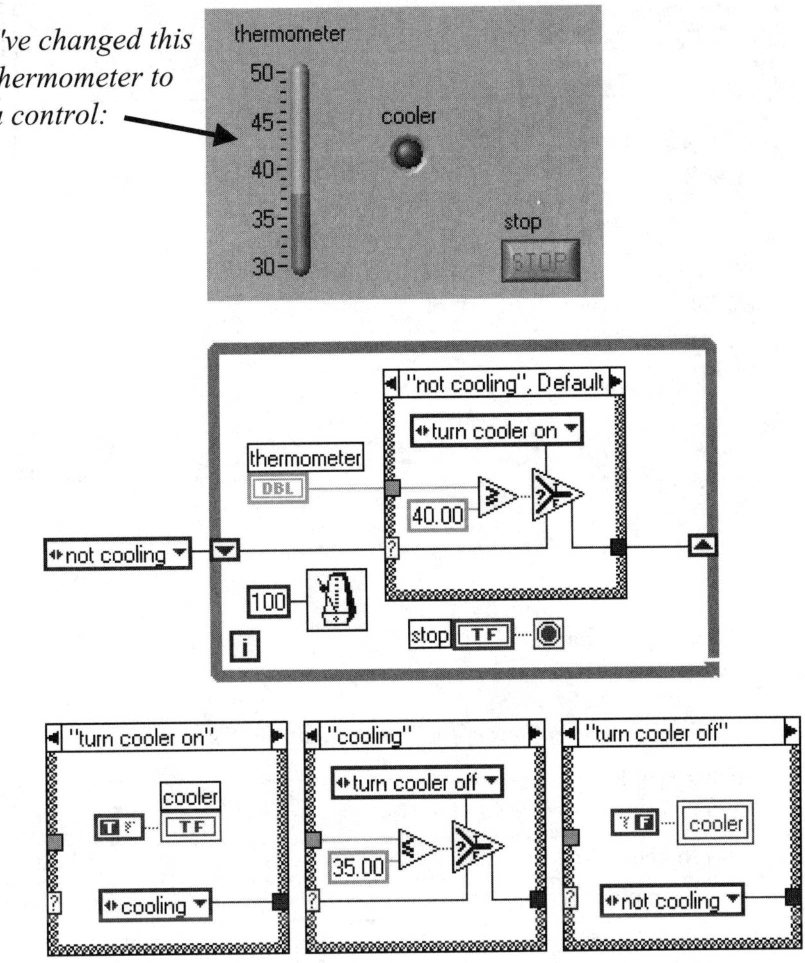

Figure 6–23
A VI performing temperature control with hysteresis could be implemented as such.

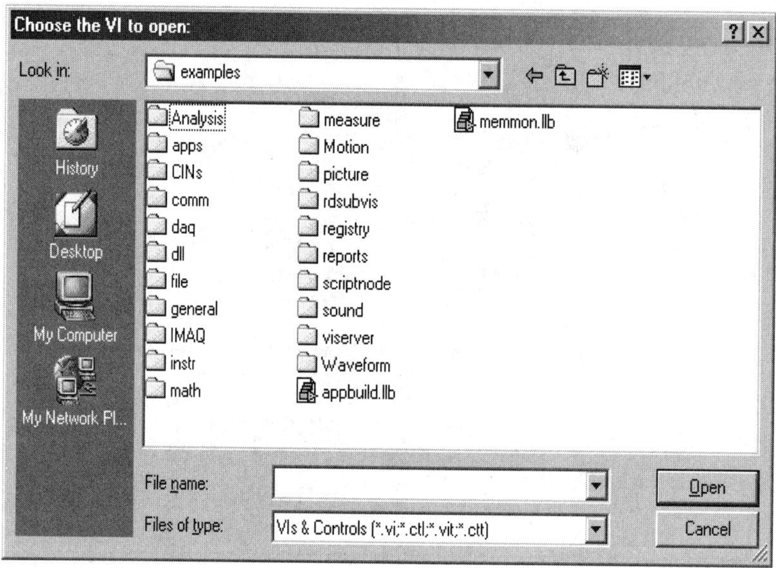

Figure 6–24
LabVIEW's `examples` folder.

Table 6.1 Detailed descriptions of the items in LabVIEW's `examples` folder

Folder	Description	DAQ Relevance
Analysis	Mathematical analysis for your data.	Often relevant.
apps	Examples of completed applications.	Somewhat relevant.
CINs	Avoid these! CINs allow you to link to external code, but use the Call Library node if you can.	Better not be relevant.
comm	Examples of interapplication communciation, such as transferring data with Microsoft Access or Excel.	Often relevant.
daq	DAQ VIs.	Almost always relevant.
dll	DLLs are external pieces of code. You may need these to link to non-NI hardware or to do non-standard things.	Sometimes relevant.
file	File I/O means reading or writing data from or to a hard drive.	Very relevant.
general	A big grab bag of useful examples.	Often relevant.
IMAQ	Image processing VIs.	Often relevant.

Table 6.1 Detailed descriptions of the items in LabVIEW's `examples` *folder* (continued)

Folder	Description	DAQ Relevance
instr	Communication with various instruments via various busses like GPIB.	Often relevant.
math	Widely varying math examples.	Often relevant.
measure	Various DAQ-related mathematical examples.	Very relevant.
Motion	Motion control for motors and servos.	Often relevant
picture	Examples using LabVIEW's picture control.	Often relevant.
rdsubvis	VIs for opening and editing "read me" files	Could be relevant.
registry	These from and write to the Windows registry; *leave these alone.* Unless you really know what you're doing, you can foul up Windows very easily.	Hopefully not relevant.
reports	Report generation.	Often relevant.
scriptnode	Execute scripts in certain other math applications, HiQ and MATLAB.	Sometimes relevant.
sound	These VIs use your computer's sound card. Looking for cheap DAQ? Here it is!	Could be relevant.
viserver	Networking VIs.	Sometimes relevant.
Waveform	VIs illustrating the Waveform data type.	Oten relevant.

Let's look into the `daq` folder now (see Figure 6–25 and Table 6.2). Take your time to browse around the examples—you don't want to reinvent the wheel!

Table 6.2 Detailed descriptions of the items in LabVIEW's `examples/daq` *folder*

Folder	Description
`Accessories`	DAQ accessories—currently has only the BNC-2120, a very fancy terminal block.
`anlog_io`	Simultaneous analog input and output.
`anlogin`	Analog input.
`anlogout`	Analog output.
`counter`	Counters.
`daqcan`	Synchronize a DAQ and a CAN board together via a RTSI connection.
`digital`	Digital I/O.

Table 6.2 Detailed descriptions of the items in LabVIEW's `examples/daq` *folder* (continued)

Folder	Description
`I_Series`	Supports a wide variety of specialized DAQ features on NI products.
`multidev`	Multiple devices synchronized with the RTSI bus.
`scxi`	SCXI devices.
`solution`	A variety of completed applications.

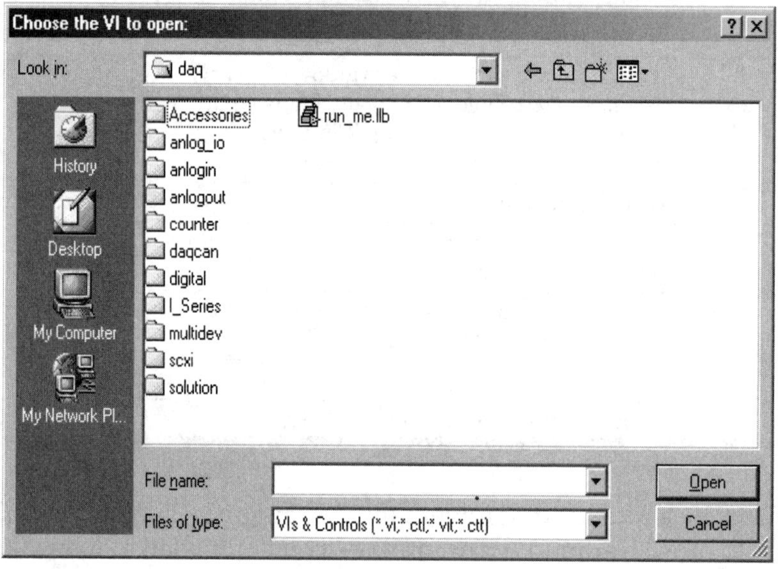

Figure 6–25
Here are some folders in LabVIEW's `examples/daq` folder.

6.1.4 Handling Errors with Dialog Boxes

Handling dialog boxes while performing DAQ (or doing anything in LabVIEW) is inherently cumbersome. Why? Given a standard LabVIEW program with one main While Loop, user input and looping are suspended while the standard dialog boxes are showing from the One Button Dialog and Two Button Dialog functions.

Open `Simple Slow DAQ.vi` and save it as `Dialog Boxes with DAQ 1.vi`. Go to its block diagram, then build the following VI, using these tips:

1. Streamline our timing code with the Select function, as shown in Figure 6–26, then select the portions shown.

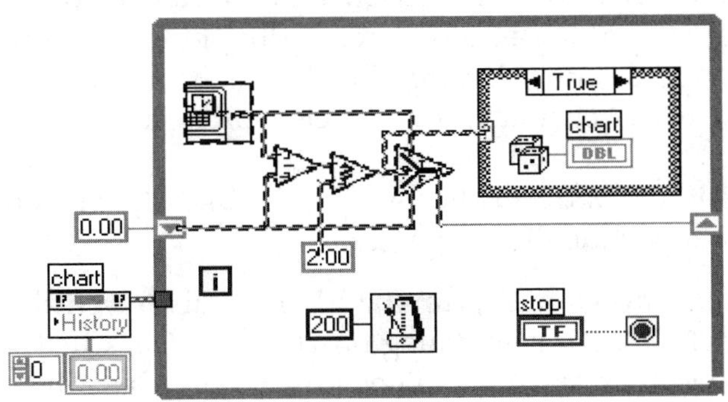

Figure 6–26
Timing functions inside a block diagram are selected to make a generic timing subVI.

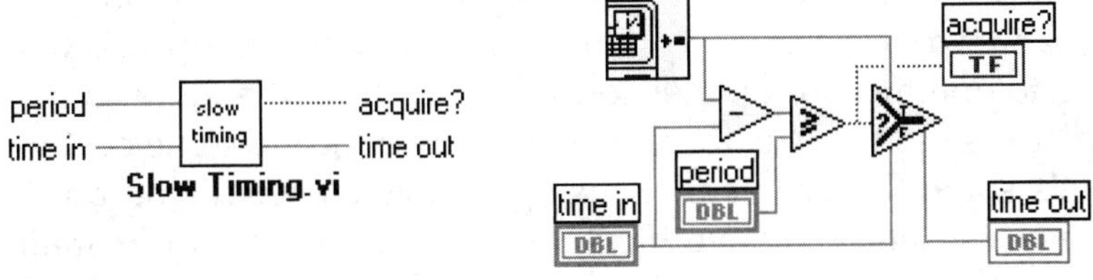

Figure 6–27
A generic timing subVI is created from the items selected in Figure 6–26.

   Create a subVI from it, and save this subVI as `Slow Timing.vi`, creating the icon/connector and block diagram shown in Figure 6–27.

2. On the front panel of `Dialog Boxes with DAQ 1.vi`, add the `data` Horizontal Pointer Slide control, as shown in Figure 6–28.
3. Replace the chart with the XY Graph, and label it `xy graph`. Pop up on this graph, and using the **X Scale»Formatting...** menu item, change the **Format:** item to **Time & Date**. Ignore the data on the graph for now; we'll create it soon enough. Each data point shown on the XY Graph is a cluster of two numbers, where the first num-

ber is the point's x-coordinate (time), and the second number is the point's y-coordinate (amplitude).

4. Modify the block diagram of `Dialog Boxes with DAQ 1.vi` as shown in Figure 6–28 by using the Bundle function, the Build Array function, and your previously created `Safe Dialog.vi`. The new constant wired to the lower left shift register is most easily created after everything else is wired by popping up on that shift register and selecting **Create»Constant**. The Build Array function automatically changes its upper element to an array type for you here.
5. Save the VI again (as `Dialog Boxes with DAQ 1.vi`) once you've successfully made these changes.

Set your `data` slide control to zero, run the VI, then raise `data` above 9.00 so you see the `out of range` safe dialog box pop up. Click its **Continue** button a couple of times, then leave the dialog box up for at least 10 seconds. Note that you cannot operate the `data` control when this dialog box is up. Finally, click **Continue** again, then stop the VI however you can.

The resulting graph might look like the one shown in Figure 6–28—see the gap between the data points? In this implementation, the dialog box interferes with the DAQ timing, thus causing the gap. We would like to continue acquiring data every two seconds, even when the dialog box is up. We would also like to able to operate front panel controls, like the `data` slide control, when the dialog box is up. Let's fix these problems by building some custom VIs.

First, let's solve the problem by using an even safer dialog box than the one we have; the one we have blocks all user input from our main screen whenever it pops up. Technically, we can no longer call this a dialog box, as a dialog box, by definition, blocks all user input from its parent window—so we'll call it a message window. Build the VI shown in FIgure 6–30 using these instructions:

1. Notice that the `message in` terminal on the block diagram has no visible control on the front panel. To make this happen, create a `message in` string control on the front panel, then hide it to the left of the window by sizing the window and using the scroll bars.
2. Create the global variable `message` as a string in your previously created `Globals.vi`.
3. The **Functions»Comparison»Empty String/Path?** function tests for an empty string, and the **Functions»Application Control»Stop** function will stop all VIs.

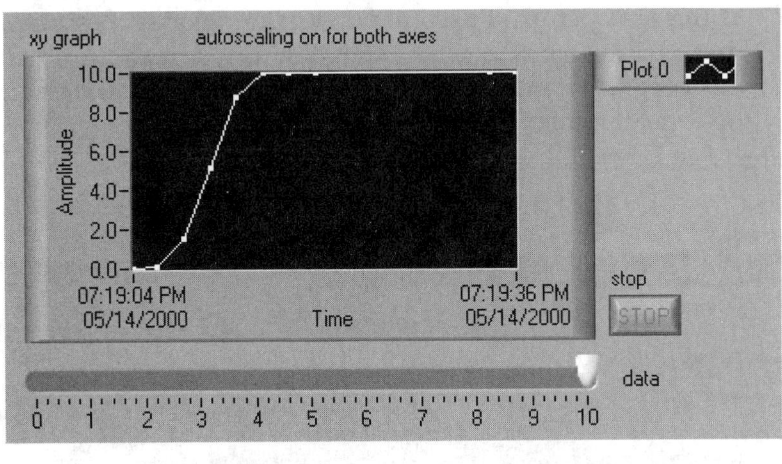

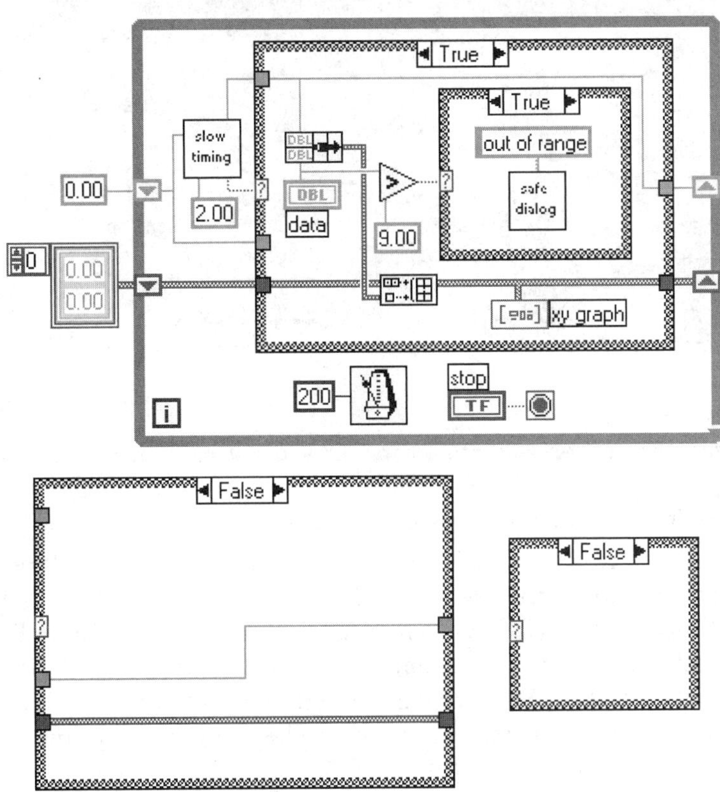

Figure 6–28
This VI has a dialog box that halts DAQ when it shows.

4. As this is a special VI used to pop up and display messages, go to **File»VI Properties...**, select the **Window Appearance** category, select the **Custom** radio button, then push the **Customize...** button to set the following screen, shown in Figure 6–29.
5. Save it as `Message Window.vi`.

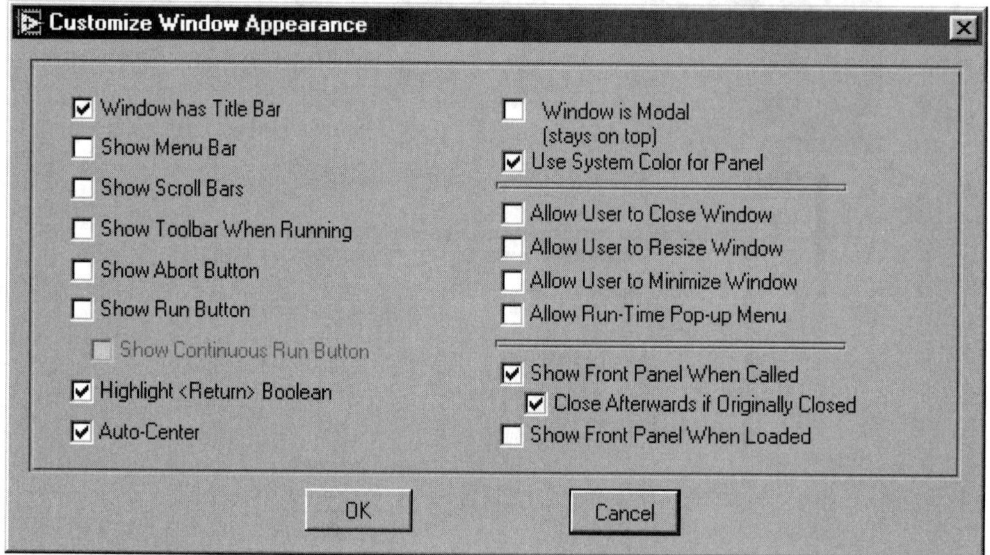

Figure 6–29
This Customize Window Appearance window lets you customize front panels.

Figure 6–30 shows what `Message Window.vi` should look like.

`Message Window.vi` is designed for use as a subVI in a calling VI. Notice how the **Show Front Panel When Called** button is checked in the **Customize Window Appearance** window. Whenever you have **Show Front Panel When Called** checked, you should usually check **Close Afterwards if Originally Closed** checked as well. Since that **Show Front Panel When Called** property is set, be sure to have `Message Window.vi` closed before you run its calling VI.

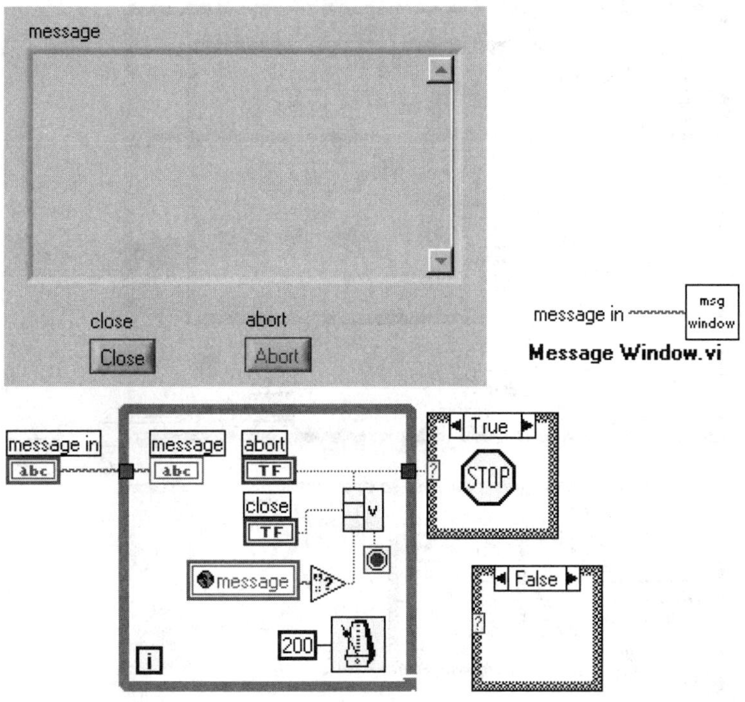

Figure 6–30
This is one way to implement a dialog box that does not halt DAQ when it is showing.

Save `Dialog Boxes with DAQ 1.vi` as `Dialog Boxes with DAQ.vi`, then modify it as shown in Figure 6–31 (I made it as simple as I could, honest), using these tips:

1. Two local variables are created from the `stop` button. Set the `stop` button's mechanical action to a non-latching value; local variables are not allowed with latching buttons. The front panel should look the same as it did, however, so we won't show it again here.
2. Create the global variable String Control `message` in `Globals.vi`, and use it as shown on the block diagram.
3. Be patient; it takes a lot of work to handle this seemingly simple situation.

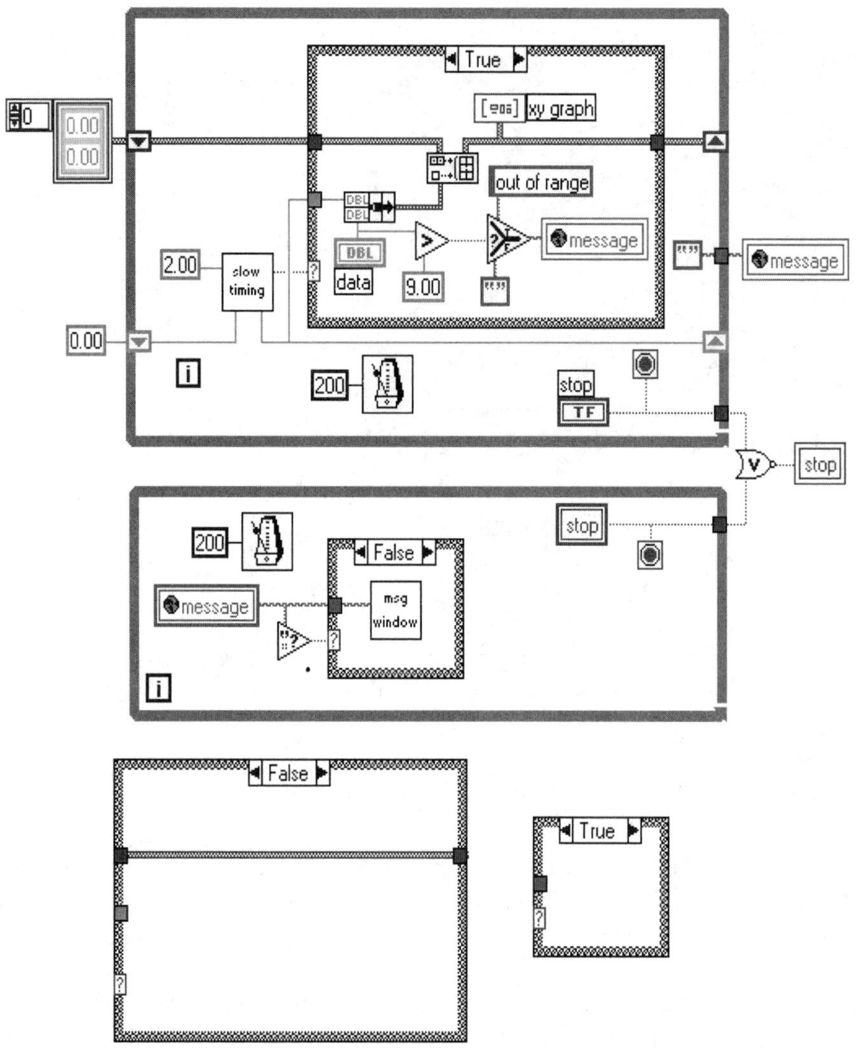

Figure 6–31
This VI illustrates a dialog box subVI that does not interfere with the main VI's execution.

The main trick here is having two While Loops running simultaneously. If we display a standard dialog box window from our upper loop, it stops and therefore DAQ stops—so we need another loop to display a message box. The upper loop acquires the data, and the lower loop displays the message box. When the upper loop discovers a data value greater than 9, it sets the global variable `message` to `out of range`. When the lower loop detects anything but an empty string, it calls `Message Window.vi`, which pops up and displays the message. Notice that the lower loop stops whenever the message window is showing, so `Message Window.vi` must monitor the `message` global internally so it can automatically close itself whenever there is no message. We set the `message` global to an empty string when the upper loop stops so the message window closes itself if it's showing.

The `stop` button uses its two local variables to stop both loops at once, and then to reset itself to its original state.

6.1.5 Monitoring Bus-Based Ports

Don't bother building the VIs in this section, unless you have a null modem cable, two computers with LabVIEW and serial ports, and the desire to actually do bus-based programming.

Many times, you might have external equipment that communicates via a bus, such as the RS-232 port. Although this might not fall under the strict definition of DAQ, it's close enough for me, and it's often used with DAQ, so this section discusses the key issues. I will be using the RS-232 port, as most computers have these, but the basic concepts herein apply to all bidirectional ports.

Suppose your computer will be connected to an RS-232 gadget that can read and write data.

Some RS-232 devices only write data, sending out a constant stream of information, but we won't discuss that in this book, as it's really a subset of the read/write scenario.

Suppose you want to acquire a piece of data from this read/write gadget; you will usually need to send the gadget a special "request" code. This code will vary per gadget. The gadget then responds with the data, which may be in binary or ASCII format. You need to read the gadget's manual to figure out how to decode this data.

Before any communication at all will work, you may need to set some low-level hardware parameters relevant to your particular port. For an RS-232 port, these parameters are usually *baud rate, number of data bits, number of stop bits,* and *parity.* Hardware control can also be tweaked, but this is usually not needed. You must make sure these parameters match your serial gadget's settings, or you will get garbled data, if any.

LabVIEW 6i introduced a new set of serial I/O VIs in the **Functions»Instrument I/O»Serial** palette, with the word VISA attached. I prefer the older-style VIs, found in the **Functions»Instrument I/O»I/O Compatibility»Serial Compatibility** palette, because they are currently more reliable. Figure 6–32 shows a comparison of the two palettes, and Table 6.3 describes the serial VIs. The left palette does have the advantage of more elegant error checking, by using the error cluster, and I would recommend using them for this reason if you can get them to work reliably.

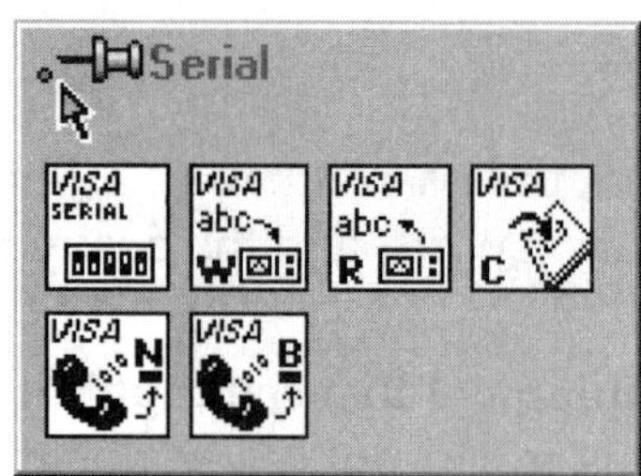

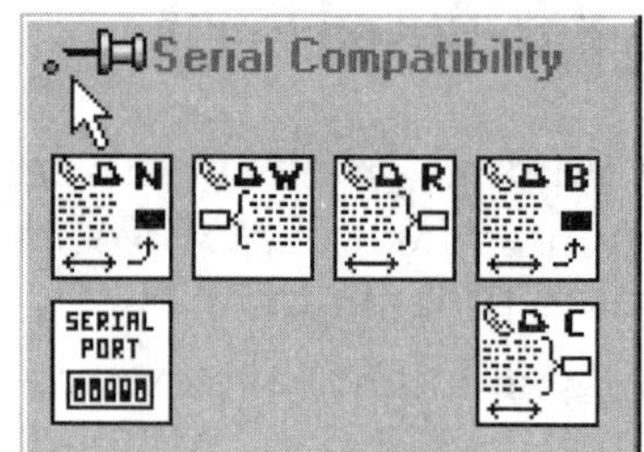

Figure 6–32
The **Serial** palette and the **Serial Compatibility** palette.

Figure 6–33 shows a very simple VI that monitors the serial port for data, and reads it into an ever-growing string.

In a real system, you would want to empty the string in the Shift Register after you've read valid data, to keep that string from getting too large. And to be more robust, you would want to check for errors on all the serial VI functions. I have foregone this error checking for the sake of simplicity here, except for `init error`, which in my experience is the only error that ever occurs (when a port is not valid for whatever reason). To implement more professional error checking, use the error cluster as in the VISA versions of serial VIs, or as in file I/O VIs, by wrapping a VI around each of these serial VIs.

The data in the `data read` indicator shows what came through the serial port. I specifically set my system up so we could read ASCII bytes, but your serial gadget may give you non-ASCII bytes. Notice the very important Concatenate Strings function. If you read the serial port in mid-message, you may only get the first few bytes of a multibyte string; the shift register saves this and tacks additional bytes, if any, onto the end of your first few bytes on the next iteration of the loop.

Table 6.3 Detailed descriptions of the items in the Serial Compatibility palette

Serial VI	Icon	Description
Serial Port Init	SERIAL PORT	Initializes any of your computers' serial ports. If the port is being used by another application, or does not exist, an error is returned. Be careful to remember that the **port number** input is really one less than the COMx designation; in other words, COM1 is **port number** 0, COM2 is 1, COM3 is 2, and so on. The hardware parameters, such as baud rate, parity, etc., are set here.
Bytes At Serial Port	N	Checks to see if there are any bytes ready to read at the serial port. If so, the number of bytes is returned; if not, zero is returned.
Serial Port Write	W	Writes a string to the serial port.
Serial Port Read	R	If the Bytes At Serial Port VI indicates that you have data, call this VI with the number of bytes specified by Bytes At Serial Port to read the data into a LabVIEW string.
Close Serial Driver	C	Before exiting, call this on your serial port. This undoes the action of the Serial Port Init VI.
Serial Port Break	B	Breaks the serial port; you'll have to buy a new one. ☺ But seriously, it sends a break on an output port for a specified period of time.

A serial gadget could have written this data. To verify this VI's operation for this book, I built a different VI running on a different computer, connected with a *null modem cable*, which wrote the data you see—I rigged the data so that every character is a readable ASCII character. A null modem cable allows two computers to connect directly to one another—on a basic 9-pin null modem cable, lines 2 and 3 are crossed, and line 5 is connected directly. The other six lines may remain unconnected unless you want hardware handshaking. Figure 6–34 shows the VI running on a second computer that wrote the data in the previous figure via null modem cable.

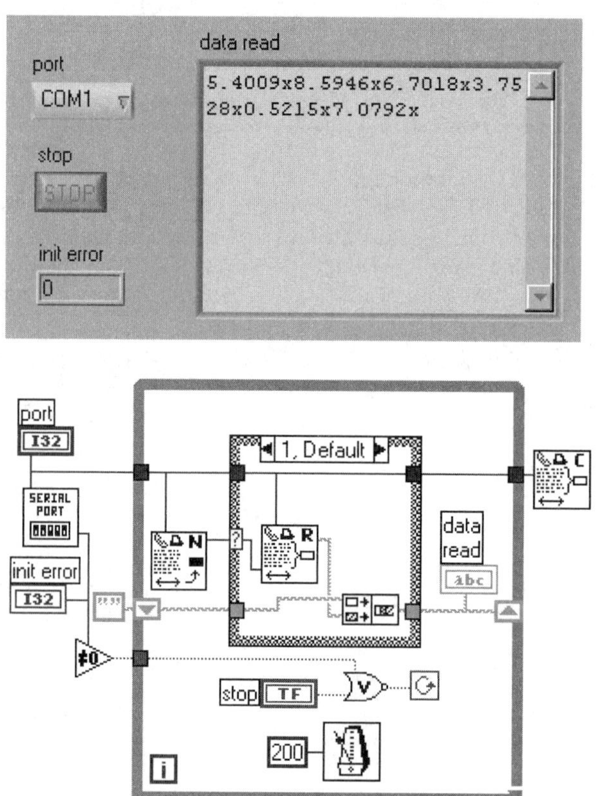

Figure 6–33
A simple serial port monitor.

Like the previous VI, you would want to check for errors on all the serial VI functions. This VI writes the bytes in `data to write` to the specified serial port whenever the `write` button is pushed. The `write` button must have a latching mechanical action.

I will now put together a VI that simulates a typical serial gadget. The VI will also double as a VI running on a host computer monitoring that serial gadget. I have here in my office two computers connected via null modem cable. Both computers have LabVIEW and a COM1 serial port. One of my computers, I will call the "gadget," as it will be simulating a serial gadget, and the other, I will call the "host" as it will simulate a computer with a real serial gadget. The gadget will transmit a random number in the range 0 to 10 to the host upon request. The magic letter here is "x" for both computers. The gadget will send an x when it is finished transmitting its random number, so that the host will know when to interpret its received bytes as a complete number. For example, if the gadget wanted to transmit a 1.9, it would really transmit 1.9x. That way, if the host sees just a 1, it knows there's more coming, and it stores the 1 in its shift register. It then gets a .9x, or at least some of it, on the next iteration of the loop. The computer requests a random number from the gadget by sending it an x. Sound complicated? Well, this is basically the way most bus-based instruments work, serial or not—they request data from an instrument by sending a command, then wait for the data, and be prepared to receive it on the next bus read *or* reads. Figure 6–35 shows `Serial Monitor.vi`, which runs on both the gadget computer and the host computer.

This block diagram is sloppy because of how we're checking the errors—in fact, we might not even know exactly where the error came from! But it is more concise than doing it the correct way, which would be to wrap each serial VI in another VI which uses the error cluster, then daisy-chain the serial VIs together. Or, use the newer version of serial VIs that have already done this for you! Here, however, we'll save much time by checking the errors in this clumsy fashion. Here is a list of the important points to recognize:

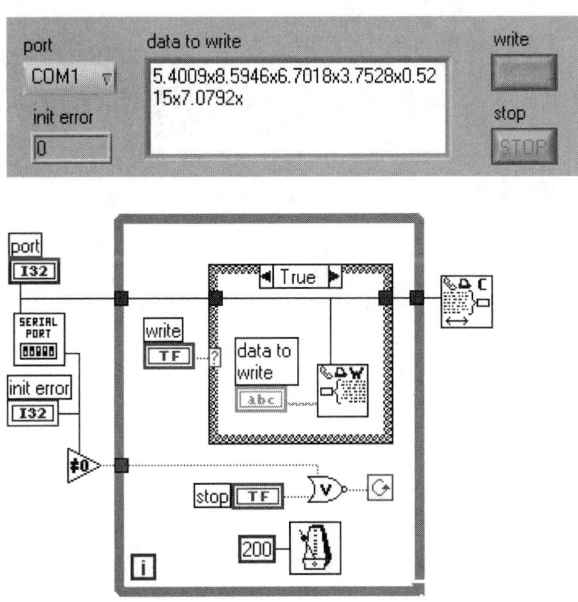

Figure 6–34
A simple serial port writer.

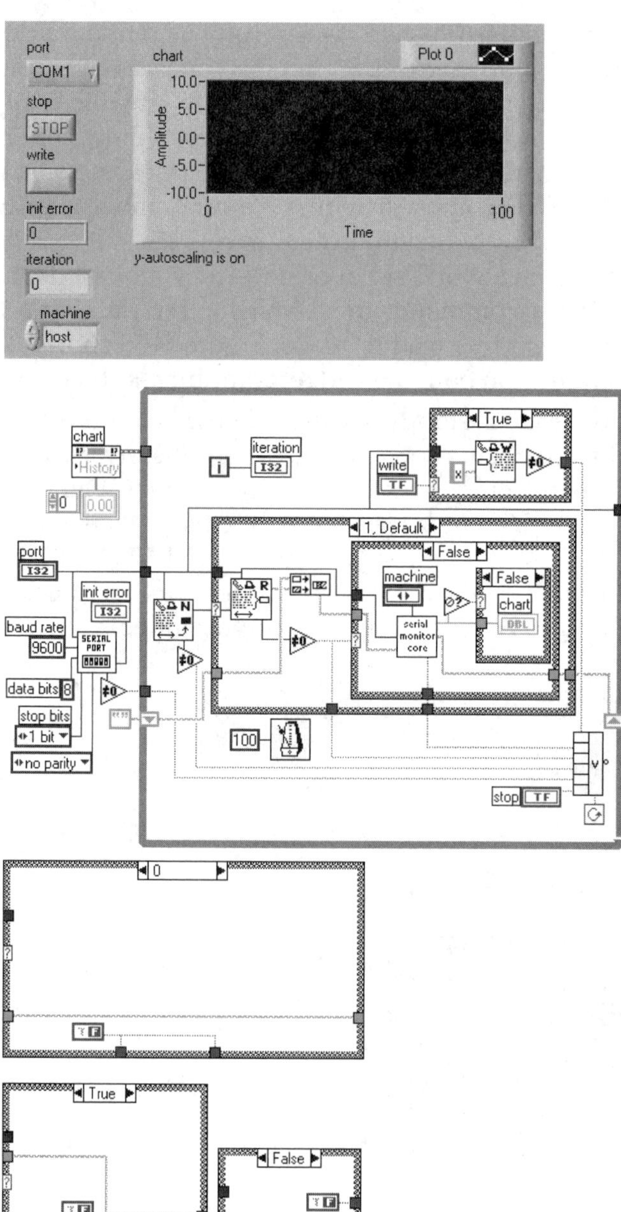

Figure 6–35
This VI works, but the error checking is very awkward and takes up much space. See your other options below.

1. This VI will run on both computers, and it continuously monitors the serial ports for data on both computers. The gadget computer is looking for a lone "x", and the host computer is looking for a number followed by "x", like "2.7183x". When data is found, it is sent to `Serial Monitor Core.vi`, which is customized to run differently on the two computers.

2. Like the simpler serial reading VI shown previously, we should use the Concatenate Strings function with a shift register to make sure that we piece together data in case it's received on different iterations of the While Loop.

3. The wire entering the largest Case Structure's selection terminal indicates how many bytes of data are waiting at the serial port to be read. By making case 1 the default case, this technique will read the data whenever there are more than zero bytes available (not *just* one byte).

4. `Serial Monitor Core.vi` returns NaN (not a number) whenever it gets numeric data that is not followed by an x. The little triangle to the right of `Serial Monitor Core.vi` is the Not A Number/Path/Refnum? function , and it uses the smallest Case Structure to write data to the front panel chart *only* when the data is valid (followed by an x).

5. The Case Structure with `Serial Write.vi` writes an x whenever the `write` button is pushed. The `write` button must have latching mechanical action.

6. The logic connected to the While Loop's conditional terminal stops the main loop whenever an error occurs or whenever the `stop` button is pushed.

7. On the front panel, I made the `port` ring a typedef and saved it to disk, so I can use this as the COM port control for all my serial VIs. Its values are COM1, COM2, COM3, and COM4, which conveniently correspond to the values 0, 1, 2, and 3, respectively. You can go higher than COM4 if need be by simply adding more items to the typedef.

Let's go down one more level and look at the guts of `Serial Monitor Core.vi`, as there are some useful concepts there. See Figure 6–36. Here are the key points of this subVI:

1. The default value of `data` is NaN (not a number), so that whenever something other than a number is written to it, this VI outputs a NaN through `data`.
2. The powerful Match Pattern function searches the incoming string for the letter x. If no x is found, the False case of the main Case Structure is executed, which simply passes any data in the shift register of `Serial Monitor.vi` back to the shift register, unchanged.
3. Data may arrive in incomplete chunks. If an x is found by the Match Pattern function *and* data is found after the x, that data is saved for the next iteration of the main While Loop, because it's part of the next number to be sent. This is a key point in any bus-based LabVIEW acquisition—don't throw away *any* data! That's the purpose of the wire going into `data out` in the True case in Figure 6–36; it will pass this incomplete data to the next iteration of the While Loop. In that same case, the wire passing through the inner Case Structure's host case contains all data before the x, which should be a complete floating-point number in string form.
4. The outer Case Structure is executed whenever an x is found on a serial port, regardless of the computer. If we're on the gadget computer, the `computer` ring will be set to gadget and a random floating-point number with four digits of precision will be written to the serial port. Remember the data

   5.4009x8.5946x6.7018x3.7528x0.5215x7.0792x,

   in our simple serial-read VI shown in Figure 6–35? This was the result of six such numbers being read in succession. The other case of the inner Case Structure executes whenever an x is seen on the monitoring computer, in which case the data before the x is interpreted as a floating-point number, and sent to the `data` indicator to be charted by our calling VI, `Serial Monitor.vi`.

In real life, you will probably never see the letter x used as a delimiter. I only used it here because it's easy to see. More often, the delimiter is a carriage return, line feed, comma, tab, or space.

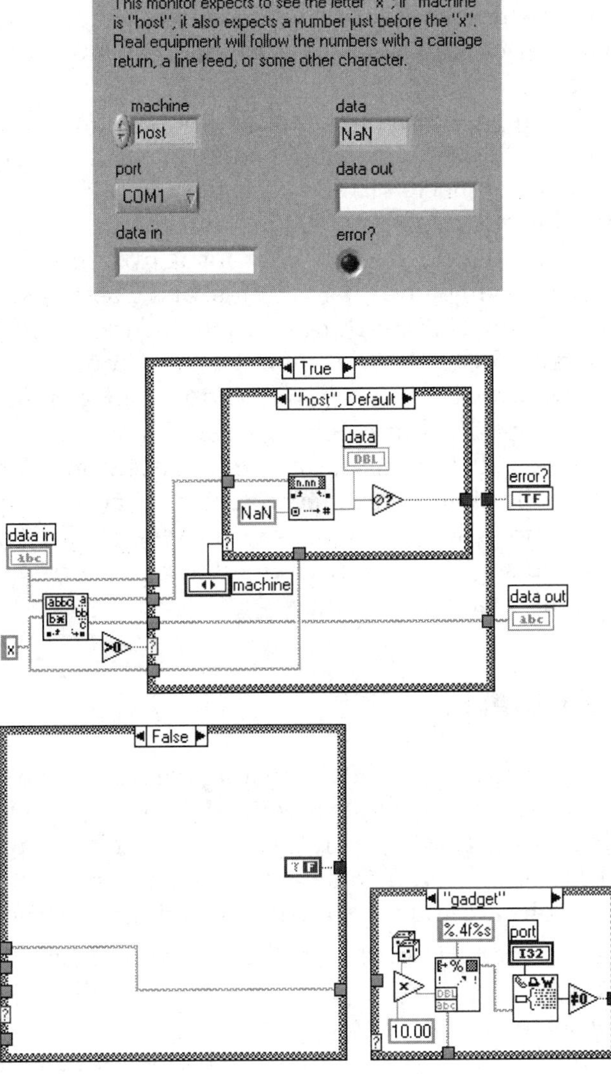

Figure 6–36
Our serial VI reads and writes formatted data.

Some serial devices output a constant stream of data without being asked, not unlike my last girlfriend. If you get such a device (a serial device), take care to not misinterpret incomplete data when you begin receiving. For example, let's pretend you started reading the stream 5.4009x8.5946x6.7018x midway into the first number, so you read 009x8.5946x6.7018x. You have no way of knowing whether that first 009 is valid, so ignore any data until you get past your first delimiter. In this case, the first number you should take seriously is 8.5946.

Before leaving this section on busses and ports, you should learn one more trick about what to do when you are monitoring multiple ports. Suppose you have eight devices connected to eight serial ports on your computer. Suppose also that each device takes about 400 ms to respond to a data request. If you request data from one device and wait for it to respond before requesting from the next, you will be waiting 3200 ms, or 3.2 seconds, to read data from all eight devices; this will introduce an annoying delay to the responsiveness of your program! Instead, you should send requests to all eight serial devices, then wait for each one to respond in a loop. Remember the string we've been keeping in a shift register for serial monitoring VIs? We will now need an array of eight strings in a shift register, one per device, as each device will be collecting data independently in its own string. This sort of LabVIEW program will grow complex, and you'll likely need to send this array of strings to subVIs, so don't forget the typedef trick mentioned in the first section of this chapter.

6.1.6 Multiplot XY Graphs

Suppose you want to acquire data for a long period of time (anywhere from a few seconds to a few years) and show the data on a graph with timing information on the x-axis. Usually, you'll want to do this with a very slow sampling rate, 1 Hz or slower. If this is what you want, this section is for you, because XY Graphs are good at showing such timing on their x axes.

Open `Dialog Boxes with DAQ 1.vi` from earlier in this chapter and save it as `Multiplot XY Graph.vi`. Modify this VI as shown in Figure 6–37, then save it again.

This VI simulates analog input of one channel of data, where the data is controlled by you from the front panel slide control `data`. Suppose you want to expand the above VI to accommodate multiple channels. I'm tempted to just gloss over this topic, and say "read the fine manual" concerning multiplot XY Graphs, and expand your data types as such. But there's some trickery involved, so we should explicitly work through a multiplot example.

If you want to find out exactly what the multiplot data types are for the XY Graph, you'll need to dig through LabVIEW's Help documentation to find it. I finally found this description in LabVIEW's User Manual:

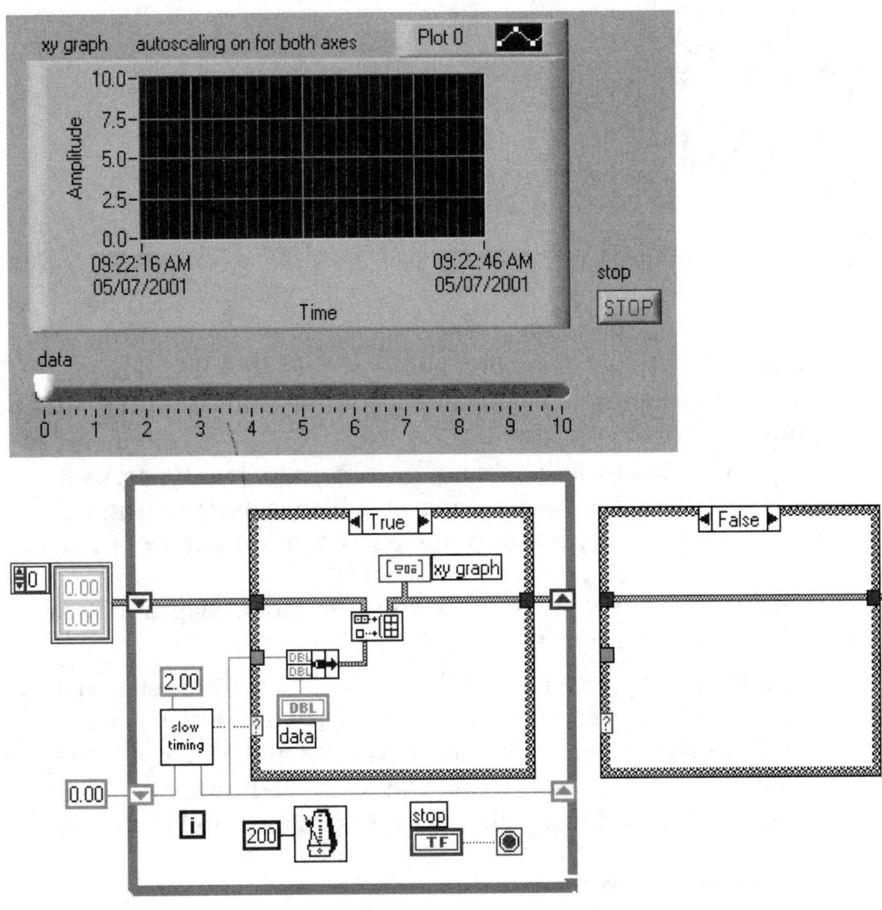

Figure 6–37
An XY Graph will be used with simulated data.

The multiplot XY graph accepts an array of plots, where a plot is a cluster that contains an x array and a y array. The multiplot XY graph also accepts an array of clusters of plots, where a plot is an array of points. A point is a cluster that contains an x value and a y value.

Refer to theXYGraph VI in the `examples\general\graph\gengraph.llb` for an example of multiplot XY graph data types.

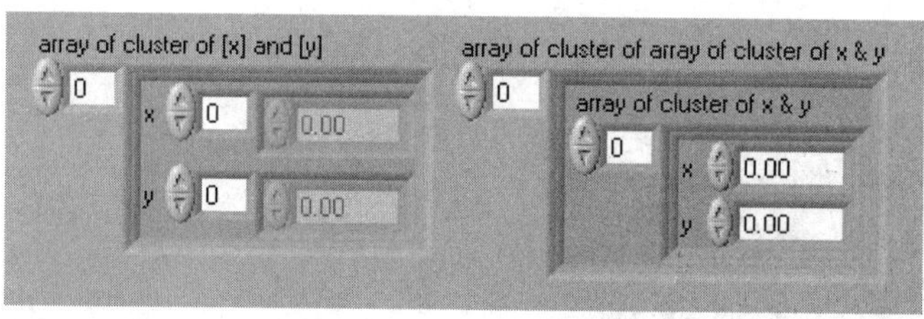

Figure 6–38
XY Graph data types.

The two data types described above would look like Figure 6–38 on the front panel (the example listed shows you the same data types on the block diagram).

I've always preferred the second of these two data types, as the first form allows differently sized X and Y arrays, which is useless to an XY Graph. The second data type neatly bundles the X and Y coordinates of each data point.

1. Delete the appropriate objects on the block diagram of `Multiplot XY Graph.vi`, as in Figure 6–39.
2. On the front panel, replace the `data` slide control with an array of three such slide controls having values of roughly 0, 1, and 2. Make this array the default value, or you'll have an empty array next time the VI is opened. While you're on the front panel, change the graph's legend to have the point styles shown in Figure 6–40.

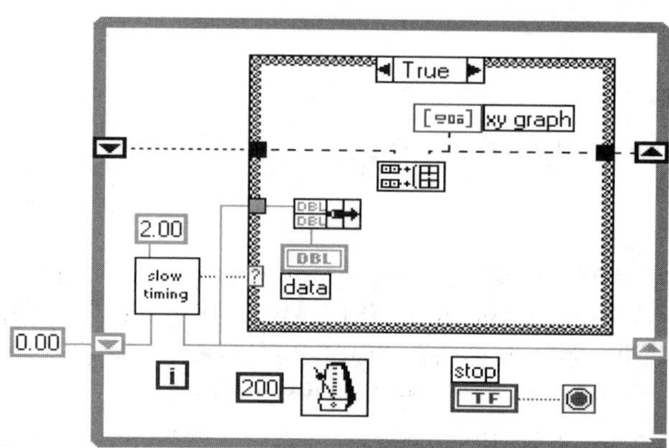

Figure 6–39
A VI using an XY graph is being built.

If you've forgotten some of Chapter 1, which is quite understandable, you need to use the Positioning tool to grow the `data` array to show three slide controls, then switch to the Operating tool to change the data to approximately 0, 1, and 2.

3. Build the block diagram shown in Figure 6–41, saving the complex-looking constant of arrays and clusters at the far left until last, at which point you can pop up on the left shift register and select **Create»Constant**. The functions inside the For Loop are the Bundle (two of them), Unbundle, and Build Array functions.

This VI appears to be set up to plot data every two seconds. Run this VI, and notice that it does not plot data at all. ☹

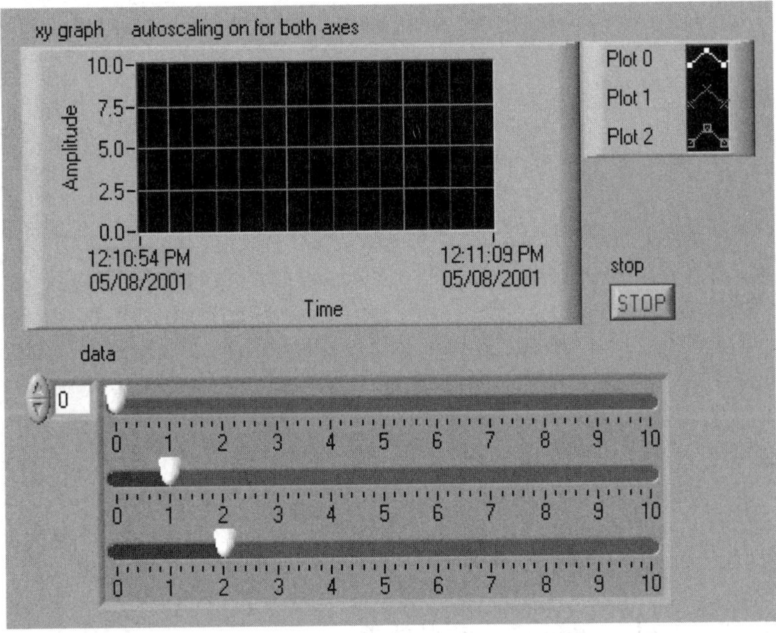

Figure 6–40
A front panel using slide controls for DAQ simulation is shown.

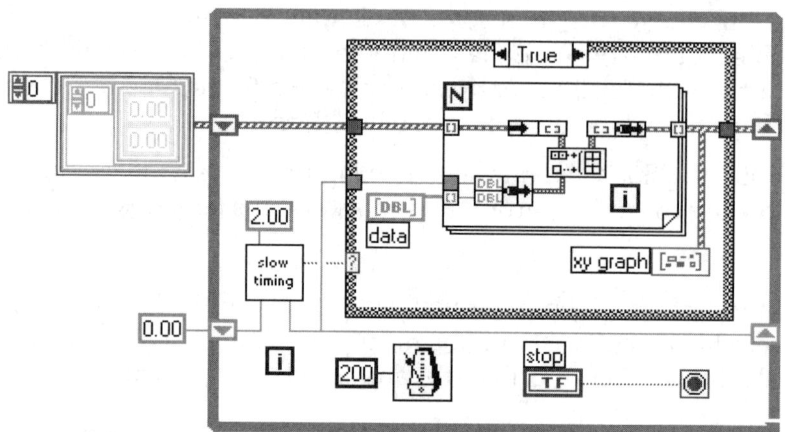

Figure 6–41
This VI might make you think it will plot data, but mysteriously, it does not.

What's going on here? Who said LabVIEW was simple? Don't look at me! If you understand why the data is not being plotted right away, maybe you don't need this book.

Let's analyze this block diagram. Every two seconds, the True case of the Case Structure executes. The For Loop should iterate three times, since we created an array of three elements on the front panel. Each iteration of the For Loop appends another data point onto the corresponding array of data points for each of our three plots.

So why won't the data plot? The data in the shift register appears to be an array of clusters of an array of clusters of X and Y data points. But in reality, the constant initializing the shift register is an *empty* array. This means the For Loop will execute *zero* times, not three, because its smallest array input has size *zero*! You can verify this by running the program with execution highlighting on and watching the block diagram. How do you correct this? You could modify that complex-looking constant to be an array of three clusters of empty arrays, but that's awkward and leaves a hard-to-read block diagram. I prefer the method shown in Figure 6–42, in which you drag the cluster out of the complex-looking array constant and use the Initialize Array function.

I've run this VI, slowly changing each slide control, to produce the graph in Figure 6–43.

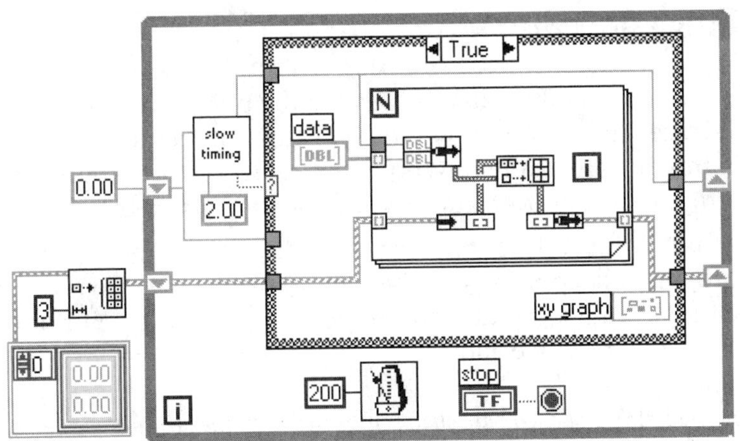

Figure 6–42
The VI from Figure 6–41 is cured.

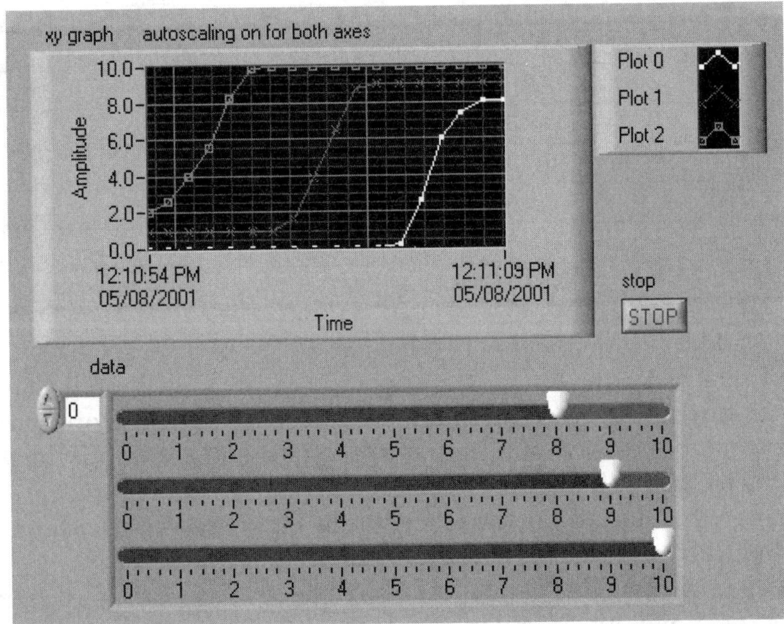

Figure 6–43
Our XY Graph VI, plotting simulated DAQ data, finally works! The plots correspond to how the slide controls were moved by the user.

6.1.7 Limiting Graph Data Size

Regardless of your scan rate, you might get to the point where the graph data becomes too large, and you only want to see the last section of the graph. A chart will do this automatically for you if you set its history length, but you'll need to programmatically control the length of any sort of non-chart.

Usually, you'll want to track the last few hundred or few thousand data scans on a graph. To keep things really simple, suppose you wanted to only show the last four scans of data, and you're acquiring one point of data per iteration of a While Loop; see Table 6.4.

Table 6.4 *Eight iterations of a loop are shown in which only the last four elements of an array are retained.*

Loop Iteration	Graph's Data Scans Without Size Control	Graph's Data Scans With Size Control
0	0	0
1	0,1	0,1
2	0,1,2	0,1,2
3	0,1,2,3	0,1,2,3
4	0,1,2,3,4	1,2,3,4
5	0,1,2,3,4,5	2,3,4,5
6	0,1,2,3,4,5,6	3,4,5,6
7	0,1,2,3,4,5,6,7	4,5,6,7

Let's add this functionality to our `Dialog Boxes with DAQ 1.vi`; save this VI now as `Last Four AI Points.vi`, and we'll modify it so we are only looking at the last four simulated scans. In the real world, you'll more likely want to see a few hundred or a few thousand scans, but four will illustrate the technique here.

Change the block diagram as shown in Figure 6–44 by adding the Array Size, Subtract, and Array Subset functions.

If you now run `Last Four AI Points.vi`, you can see how the Array Subset function is always taking the last four points of data stored in the shift register. If you're really sharp, you will have noticed that when the first point is added to the shift register, the Array Subset function gets a -3 in its **Array Index** input and a 4 in its **Array Length** input. Luckily, the Array Subset function is smart enough to not crash your computer, which might have happened had you been programming in a text-based language! Instead, the clever Array Subset function internally changes the -3 to a 0, and the 4 to a 1, thus doing the right thing. All LabVIEW functions that manipulate arrays are similarly clever—yet another reason to use LabVIEW!

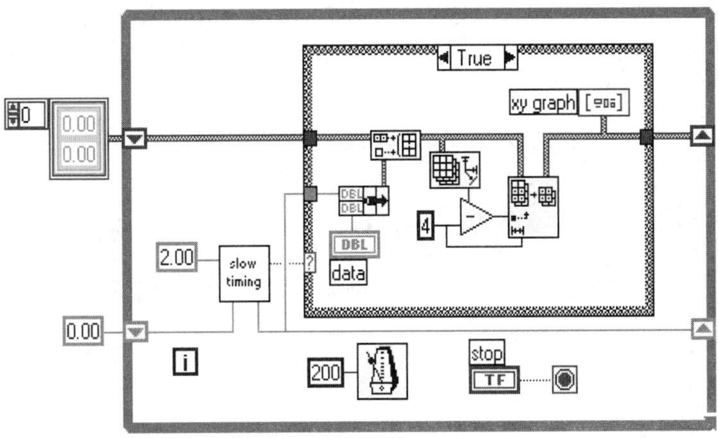

Figure 6–44
This VI looks at only the last four data samples by exploiting the Array Subset function.

6.2 Analog Waveform Analysis

Suppose you are running a test and must analyze some analog data. If acceptable, collect all of your data first in an array or file, then once the test is complete, analyze your data as a LabVIEW array. But if you need to analyze the data while the test is occurring, it can be tricky. In this section, we will demonstrate analysis occurring *while* data is being acquired continuously.

Two different buffering methods will be presented in this section. These buffers are not the hidden buffers that NI-DAQ uses during continuous acquisition; we will explicitly create these buffers with LabVIEW arrays. Suppose you are continuously acquiring analog data at a constant rate, with four scans per iteration of a While Loop. Figures 6–45 and 6–46 illustrate two buffering mechanisms you could use. These diagrams illustrate how data is buffered per iteration of the acquisition loop. The bold numbers towards the left indicate the iterations in which an analysis takes place—whenever our 16-sample buffer is filled with new data. For this example, let's pretend we're taking 20 samples per second (20 Hz), but acquiring only four data points at a time, although real applications often involve much higher data rates and correspondingly many more data points sampled at a time.

Open `Simulate AI Waveform.vi` which you should have built in Section 4.1.1, save it as `Continuous AI Analysis.vi`, and modify it as in Figure 6–47, using these tips:

1. Give the `sgl` Numeric Constant an SGL data type via its **Representation** pop-up menu.
2. The Initialize Array function is used.

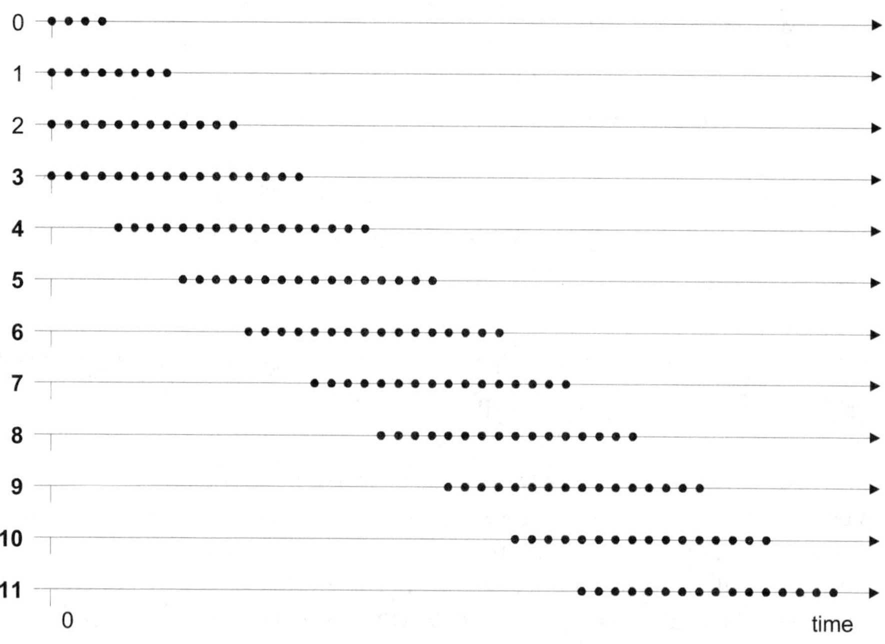

Figure 6–45
Buffering during continuous analog acquisition is illustrated where data can be analyzed every iteration once the buffer is filled.

Figure 6–46
Buffering during continuous analog acquisition is illustrated so that data can be analyzed less frequently than in Figure 6–45.

Create a subVI by selecting the Case Structure in Figure 6–47 then using the **Edit»Create SubVI** menu item. Save this new subVI as `Analyze Analog Waveform.vi`. The block diagram of `Continuous AI Analysis.vi` should like Figure 6–48.

Modify `Analyze Analog Waveform.vi` as shown in Figure 6–49 using these tips:

1. Make sure the `iteration` terminal is the one connected to its calling VI's While Loop's iteration terminal (shown in Figure 6–48).
2. In the short fat Case Structure, the Replace Array Subset function and the Build Array function are in the **Functions»Array & Cluster** palette. You will need to pop up on the Build Array function and check its **Concatenate Inputs** menu item—otherwise it will produce a 2D array rather than the 1D array we want.

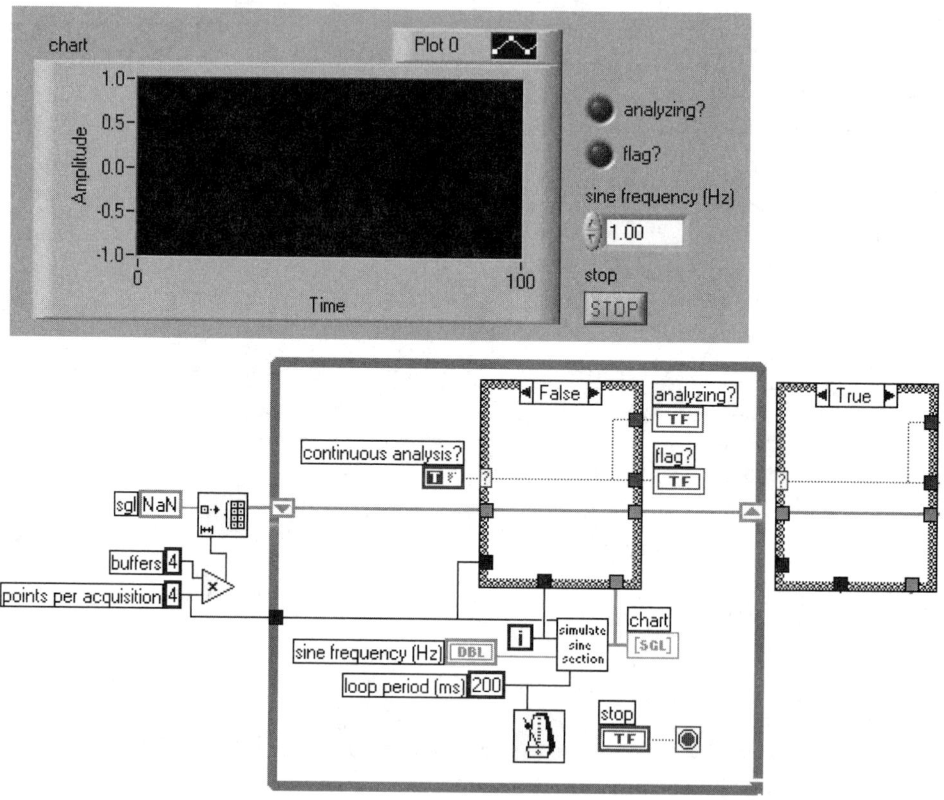

Figure 6–47
`Continuous AI Analysis.vi` is prepped for continuous analog acquisition.

3. In the tall skinny Case Structure, the In Range and Coerce function is in the **Functions»Comparison** palette. `Mean.vi` is found in the **Functions»Mathematics»Probability and Statistics** palette.
4. If you replace the numeric elements of `buffer in` and `buffer out` with those shown (from the **Controls»Classic Controls»Numeric** palette), your array elements will line up with one another.

Enter the front panel data shown. Repeat the following procedure at least a dozen times or until you can see how this implements the buffering scheme shown in Figure 6–46.

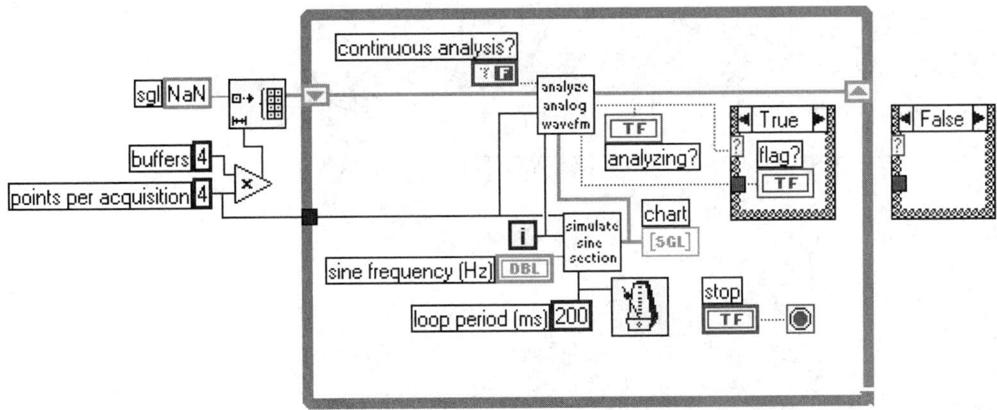

Figure 6–48

`Continuous AI Analysis.vi` is prepped further for continuous analog acquisition.

1. Run the VI with the <Ctrl-R> key.
2. Look at where the data appears in the `buffer out` array.
3. Click the up arrow on the `iteration` numeric control so its data is incremented.

Set the `iteration` numeric control back to zero, set the `continuous analysis?` control to True on the front panel, then repeat the recent three-step procedure at least a dozen times or until you can see how this implements the buffering scheme shown in Figure 6–45.

Let's explain what's going on here with respect to the block diagram. As you can see from Figures 6–45 and 6–46, the first four iterations of the two buffering mechanisms are identical, in which case the False case of the short fat Case Structure of Figure 6–49 is executed. The Replace Array Subset function in that False case clearly replaces a subset of the array. After the fourth iteration, the two buffering mechanisms differ. The mechanism of Figure 6–46 will continue to use the False case. The mechanism in Figure 6–45 will use the True case of the Case Structure, which simply strips off the first four points of the buffer and appends the new four points, exactly as Figure 6–45 suggests.

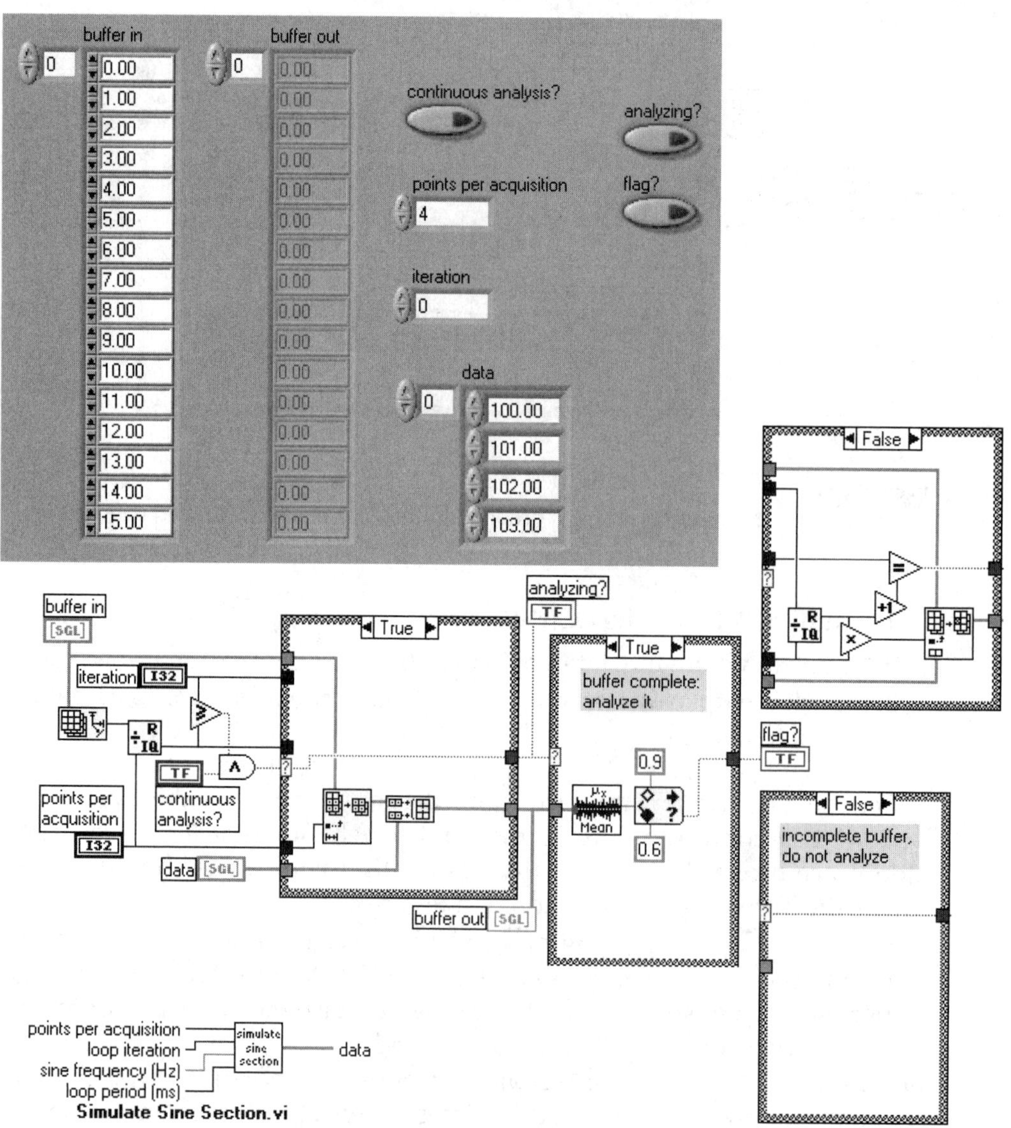

Figure 6–49

`Analyze Analog Waveform.vi` is shown.

Run Continuous AI Analysis.vi with its sine frequency (Hz) numeric control set to 0.1. Try other frequencies, never more than about 4 Hz, and experiment with the continuous analysis? constant on the block diagram of Continuous AI Analysis.vi. Once you know that the flag? indicator should be true whenever the last buffer analyzed has an average in the range of 0.6 to 0.9, and if you keep in mind Figures 6–45 and 6–46, everything should make sense. If your frequency is too high, the flag? indicator will not light.

Notice that the continuous analog input acquisition examples that ship in LabVIEW's examples folder (mentioned in Chapter 2) output 2D arrays of data, whereas our example here was working with a 1D array. In general, our VIs in this section could be modified to work with 2D data by adding a dimension to the 1D arrays and dragging For Loops around the functions connected to your new 2D wires. You will run into a few other array-related issues depending on how your VIs are built, but that's the basic approach for converting from 1D math to 2D math.

6.3 File I/O for DAQ

This section will discuss a few diverse file I/O topics pertinent to DAQ.

6.3.1 Binary data files

In order to understand this section, you must not only understand bits and bytes, but you must understand LabVIEW data types and the difference between binary and ASCII data. These topics are covered in Appendix A and Chapter 1.

Let's use one of the VIs from Chapter 3 that saved ASCII data, and we'll modify it to save binary data. Open AI Single Points and Chart.vi along with our recently created Multiplot XY Graph.vi. Since we're not using any real DAQ devices in this chapter, save AI Single Points and Chart.vi as AI Single Points and Chart (sim).vi. Modify your new VI as in Figure 6–50, using these tips:

1. Copy the data array of slide controls from the front panel of Multiplot XY Graph.vi to the front panel of your new VI.

Empty this array with its **Data Operations»Empty Array** pop-up menu item, then make it a two element array having values 0 and 1. Make this the array's default value. You may close `Multiplot XY Graph.vi` now, as we only wanted it for its slide controls.

2. Replace all the real DAQ stuff on the block diagram with the `data` terminal from your array of two slide controls; we're simulating all data in this chapter.
3. Enter a valid path to a file we'll create, called `ascii4.txt` in the `file path` control. Your path will likely be different from mine, of course.

Run the VI for about 10 seconds, changing the slide controls as you wish, so you get some changing data on your chart. For example, I get the data shown in Figure 6–51.

Now, open your `Read and Display CSV Data.vi` and copy the path to your `ascii4.txt` file into its `file path` control. Run the VI, and you should see the same data you created. This data was written to disk in ASCII format.

We will now write similar data to disk in binary format. Save `AI Single Points and Chart (sim).vi` as `AI Binary Save (sim).vi`. From its block diagram, open `Write 1D Ascii Data.vi`, and save it as `Write 1D Binary Data.vi` with the **File»Save As...** menu item, which effectively replaces itself in the block diagram of `AI Binary Save (sim).vi`. This will be more apparent when you update these VIs' icons to reflect their new file names. Modify your new `Write 1D Binary Data.vi` as follows, using these tips:

1. On the front panel, remove the `data string` control.
2. That new oddly shaped function (see Figure 6–52) is the Type Cast function, which converts binary data to string data. It is found in the **Functions»Advanced»Data Manipulation** palette. The file I/O VIs accept string data as their default data type at the lowest level. There are specialized VIs that could have saved this floating-point data without explicitly using Type Cast function, but this is a more powerful technique in that it can be easily modified to many data types.
3. Make the icon look like this: write 1d bin. data
4. Your block diagram for `AI Binary Save (sim).vi` should look much like Figure 6–53, if you've followed the instructions carefully.

5. On the front panel of `AI Binary Save (sim).vi`, change your path so that you're writing to `ascii4.bin` rather than to `ascii4.txt`, making it the default value as well.

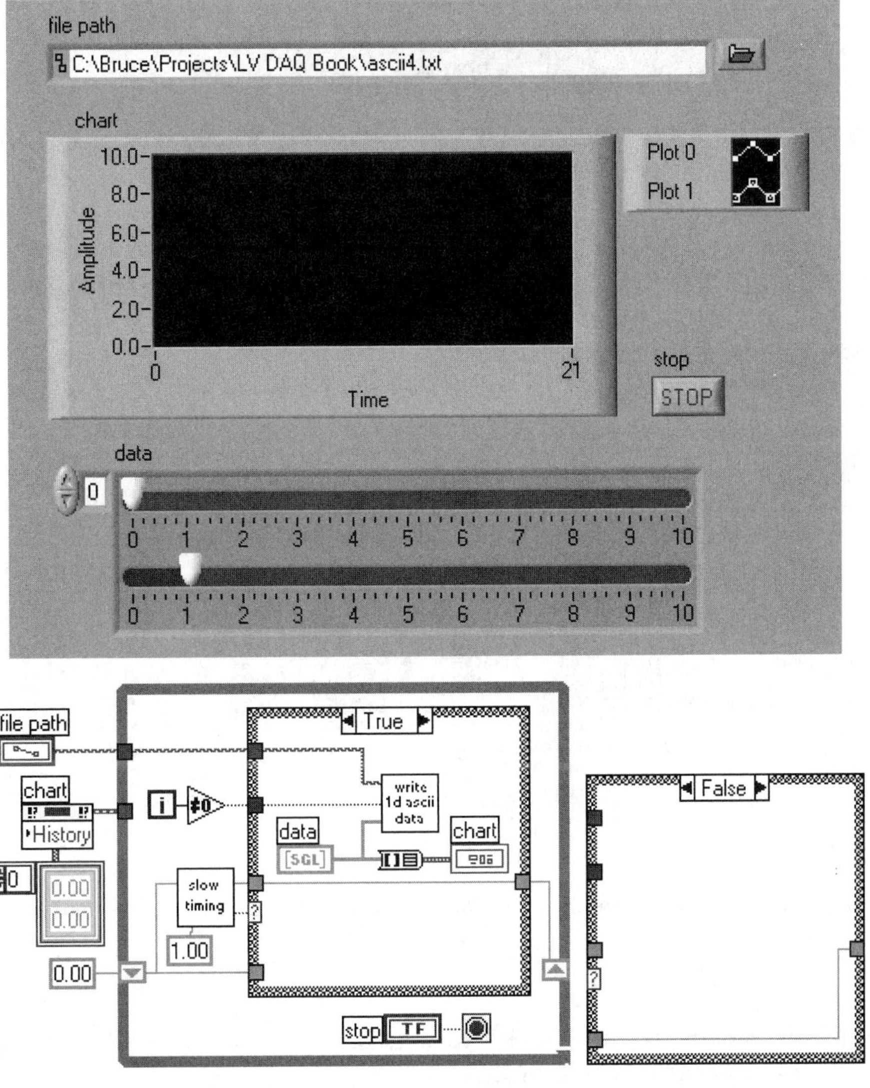

Figure 6–50
This VI writes simulated DAQ data to disk in ASCII format.

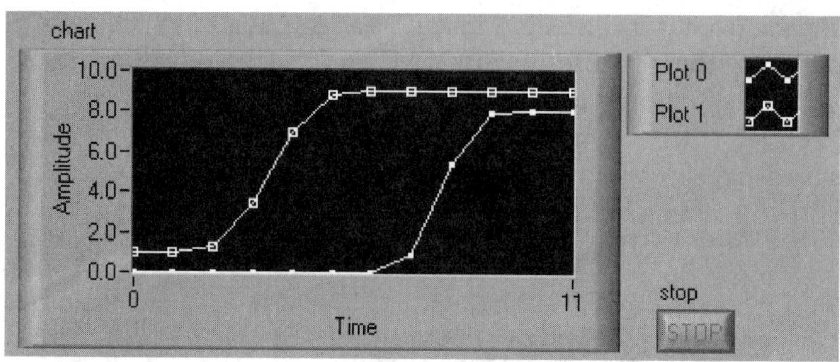

Figure 6–51
These plots represent data corresponding to the movements of the slide controls in Figure 6–50.

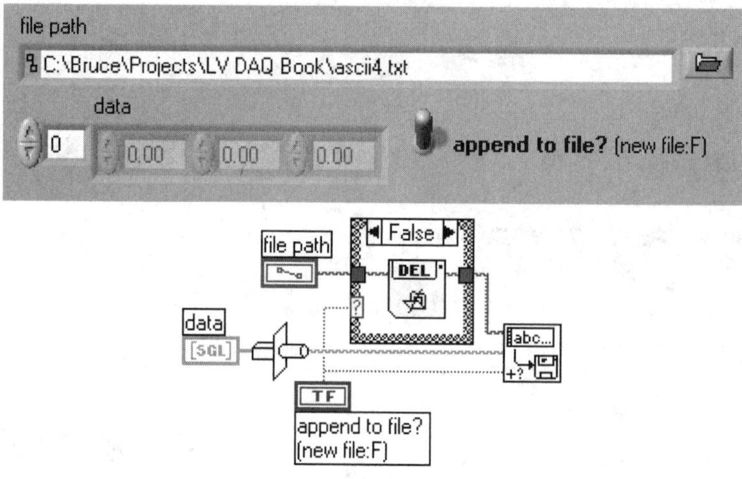

Figure 6–52
This snippet of code uses the Type Cast function to convert data to binary format then write it to a file.

6. Run your VI and change your data slide controls while it's running, so you see some data looking something like the plots in Figure 6–54.

First, understand what your data "looks" like on disk if you happened to get exactly 12 samples (at two data points per sample). See Figure 6–55.

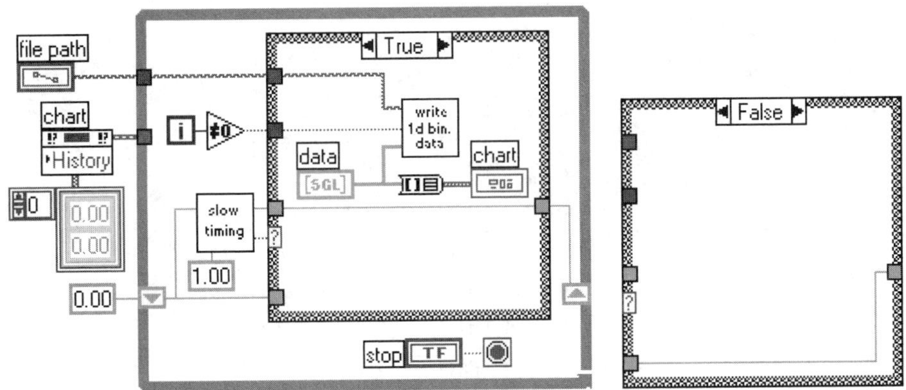

Figure 6–53
This block diagram is a modification of an earlier VI so that is saves data in binary format rather than ASCII.

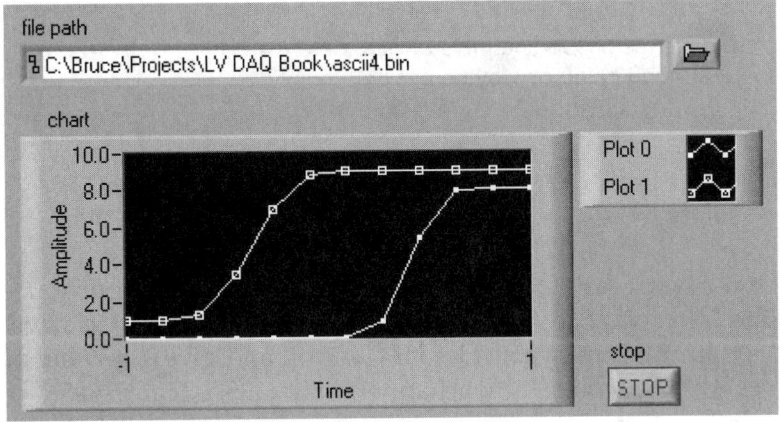

Figure 6–54
This familiar front panel is used for saving binary data, not ASCII as before.

Here's how to read that data. Open your `Read and Display CSV Data.vi`, if it's not already open, and save it as `Read and Display 2D SGL Data.vi`. Now, make it earn its name by modifying it as follows (and saving it afterwards):

1. Copy your `ascii4.bin` path's text to the `file path` control of this VI, making it the default value.

sample #	4 bytes	4 bytes
0	channel 0 data	channel 1 data
1	channel 0 data	channel 1 data
2	channel 0 data	channel 1 data
3	channel 0 data	channel 1 data
4	channel 0 data	channel 1 data
5	channel 0 data	channel 1 data
6	channel 0 data	channel 1 data
7	channel 0 data	channel 1 data
8	channel 0 data	channel 1 data
9	channel 0 data	channel 1 data
10	channel 0 data	channel 1 data
11	channel 0 data	channel 1 data

12 samples of two channels of SGL data (32-bit floating point, or 4 bytes per data point), occupying 12 x 2 x 4 = 96 bytes.

Figure 6–55
Binary data is illustrated.

2. Given our very raw form of binary data, we must tell our data-reading VI how many columns, or channels, of data we have. Add a `channels` numeric Digital Control, and give it a value of 2, since we just acquired two channels of data. See Figure 6–56.
3. Now, let's fix the block diagram so it is expecting a simple 2D array of SGL (single precision, or 32-bit, floating-point data). See Figure 6–57.

This VI takes a string of bytes and breaks it into substrings such that the length of each substring corresponds to each scan. The Type Cast function, which just recently changed binary data to string data, now does exactly the opposite—it changes string data to binary, because we wired its center terminal. You might ask, why didn't we just skip the For Loop altogether and wire a 2D array of SGL data? Answer: The Type Cast function can only deal with very simple types of data. For more complex types of data, you could use the functions described in the next paragraph.

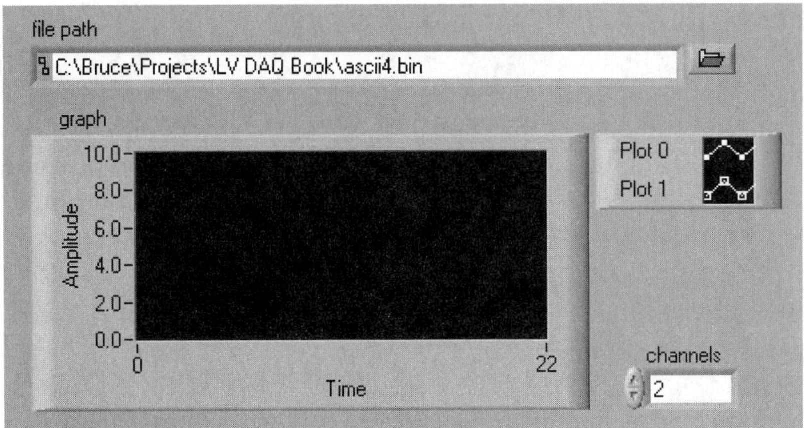

Figure 6–56
This front panel will display our recently created binary data.

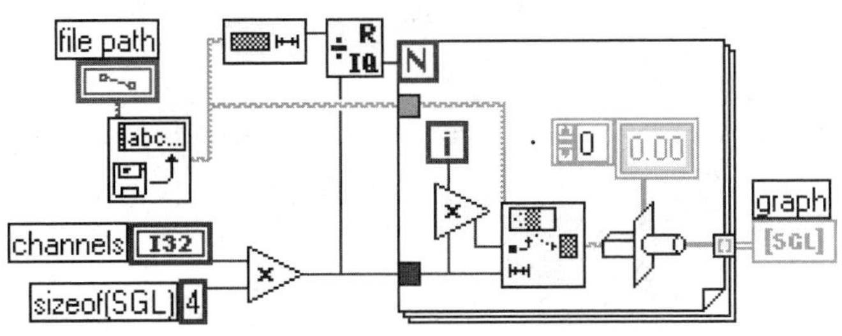

Figure 6–57
This block diagram is used to display our recently created binary data.

Suppose that you saved all your DAQ data in memory, rather than writing it to disk one piece at a time. There's a really simple way to write *any* data, even very complex data, to disk in LabVIEW—the Flatten To String function. Like the Type Cast function, it is in the **Functions»Advanced»Data Manipulation** palette. The binary data from the Flatten To String function will have some header information at the beginning, which helps decode the data format later, but other than that, this data looks similar to data created by the Type Cast function. To display the data, use the Unflatten To String function, passing the exact same data type to one of its inputs (this input ignores the data, but just looks at the data type), and you have your complex data.

6.3.2 Automatic Data File Organization by Date

You might not agree with my strategy in this section when you first hear it (most people don't at first), but I've been using this trick for many years now, and this turns out to be a very effective strategy. Until Windows 95 came out, we didn't have the luxury of using long file names and folder names on the PC; but now we do, so let's use them!

The VI in Figure 6–58 simply creates a new path for your test data, based upon the time and date.

Here are its advantages:

1. Unless you're creating a file more than once within a second, this VI automatically creates unique file names for you. If there's a chance you'll write more than once within one second, you may want to append an extra number, such as an "iteration" number, to the name.
2. If you sort your file names alphabetically, they'll also be sorted chronologically.
3. The `file subname` control allows for data-dependent customization.
4. You can categorize according to other criteria, such as your project's name, by using the base folder.

I have yet to see any legitimate disadvantages to using this scheme.

6.3.3 When to Flush

When you're saving data to disk, and using the standard file I/O functions, the data may not be physically written to the disk until you flush, or close, the file. If your application continuously saves data to disk, your operating system sneakily buffers this data in memory, without telling you, until it decides to physically write to disk. This can result in noticeable delays if the buffer happens to be too large. To get around this, you can use LabVIEW's Flush File function, which physically writes any and all file data from the sneaky buffer to the disk as illustrated in Figure 6–59.

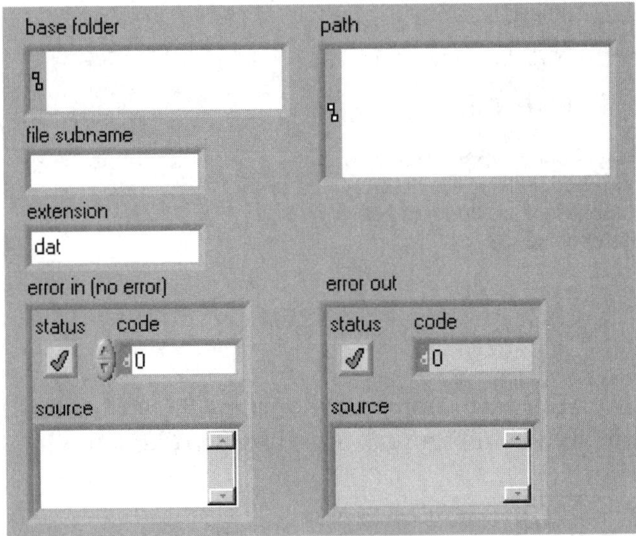

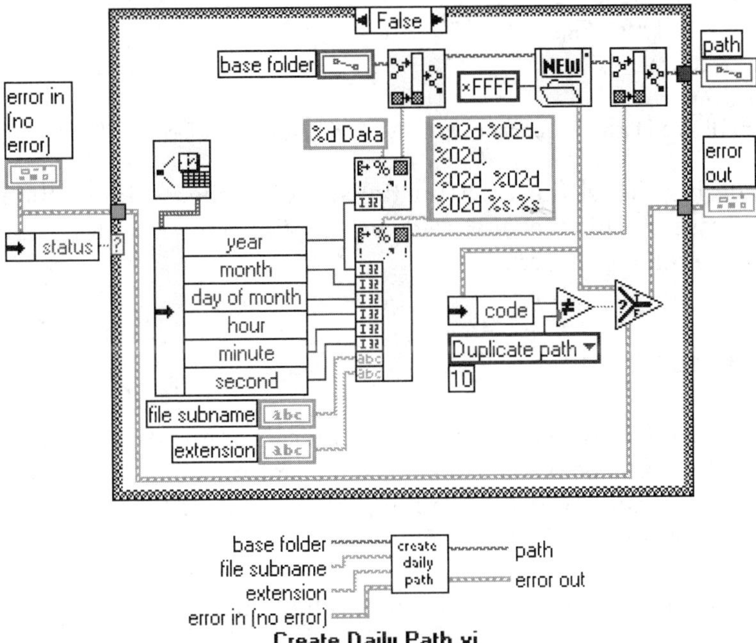

Figure 6–58
This VI creates a path that contains the time and date. For repeated testing, this is quite useful for organizing data.

Flush File

Writes all output buffers associated with refnum to disk and updates the directory entry of the file. File stays open and its refnum stays enabled.

Figure 6–59
LabVIEW's Flush File function.

The LabVIEW team immediately canned my idea for the Flush File function's icon, shown in Figure 6–60, when I worked at NI.

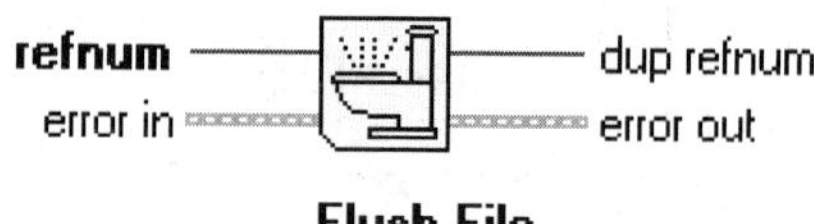

Flush File

Figure 6–60
Not LabVIEW's Flush File function, just my quickly-rejected idea.

Some high-level file I/O functions open a file, write it, then close it, in which case flushing is not relevant, as file data is automatically flushed whenever a file is closed.

6.3.4 Databases

NI offers a SQL Toolkit (Structured Query Language; that acronym is pronounced *sequel*), which allows direct communication with most databases, such as Microsoft Access. Copying directly from NI's Help window, here's a summary of the SQL Toolkit's features:

- Direct interaction with a local or remote database
- Connection to most popular databases
- High-level, easy-to-use VIs for common database operations
- Complete SQL operation
- Low-level VIs for direct access on columns, records, and tables

If timing or speed is not an issue, skip this paragraph. Be warned that SQL functions can take much more time to execute than LabVIEW's file I/O functions. Also, it's rather easy to use the SQL Toolkit in an inefficient manner, so make sure you are familiar with SQL speed issues independently from LabVIEW. Another way to speed up database acquisition is to write your data to your own custom database while acquiring, then convert it to your target database format later, when timing is not critical.

If your database does not support SQL, look in the **Functions»Communication** palette to see if another means of communication is applicable.

6.4 AVERAGING

Averaging is a technique for reducing noise in signals. You should try to reduce the noise via hardware techniques, as described in Chapter 1. There are software filters other than averaging built into LabVIEW, which you may find more useful, particularly if frequency response is an issue. For a full technical description on the effects of averaging, see NI's Application Note 152, *Reducing the Effects of Noise in a Data Acquisition System by Averaging*. The brightest analog engineer I know, who also happens to have the shortest email address I've ever seen (coincidence?), wrote this app note; here's his summary of averaging at the end of said app note, where *i.i.d.* means "independent and identically distributed."

> Underlying the application of averaging or filtering is a trade-off between the degree of certainty achieved and the number of samples that must be taken (and the time it takes to obtain them). When samples are independent and identically distributed, averaging a collection of samples reduces measurement uncertainty by a predictable amount. If σ is the amount of rms noise in a set of n i.i.d. samples, the rms noise of the average taken over the samples is
>
> $$\sigma_{avg} = \sigma \sqrt{1/n}$$

Suppose the your rms (root mean square) noise on a 0 to 10 volt signal is 0.1 volts. By taking the average of 100 "independent and identically distributed" samples, which is easy to do with the DAQ VIs, you will have reduced your noise by a factor of 10, so your rms noise is effectively 0.01 volts.

Suppose you have the following data samples:

1, 2, 3, 4, 5, 6, 7, 8, 9, 10, 11, 12

Next, suppose you are taking the average of four points. To get "independent" data, you must average "1, 2, 3, 4" "5, 6, 7, 8" then "9, 10, 11, 12" as opposed to "1, 2, 3, 4" "2, 3, 4, 5" then "3, 4, 5, 6". In other words, you should not use any of the same data from one averaged sample to the next, as this is not independent information.

LabVIEW ships with an averaging VI called `Mean.vi` (*mean* means average), to be found in the **Functions»Mathematics»Probability and Statistics** palette (see Figure 6–61).

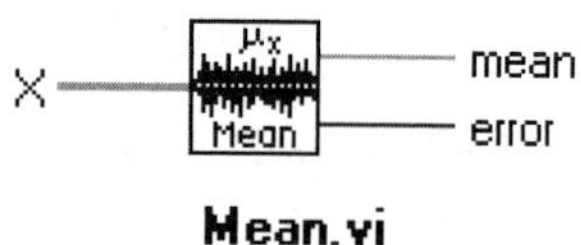

Computes the mean(average) of the values in the input sequence X. The mean is computed using the following formula:

mean = Sum{ X[i] } / n.

Figure 6–61
LabVIEW's `Mean.vi`, which simply averages data in an array.

For some reason, a homemade averaging VI is much faster than `Mean.vi`. If you don't have `Mean.vi`, or if you need the speed, it's not rocket science to build your own version of `Mean.vi`. If you want, you can build the following VI and save it as `SGL 1D Mean.vi`, using these tips:

1. The Add Array Element function is in the **Functions»Numeric** palette.
2. That's the Array Size function near the bottom of the block diagram in Figure 6–62.

You'll probably want a 2D version of this as well, which you had better know how to build by this point. To create it, save the above VI as `SGL 2D Mean.vi`, make **X** a 2D array, then make **mean** a 1D array. You may want to throw in an array-transposing option as well.

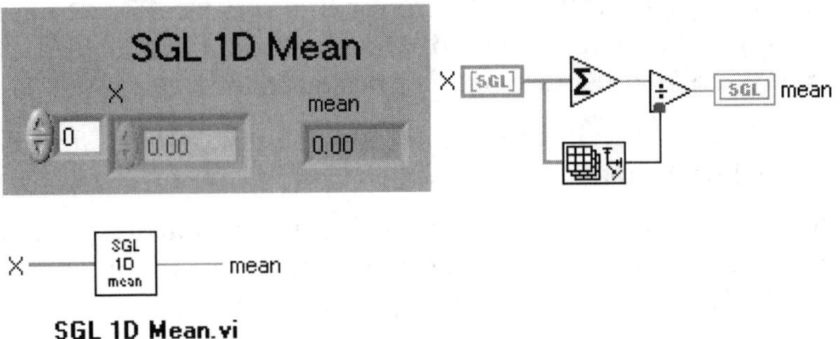

Figure 6–62
A homemade averaging VI is much faster than LabVIEW's, but I'm not sure why.

6.5 Speed of Execution

When your program is running, many factors affect its speed. There's no way I can completely cover these topics, as speed is affected by how you write your program, so I will cover the more important points.

1. Graphical operations on your front panel can take a long time to display.
   - Larger graphs, charts, and picture controls draw more lowly than smaller ones.
   - You can update your graphs and charts every few iterations of the loop they're in, rather than every iteration, by using the Case Structure. Bear in mind that those iterations that perform the update will generally take longer than those that do not.
   - Do not overlap front panel objects; this takes longer than you might think.
2. Objects requiring large amounts of memory can take a long time.
   - Graphical operations, as described in item 1, can easily fall into this category.
   - Arrays and strings can take up a large amount of memory.

- If you have an array on the front panel, an extra copy of the array's data is kept in memory for that front panel object, regardless of whether it is a control or an indicator.
- Read Globals and Read Locals (as opposed to Write Globals and Write Locals) make a copy of their data.

3. Complex data types, such as arrays of clusters of arrays, can take a long time to process. Try to keep your data types simple, particularly those which are being manipulated.
4. Other applications running on your computer can slow down your application.
5. File I/O and networking operations (from any application running on your computer) can result in low-level delays that cannot be interrupted by other processes or threads.
6. Be careful not to build very large arrays one element at a time with the Build Array function. If you must build a large array, either build it with an indexed loop tunnel, or first preallocate the entire array in a shift register, then use the Replace Array Subset function to replace each element of the large array.

6.6 Display Techniques

6.6.1 Screen Resolution

If you're developing LabVIEW programs with a screen resolution of anything less than 800 × 600, you're wasting your time. See Appendix E, item 7 for more details.

If you open up a VI from a system with a larger screen resolution, your front panel or block diagram may be partially or totally off your screen! To find your front panel and/or block diagram if it's off-screen, use <Ctrl-T> (the tiling function). Alternately, in Windows operating systems, your window may allow the use of <Alt-Space> to begin movement or sizing. I learned this trick from a National Instruments LabWindows/CVI developer, all of whom seem to be keyboard wizards.

6.6.2 Displaying Numerous Numerics

Suppose you wanted to display hundreds of numbers simultaneously from your DAQ experiment. You could use the Multicolumn Listbox or Table indicator (from **Controls»List & Table**) if you really want to maximize your use of front panel real estate. To build a grid of numbers in which you can control the color of each number in order to indicate an alarm condition, create the VI shown in Figure 6–63 using these tips:

1. First, drop only one Simple Numeric control from the **Functions»Classic Controls** palette—I chose this one to optimize screen area. Change it to an indicator. Label it "n 0" so that when we begin cloning, the numbering will happen automatically—this would not work if we had used the label "n0".
2. Create a Property Node for "n 0" and change its property to the one shown in the figure (use its **Properties»Numeric Text»Text Colors»BG Color** pop-up menu item). Make this Property Node a "write" node with its **Change All To Write** pop-up menu item.
3. To most efficiently create 11 copies of our numeric indicator *and* its Property node, we cannot clone them. Instead, select both nodes on the block diagram, then copy and paste them repeatedly (<Ctrl-C>, click where you want the copy, then <Ctrl-V>).
4. The two boxes connected to the Select function are Color Box constants from the **Functions»Numeric»Additional Numeric Constants** palette. Using your Coloring tool and Color Copying tool , color the top one the same gray color as the front panel, and the bottom one white.
5. The Case Structure has 12 cases wired just like the three shown in Figure 6–63.

Run the VI and see how the alarm conditions are shown.

6.6.3 Speeding Up Graphs (or Charts)

Compared to many other common front panel items, graphs and charts take a long time to display. Here are a number of techniques for speeding them up:

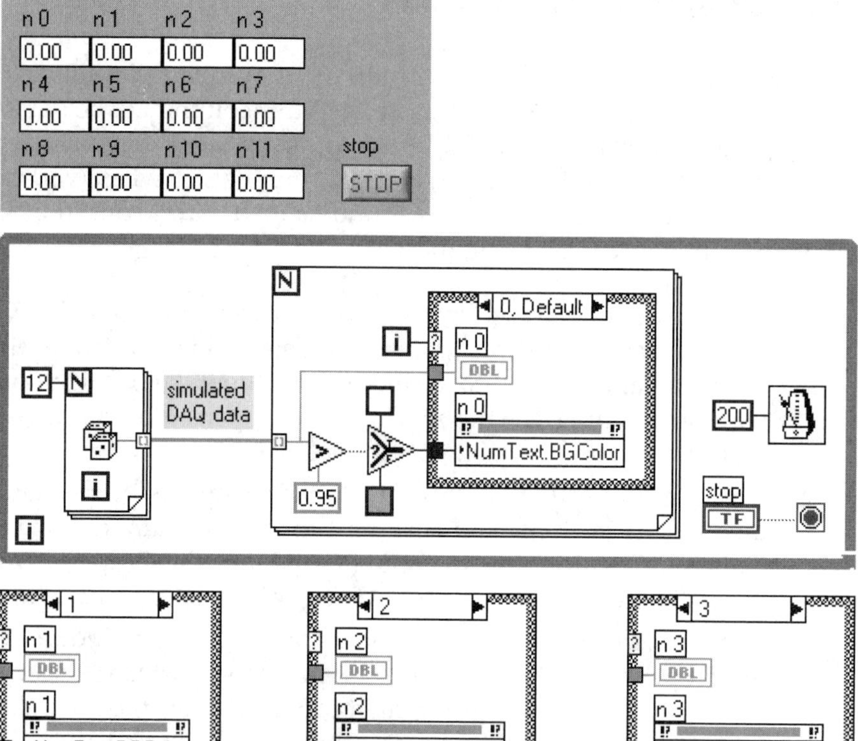

Figure 6–63
Many numeric values are shown.

1. Get a faster computer or a faster video card.
2. Make your graph smaller. The time required to update a graph is roughly proportional to the area of the graph.
3. If the graph is a chart, you can set its update mode with its pop-up menu item **Advanced»Update Mode**. The slowest update mode by far is **Strip Chart**, and the fastest is **Sweep Chart**.
4. Do not have other front panel objects overlapping the graph.
5. Turn off autoscaling if it's not needed.

6. Only update the graph every N iterations of its enclosing loop with a Case Structure. The graph will still take just as long to draw whenever it does.

7. If the technique in step 6 interferes with your DAQ timing whenever the graph is drawn, you can use two While Loops running simultaneously. One loop has the graph, and the other has the DAQ operations implemented as a subVI with a high priority (via **File»VI Properties»Category»Execution»Priority**). This priority business is a very inexact science, so you'll need to experiment with it! You'll also need to use a local or global to get your data from the DAQ loop to the graph loop, and this can be tricky.

6.7 Alarms

Visible or audible alarms will let you know if something is going wrong with your DAQ project, and there are many ways to implement them.

6.7.1 Visible Alarms

The simplest way to create visible alarm on a front panel is to set a big Boolean indicator to True, where True is a noticeable color. Flashing this indicator makes it even more noticeable. The second simplest method, if your front panel is already crowded (as most are), is to change the color of an existing very large control (or controls). The third simplest method is to show a dialog box. If you must still perform DAQ when this dialog box is showing, things become less simple; refer to Section 6.1.4 if you want to try this.

There is *still* no way to programmatically change the color of your front panel. You can, however, create a giant Boolean indicator that covers the entire background of your front panel to simulate such a thing. Another more effective approach is to use a digital output on your DAQ device to drive a relay that drives a bright flashing light. This is appropriate for noisy industrial environments or whenever the operator may not be able to see the monitor.

6.7.2 Audible Alarms

The simplest audible alarm is to use the computer's sound card to play WAV files, or some other sound file, through common amplified speakers connected to your computer. The **Functions»Graphics & Sound** palette has functions that can handle this. There are a number of disadvantages to this sound card approach:

1. Your computer's sound driver might become disabled at any time without your knowledge. For example, if somebody opens up another application that uses the sound driver, or sometimes even if your operating system is running very low on memory, LabVIEW might be unable to use it.
2. Somebody might turn down the speaker's volume.
3. Somebody might switch off the amplifier's power supply.

A more reliable approach is to use a digital output on your DAQ device to drive a relay that drives another audible alarm, such as a buzzer or loud siren. See Chapter 8, Section 8.2.2 if your DAQ device doesn't have any spare digital I/O.

6.7.3 Paging Techniques

There exist a number of modem-driving utilities on various Web sites that you can use to send a message to a pager. Sending an alphanumeric message can be especially helpful, as text is more descriptive than numbers. Sometimes you cannot connect to an analog telephone (for modem driver reasons), or you might have no access to any phone line at all (for security reasons). It is possible to use LabVIEW to send an email that will relay a message to a pager. You're on your own when it comes to implementing this, as this will vary per company, but NI's Internet Developers Toolkit (G Web Server) makes emailing from LabVIEW easy. If you work at a company with network-savvy staff, they will already know how to reliably send an email that will activate a pager.

6.8 Power Losses

Your computer may lose power for any number of reasons. Since this could happen while your DAQ experiment is running, you may want to consider taking the necessary steps to automatically continue your experiment in the event of power loss. Here are issues to consider:

1. For lengthy experiments, you will need to write your program so that it expects to automatically resume upon power up. To do this, you will need to occasionally write the status of your test to disk while the experiment is running, as the disk will likely be the only place with any data remaining when the power comes back on. Everything in memory, like shift registers and global variables, not to mention your entire LabVIEW program, is immediately terminated when your computer powers down.
2. If you're running Windows, you can place a shortcut to your VI in Windows' startup folder so it will start itself upon power up. This folder is typically something like `C:\Windows\Start Menu\Programs\StartUp`. A similar mechanism can be implemented on all non-Windows LabVIEW-compatible platforms.
3. You will want to use item 15 in Appendix E.
4. If your computer has dialog boxes that pop up and prevent LabVIEW from running during power up, you will need to figure out some way to get rid of these boxes. The usual such dialog box is the networking login box. If you're running Windows, Microsoft was kind enough to offer TweakUI, which can automatically log you in when this networking box shows up. Search the Web using a search engine to find TweakUI; it has a habit of moving around frequently, as does everything else on Microsoft's Web site. This has security consequences, of course.

Transducers

7

With DAQ, the signals you are measuring or controlling will ultimately be translated into voltages compatible with your DAQ device. Even a current signal is converted to a voltage at some point; with current-reading DAQ devices, this conversion will occur internally. The devices that convert any real-world parameter to or from an electrical signal, often a voltage, are called *transducers*. NI has application notes devoted to many of these transducer types—this chapter is simply meant to give you a quick overview of the most common transducers. For an entire book on LabVIEW and transducers, read *Sensors, Transducers, and LabVIEW: An Application Approach To Virtual Instrumentation*, by Barry E. Paton, a physics professor. For a list of transducer manufacturers, go to *www.LCtechnology.com/daq-hardware.htm* and see the Transducers section.

For a few very common types of transducers, LabVIEW has customized hardware and software. These will be discussed on a per-transducer basis throughout this chapter. When you see *SCXI*, this is a type of NI hardware. Notice that you cannot spell "**excessi**vely **ex**pen**si**ve" without SCXI. SCXI DAQ devices are very rugged and are housed in beautiful aluminum chassis. SCXI devices are designed for industrial applications—but there are often less expensive alternatives from NI as well. An example of an SCXI system is shown in Figure P–4 of the preface.

Following is a categorization of the most common types of transducers:

Temperature
- *Thermocouples*
- *RTDs*
- *Thermistors*
- *Other Devices*

Force and Pressure
- *Strain Gauges*
- *Other Devices*

Flow Rate

Position

Other Transducers

7.1 Temperature Transducers

This section presents a wide range of temperature transducers, including thermocouples, RTDs, thermistors, and specialty devices.

7.1.1 Thermocouples

A thermocouple is perhaps the simplest type of transducer in existence; it converts temperature into a voltage. Simply twisting together any two pieces of wire, each made of a different type of metal, produces a thermocouple! Iron and constantan (a copper-nickel alloy) are two metals that will produce a J Type thermocouple. Other types of thermocouples are E, K, R, S, and T, each with their own two types of metals, and each useful over a different range of temperatures. Thermocouples can withstand higher temperatures than can most other temperature transducers, but they do have a couple of unique downsides: (1) they produce a very small voltage, and (2) they require a second temperature transducer called a CJC (cold-junction compensation sensor). The CJC and its physical connection to the thermocouple usually limit the thermocouple's measurement accuracy. With NI's present hardware, this accuracy ranges from +/- 0.9° C to +/- 1.4° C.

Figure 7–1 shows a simple thermocouple, which is simply two wires of different metals connected.

Figure 7–1
A simple thermocouple.

There are many packages for thermocouples involving connectors, shielding, and various other bells and whistles, but the essence is always a pair of metal wires inside.

The following table lists some common thermocouple types, as well as their temperature ranges. ·

Thermocouple Type	E	J	K	R	S	T
Temperature range, degrees Celsius	0 – 1,000	0 – 760	0 – 500	-50 – 250	-50 – 250	0 – 400

Some of NI's DAQ devices have their own CJCs built in. For example, a few of the SCXI terminal blocks have a CJC, like the SCXI-1328, and many of the SCXI modules, like the SCXI-1120, are recommended for thermocouples. If you are a good electrical engineer, you can supply your own CJC, such as the National Semiconductor LM-35CAZ. You can use any generic DAQ device with a high gain, with *differential inputs*, and with a +5 volt power supply (for the CJC) to measure thermocouples over a wide temperature range. You'll need an extra analog input for the CJC. Of course, you'll also need to read the spec sheet on your CJC to hook it up correctly, and you'll want to thoroughly understand the application note mentioned at the end of this section.

With thermocouples, the relationship between voltage and temperature is highly non-linear, and an equation is required. LabVIEW provides thermocouple equations in the **Functions»Data Acquisition»Signal Conditioning** palette's `Convert Thermocouple Reading.vi:` . For more detailed information on thermocouples, read NI's Application Note 043, *Measuring Temperature with Thermocouples—A Tutorial.*

7.1.2 RTDs

An RTD, or resistance-temperature detector, is a device whose resistance increases with temperature. In practice, an RTD is typically either a wire coil or a deposited film of pure metal. The most common type of RTD is platinum, which has a nominal resistance of 100Ω at 0° C. RTDs are very accurate, some of them having an accuracy as high as 0.026° C at 0° C. RTDs come with two-, three-, or four-wire connections, depending on their purpose, but no matter the number of wires, each RTD is just one resistor. Figure 7–2 shows some typical RTDs.

Figure 7–2
RTDs for temperature measurement.

Many of NI's DAQ devices are appropriate for RTDs. As with any resistive device, an excitation voltage is required, and certain types of DAQ hardware provide this—for example, the SCXI-1121.

With RTDs, the relationship between voltage and temperature is fairly linear, but not quite, so an equation is required. LabVIEW provides RTD equations in the **Functions»Data Acquisition»Signal Conditioning** palette's `Convert RTD Reading.vi:` . For more detailed information on RTDs, read NI's Application Note 046, *Measuring Temperature with RTDs—A Tutorial*.

7.1.3 Thermistors

Like RTDs, thermistors are thermally *very* sensitive resistors, but they are semiconductors made from metal oxides. The two types of thermistors are negative temperature coefficient (NTC), in which the resistance decreases with temperature, and positive temperature coefficient (PTC), in which the resistance increases with temperature. Figure 7–3 shows the basic element of a thermistor.

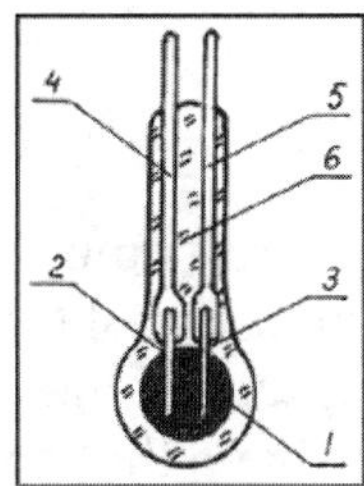

1. Semiconductor power compound
2, 3. leads contacting compound
4, 5. leads for outside connection
6. Glass hermetic seal

Figure 7–3
A thermistor.

The main advantage of thermistors is their sensitivity. As should be obvious, the tradeoff is that their temperature range is smaller than that of RTDs or thermocouples. Figure 7–4 shows temperature-resistance plots of a typical NTC thermistor versus a typical RTD.

Many DAQ devices are appropriate for thermistors. As with any resistive device, an excitation voltage is required, and certain types of DAQ hardware provide this—for example, the SCXI-1123. Thermistors' relationship between voltage and temperature is highly nonlinear, as shown in Figure 7–4, so an equation is required. LabVIEW provides thermistor equations in the **Functions»Data Acquisition»Signal Conditioning** palette's `Convert Thermistor Reading.vi:` . For more detailed information on thermistors, read NI's Application Note 065, *Measuring Temperature with Thermistors—A Tutorial*.

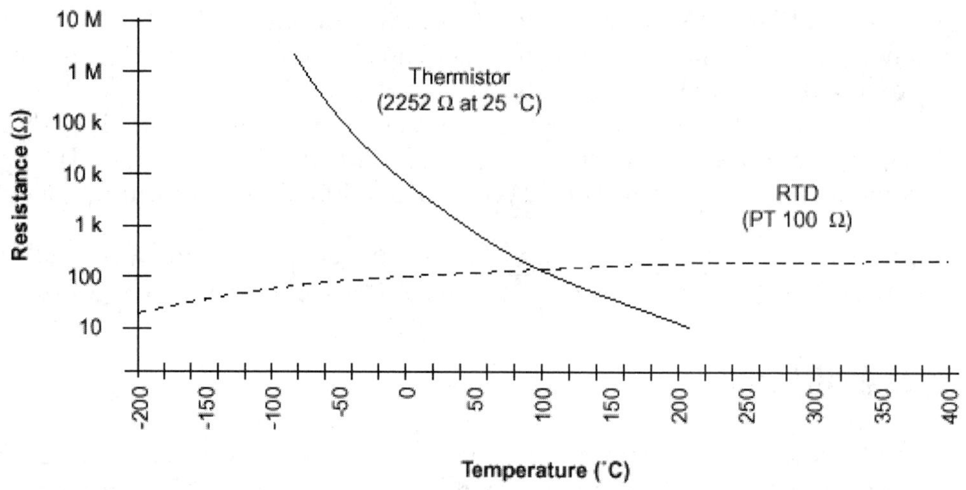

Figure 7–4
Thermistor versus RTD curves.

7.1.4 Other Devices

There are a few other types of temperature devices out there, some of which produce either a current (from Omega) or an RS-485 digital signal (National Semiconductor has some really nice chip-level devices). Either of these signal types is highly immune to noise, so consider them if they have the range and the accuracy you need. Concerning temperature range, you might wind up with some melted transducers if they can't handle the heat!

7.2 Force and Pressure Transducers

Transducers that measure force and pressure will be briefly presented in this section.

7.2.1 Strain Gauges

Strain, in the context of DAQ, is the "deformation of a material body under the action of applied forces," according to my dictionary. Since the physics behind strain gauges is fairly cumbersome to describe, I won't attempt it here; instead, please refer to the application note mentioned at the end of this section. Although there are a number of devices to measure strain, the most common is the *strain gauge*, whose electrical resistance varies with the amount of strain in the device. In other words, the more you bend the strain gauge, the greater its electrical resistance. A strain gauge can be a wire, but more often, it's a piece of plastic or similar material with metallic foil bonded to it in a grid pattern. Figure 7–5 shows an example of a common strain gauge layout, which is a metal foil bonded to a slightly flexible, electricity-insulating base, like plastic.

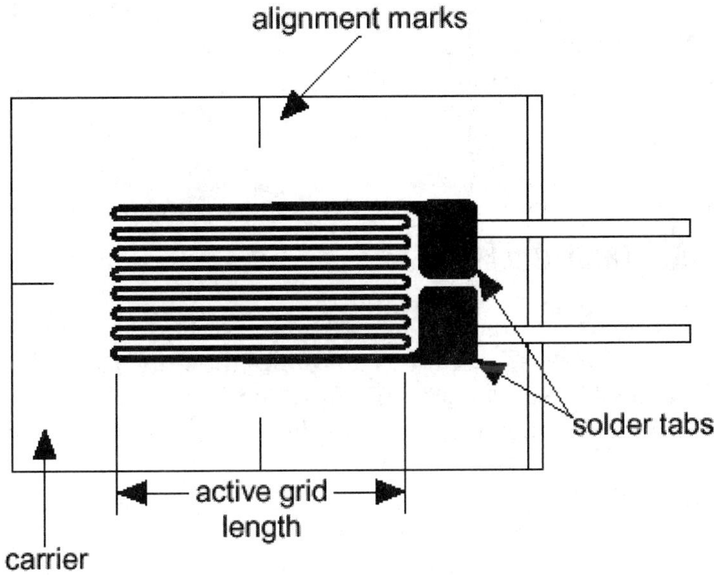

Figure 7–5
A strain gauge's basic parts.

The grid maximizes the length of wire exposed to the strain, as compared to a single wire. This entire strain gauge is then glued with epoxy (or some similar substance) to the device being measured. For accuracy reasons, multiple strain gauges with multiple wires are often used to measure one point of strain. As with any resistive device, an excitation voltage is required, and certain types of DAQ hardware provide this—for example, the SCXI-1121 and SCXI-1122 are designed for strain gauges.

Strain gauges have fairly complex equations associated with them. LabVIEW provides strain gauge equations in the **Functions»Data Acquisition»Signal Conditioning** palette's `Convert Strain Gauge Reading.vi`: [STRAIN CONV]. For more detailed information on strain gauges, read NI's Application Note 078, *Strain Gauge Measurement—A Tutorial.*

7.2.2 Other Force and Pressure Devices

Other devices for measuring force and/or pressure are *load cells, accelerometers, torque cells,* and *pressure switches,* all of which are often based upon strain gauges. Specialized transducers exist for low pressures (usually gasses, as in a vacuum) and high pressures (often for liquids or gasses).

7.3 Flow Rate Transducers

Some common names for flow rate (of gasses or liquids) transducers are *flowmeters, mass flowmeters,* or if they can *control* the flow, *mass flow controllers.* In many flow rate transducers, temperature sensors are used as the substance is flowing through a tube; in some cases, heat is applied to such a tube as well. Magnetic flowmeters have no moving parts (thus are more rugged), but work only with electrically conductive fluids by passing an electromagnetic field through said fluid flowing through a tube. If a fluid has a minimal amount of suspended particles or bubbles, ultrasonic flowmeters work by passing sound waves through the fluid as it flows through a tube. Figure 7–6 shows a mass flow controller (MFC) attached to some equipment I was recently testing.

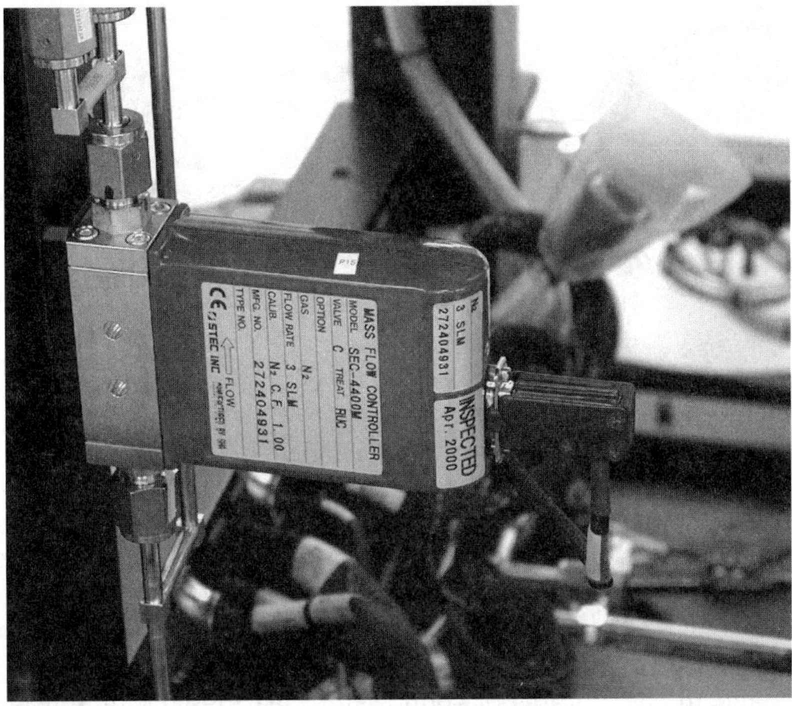

Figure 7–6
A mass flow controller used to test semiconductor-making equipment.

Of course, output is a part of DAQ as well. There exist hordes of valves that can control the flow rate of almost any gas or liquid.

7.4 Position Transducers

Figure 7–7 is a photo of a Daytronic LVDT (Linear Variable Differential Transformer).

Figure 7–7
An LVDT.

An LVDT is usually a piston that is mechanically attached to the specimen whose position is to be measured. It has magnetic coils inside that produce an AC signal. The LVDT's output is usually an AC or DC voltage signal.

Another common position sensor is the ultrasonic transducer. This transmits and receives pulses of sound, and measures the time it takes for the sound to bounce back from the object being measured. You could think of it as a little radar. Bats come with these devices organically preinstalled, although their frequencies dip into the audible range.

Ultrasonic transducers aren't much to look at, hence the bat illustration in Figure 7–8. However, bats are difficult to interface to LabVIEW programs.

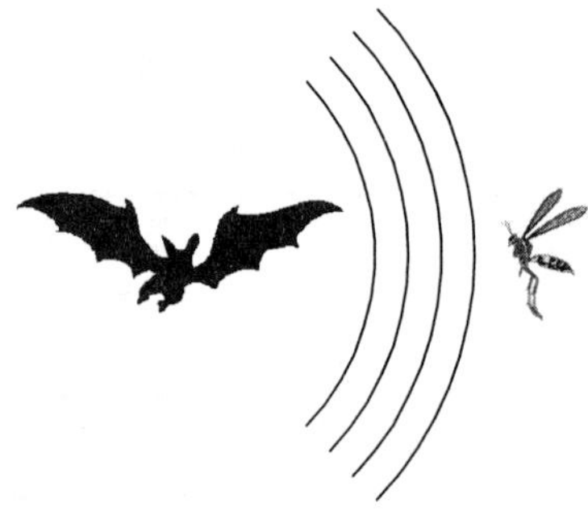

Figure 7–8
An organic ultrasonic transducer.

7.5 OTHER TRANSDUCERS

Almost any physical parameter that can be measured has some form of transducer available. Light intensity, pH, humidity, conductivity, all frequencies of electromagnetic radiation (radio waves, microwaves, infrared, visible light, ultraviolet, x-rays, gamma radiation), and so on all have specialized transducers. When the physical parameter becomes very specialized, such as measuring small concentrations of particles in a high-purity gas, expect a specialized transducer. Particularly, expect to pay lots of money for this specialized transducer, and to get some sort of bus-based interface to the computer instead of a simple voltage or current.

8

Non-NI Hardware Alternatives

This section is guaranteed to rile you readers who have been working at NI, as it mentions many direct competitors to NI. But I'm keeping this section in the book at the risk of getting evil looks from my NI contacts, as LabVIEW can be used with any sort of hardware. However, you are going to spend the least amount of time with DAQ in LabVIEW if you stick to NI hardware products. The obvious reason is that one company, NI, makes both the hardware and the software, and they're designed to work together. If you choose hardware from another company for use with LabVIEW, you are well advised to make *sure* the hardware has LabVIEW drivers. Unless you have lots of spare time, you don't want to write your own DAQ drivers!

In this chapter, we discuss non-NI hardware that could be used for DAQ. We discuss homemade hardware, including sound cards, relay interfaces, and using the parallel port for DAQ. Finally, we mention PLCs.

8.1 Non-NI DAQ Hardware Companies

Check out *www.LCtechnology.com/daq-hardware.htm* and see the DAQ Companies section for current information. These companies change all the time, so the above link should help. There are many other companies that make DAQ products, however. For unbelievably low-cost DAQ products, get the current copy of *Circuit Cellar* magazine. I cannot vouch for these companies' technical support, so be wary. Table 8.1 provides a listing of some of the more popular DAQ device suppliers.

Table 8.1 *DAQ Device Device Suppliers*

Supplier	Web site	Description
NI (National Instruments)	*www.ni.com*	Leading DAQ supplier; also makes LabVIEW.
Computer Boards	*www.computerboards.com*	Very low-cost DAQ equipment.
Keithley Instruments	*www.keithley.com*	General DAQ equipment.
ADAC	*www.adac.com*	Very low-cost DAQ equipment.
IOTech	*www.iotech.com*	General DAQ equipment.
LABTECH	*www.labtech.com*	General DAQ equipment.
ConnectTech	*www.connecttech.com*	Very reliable RS-232 and RS-485 equipment; can use Windows drivers, thus enhancing reliability.
B&B Electronics Manufacturing Company	*www.bb-elec.com*	Very reliable RS-232 and RS-485 equipment; converters from RS-485 to RS-232 allow you to use the PC's built-in RS-232 ports for RS-485.
Agilent	*www.agilent.com*	Makes high-quality lab instruments, not DAQ boards, but they can be used for DAQ purposes.
Tektronix	*www.tek.com*	See Agilent's description.

There are a number of DAQ-related publications, such as *Test And Measurement World*. See *www.LCtechnology.com/daqpubs.htm* for a current list of these publications.

8.2 Homemade DAQ Hardware

Are you the do-it-yourself type? I sure am! If so, this section is for you.

8.2.1 Sound Cards (Analog I/O)

A common sound card can be used for limited analog I/O. Sound cards have built-in DACs and usually ADCs, and as a bonus, they can use DMA for both input and output! Sound cards work in the 20 Hz to 20 kHz range, which is the frequency range of human hearing. Sound cards vary widely in quality, but with the advent of MP3, they went from low quality to high quality right away. Signal-to-noise ratios are now commonly above 90 dB (15 bits), which is good enough for most DAQ work. LabVIEW has VIs in its **Functions»Graphics & Sound** palette that can interact with sound cards. Another option is shown in Figure 8–1, the current version of a free function generator that works very well with common sound cards.

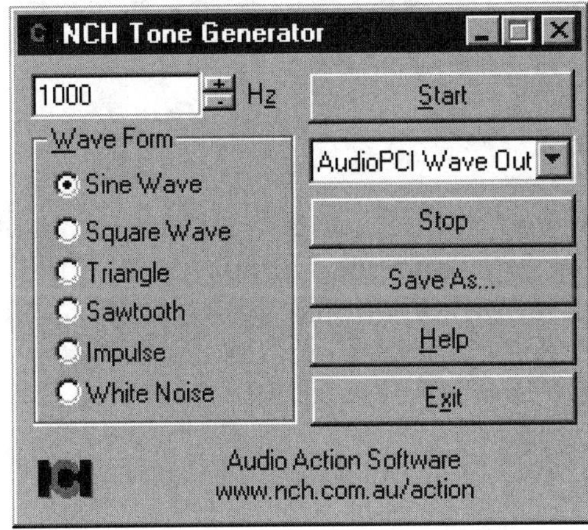

Figure 8–1
A freeware function generator.

LabVIEW's functions can be used with the sound card's inputs, and there are usually two such inputs: *line in* and *microphone*. There is usually one output, and this actually works better than many of NI's simpler DAQ boards, in that DMA is used so you will get an accurate waveform out, not to mention the greatly decreased cost. The disadvantages are (1) that the amplitude is usually limited to about three volts peak-to-peak, and (2) you will need to come up with your own wiring and connections scheme.

As luck would have it, my HP function generator just died about two days ago (it's older than I am), so I'm using a sound card as a function generator for one of my circuit design projects right now! Figure 8–2 is a graph of the sound card's actual output (a sine wave), being generated as I type by the software shown in Figure 8–1.

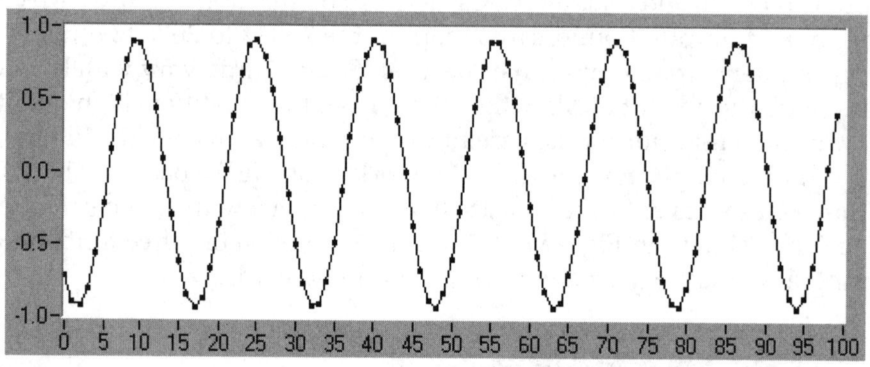

Figure 8–2
Actual output from a standard sound card.

Dell has high-quality refurbished sound cards; you can probably find some on their Web site, *www.dell.com*. Concerned about signal quality? As a rule (all these numbers are close approximations), every 6 dB of signal-to-noise ratio (SNR) corresponds to one bit of accuracy; 84 dB gives you 14 bits, 90 dB gives you 15 bits, 96 dB gives you 16 bits, 102 dB gives you 17 bits, and so on.

8.2.2 Parallel Port (Digital I/O)

A PC's parallel port can be used as a digital I/O DAQ device. It has TTL-level digital I/O lines—eight output lines and five input lines. Using the parallel port can be very tricky, especially if your printer driver is already using this port. The printer driver *may* do its own reading and writing in the background, thus interfering with your DAQ! For this reason, and to avoid frying your computer's built-in parallel port if you make a wiring mistake, you should buy a separate parallel port board. They are incredibly cheap, especially if you shop around. Accessing the parallel port can be quite tricky, but once you do access it, it's very easy to use. For useful information on this topic, see the Parallel Port section of *www.LCtechnology.com/daq-hardware.htm*.

8.2.3 Relay Interface

There are two general types of relays, solid state and mechanical. Solid state relays have the advantage of having no moving parts, so they typically last longer. The mechanical relay is an electromagnet with a coil that opens or closes a mechanical switch. Often, solid state relays require more current than many DAQ devices can provide from their digital outputs. Even more often, mechanical relays pull too much current for a standard DAQ device's digital output line, but worse, they can produce very high reverse voltages that could fry your DAQ device!

NI makes many relay boards, using either solid state or mechanical relays. Quite often you will find these economical. But if you want to build your own, Figure 8–3 provides a schematic.

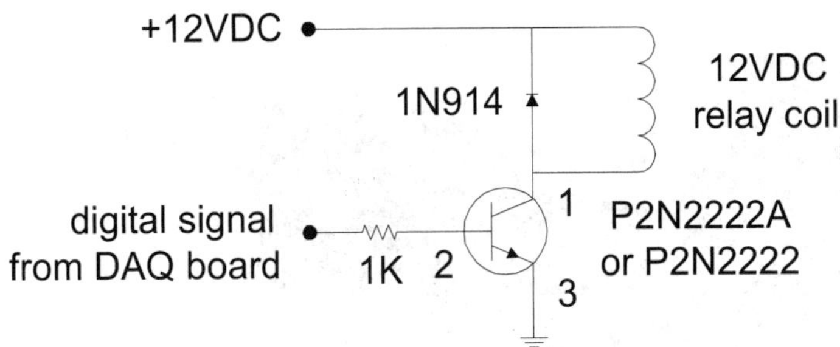

Figure 8–3
A very inexpensive, homemade relay driver is shown here; the diode is needed for mechanical relays with a coil, but may be omitted for solid state relays.

The relay's contacts are not shown, as they will vary from relay to relay. The 1N914 diode protects your circuitry against the high reverse voltage that can be generated by a mechanical relay. Your 12 volts can be unregulated, coming from a "wall wart"; just make sure it has enough current ability to drive your relay, especially if you're driving multiple relays. You could substitute other NPN transistors—just make sure their voltage rating is well above 12 V and their current-driving ability is enough for your relay. The circuit in Figure 8–3 will also work if you supply 5 volts to a 5-volt relay.

If you really want to build a cheap relay-driving system, you could use the parallel port pins to drive up to eight relays by using eight of these drivers! I have built such systems for testing motherboards.

8.2.4 Microcontrollers

If you are a real digital hardware hacker, you are already familiar with microcontrollers. If not, you may want to skip this section. If you want to learn about microcontrollers, you'll need an understanding of digital hardware. Microchip is the best company for learning microcontrollers at *www.microchip.com*. You can get a starter kit from Microchip, and with the help of a good book like *Easy Pic'N: A Beginners Guide to Using Pic16/17 Microcontrollers*, by David Benson, published by Square 1 Electronics, you will be on your way. Microcontrollers allow real-time, fast manipulation of digital signals, and with a bit of work, you can acquire analog signals and communicate with the computer's serial port.

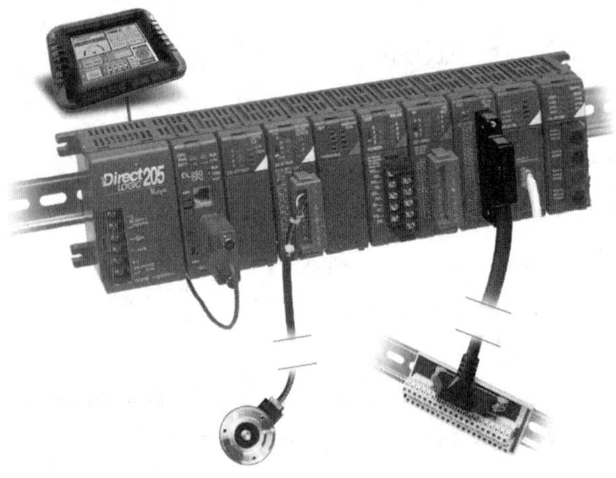

Figure 8–4
A PLC component.

8.2.5 Industrial Automation

A different type of DAQ has been traditionally used in industrial and manufacturing environments. One of the more popular devices is the PLC (Programmable Logic Controller). PLCs are very rugged and reliable devices used in industrial environments for process control. Two of the larger industrial automation manufacturers are Allen Bradley *(www.ab.com)* and GE Fanuc *(www.gefanuc.com)*. Figure 8–4 is a photo of an Automation Direct PLC controller (*www.automationdirect.com*).

These modules are usually stacked together side by side in a rack so that you can configure the PLC however you like. You can have industrial-strength analog and digital outputs that are compatible with nearly all devices used in an industrial or manufacturing environment. If you were to get an Ethernet module, you could use LabVIEW, along with NI's IA Server software, to communicate to the PLC via a standard network. The PLC has its own microprocessor inside, so you could program it as a standalone control system if you wanted. This way, a PLC can operate in a fairly real-time and reliable fashion without the insane overhead and tendency towards crashing found in a certain popular operating system.

Real-Time Issues

9

Newcomers to DAQ are very seldom aware of real-time issues until they build a system that doesn't work because of these issues. Luckily, the issues are rather easy to explain and understand, as will be done in this chapter. We will discuss real-time issues in operating systems, then we will discuss hardware and software techniques to handle these issues.

There are many definitions for the term *real-time,* so I will define it for this chapter: *a real-time system is one in which certain tasks can be guaranteed be complete within a specified time span.* This is sometimes called *deterministic.*

9.1 Timing in Operating Systems

All major operating systems are not real-time. Windows is very non-real-time. In simple terms, this means that any task you perform could take an arbitrary amount of time. For example, there is no guarantee that any Microsoft Word document can be opened in less than five seconds. In fact, there's technically no guarantee that your tasks will *ever* finish, as your entire system (particularly with Windows) could crash or hang at any time! Why are operating systems not real-time? There's a good reason. They must be flexible enough to accommodate any number of programs, as processor speed, memory, and overall computer power increase every year. Unlimited flexibility and real-time ability are opposing forces in operating system design, so very few operating systems are real-time.

The computer's processor (there may be more than one, but we'll use the singular term in this chapter) must switch between tasks, as it usually has several tasks virtually occurring at once. Disk activities typically take a higher priority than any other activities, so if you write a large file to a floppy disk (or read such a file from it), everything else that is processor-dependent will come to a halt. This includes DAQ, and that is exactly why we have DMA and FIFO, as discussed in Chapter 2: DMA and FIFO are not processor dependent. They allow a smooth, continuous stream of data points into or out of the computer, even when the processor is totally ignoring all DAQ hardware and software. You can get a good idea of what kind of delays to expect from your operating system by building the VI shown in Figure 9–1.

That's a String Indicator on the front panel with the scrollbar shown. The block diagram measures the time between consecutive iterations of the While Loop, and stores the maximum delay in a shift register. Every time a delay comes along that is longer than the previous maximum, the length of that delay is reported on the front panel, along with the time the delay occurred. The 5.00 second delay shown in the figure was intentionally caused by starting Windows' ScanDisk, a disk-scanning utility.

Suppose you are designing a control system for a fighter plane's wing movement during flight. First of all, you would *not* be using LabVIEW for this, but let's just pretend that you are. Suppose the inputs to your control algorithm come from the pilot's manual input and from the plane's instrumentation. Suppose the algorithm's output controls the wing's position. Let's also suppose that if the wing's position is not adjusted four times per second, your fighter plane will crash. Unfortunately, your plane is going to crash unless you have LabVIEW RT (to be discussed soon), because no amount of DMA or FIFO can help in this situation! Unlike data streaming into a computer, control loops involving feedback are less immune to a processor's absence, because control loops *require* actual processor interaction in order for the loop to operate properly. What is a control loop? It involves input and output, where the output must respond to the input within a certain time. Since a processor is required for such a response, a real-time operating system may be required, as is described in the next section. In general, if your control loop cannot withstand a processor delay of more than about 10 seconds, it will not be reliable. If this seems insanely long (it is), read the next paragraph.

My experience shows me there are two common situations in which lengthy operating system delays might occur: disk access and network access. Once, while running ScanDisk (a system tool that scans all disk drives for surface errors and various file I/O errors), I noticed a 14-second delay! This is a record for me, though, and I admit I was intentionally trying to see how big a delay I could create. Depending on what software and/or hardware drivers you have on your computer, you will see maximum delays that vary widely. If you have a delay-sensitive DAQ application, beware of disk and network activities. If you must read or write to disk, flush often (see Chapter 6, Section 6.4.3), as this will minimize your maximum disk-related delay.

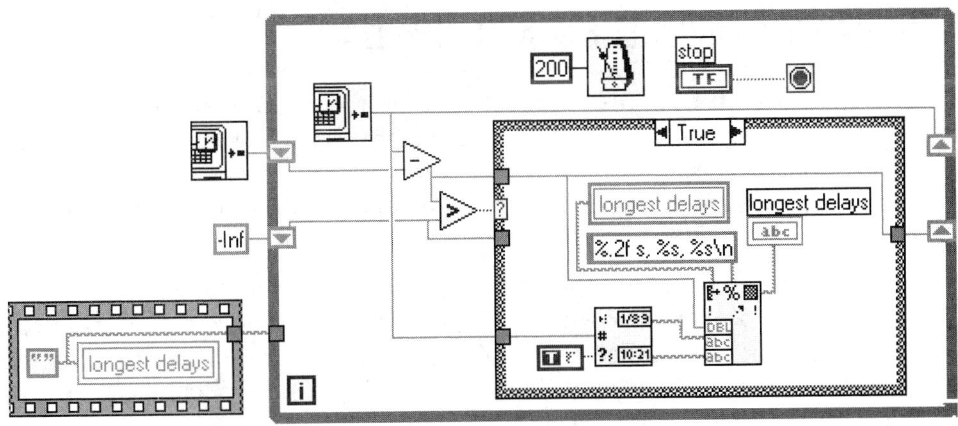

Figure 9–1
Observing your operating system's delays.

9.2 Hardware Techniques for Real-Time DAQ

Remember that DMA is not processor-dependent, so if your DAQ involves DMA, it will be unaffected during the processor's vacation. DMA and FIFO are fully described in Chapter 2, but to reiterate the key points, they can keep DAQ data streaming into or out of your computer when the computer's processor is ignoring DAQ. They do not provide truly real-time capability, however, as they can only handle processor absences of a finite length of time—all they really do is keep data streaming constantly in one direction or the other.

Probably the most reliable way to perform real-time DAQ using LabVIEW is with LabVIEW RT, where RT stands for real-time, as you might have guessed. NI has special DAQ devices (RT Series DAQ boards) that are required with LabVIEW RT. Currently, these devices are DAQ boards that live inside the computer and have their own processors; they are, effectively, embedded controllers. LabVIEW RT compiles real-time code specifically for these DAQ boards' processors, so that deterministic real-time performance is achieved; you can even warm boot your main computer, and LabVIEW RT will theoretically run unaffected.

9.3 Software Techniques for Real-Time DAQ

There is no such thing as true real-time performance using software timing alone in any non-real-time operating system (such as Windows or any other major operating system). However, there are a number of software techniques that greatly reduce the delay time encountered in real-time operating systems.

The best way to reduce the timing of your DAQ functions, other than NI's RT products, would be to write a device driver. A device driver can provide timing with much smaller delays, but it is still not truly real-time. Currently, the easiest way to develop a device driver for Windows is by using utilities from BlueWater Systems, *www.bluewatersystems.com*. These utilities exist for all of the popular versions of Windows. If you have months of time to kill, you could try writing a device driver for Windows without such utilities. No matter how you choose to write a device driver, its linkage to LabVIEW will be difficult at best. But remember, device drivers are not truly real-time, as they cannot guarantee a maximum delay in your program, but they can greatly minimize this maximum delay.

A completely different approach would be to use a real-time operating system. As of this book's writing, LabVIEW does not run under any such operating system, other than LabVIEW RT, which requires special hardware.

If you're good with microcontrollers, they can easily be programmed for real-time control. You could pass the data to LabVIEW via RS-232, but you need to be good with microcontrollers to even think about this.

DOS is very close to a real-time operating system, in my opinion. LabVIEW doesn't run under DOS, and NI doesn't support any DOS issues, but some of their older hardware and non-LabVIEW software products work very well under DOS. LabWindows for DOS, the predecessor of today's LabWindows/CVI product, is a wonderful, albeit simple, C-based development environment—if you want to use it, it helps if you know the C programming language. NI does not advertise LabWindows for DOS, it does not run on computers with more than 32 Mbytes RAM, and it is increasingly unsupported. With the need for real-time systems, it is beyond me why DOS seems to be disappearing to such an extent.

DAQ at a Distance—Networked and Distributed Systems

10

The term *intranet* refers to a local area network (LAN) of computers, in contrast to the Internet, which consists of all computers connected together worldwide. This chapter will discuss general networking issues, Internet DAQ, intranet DAQ, and finally locally *distributed* DAQ systems (connected by non-Ethernet means).

10.1 General Networking Issues

First, expect trouble if you're working with networks. Here's the standard solution to any problem from Windows' network troubleshooting wizard: "Contact your system or network administrator..." You will likely need access to a network expert to build a functional network. Once the network works reliably enough for you, it can then be used for DAQ. I have yet to experience a 100-percent reliable network.

Within any company, and even on my own home LAN, there exist firewalls for networks. A firewall is a barrier between an intranet (LAN) and the

Internet. There are often technical difficulties when trying to perform network activities through firewalls, so be prepared for some headaches.

You can set up your DAQ system so that if you should lose power, then regain power later, your system will reboot and your LabVIEW DAQ software will start running again. If you have networking on your DAQ system, the computer might prompt for a user password as it boots, which could prevent your DAQ software from automatically starting up. Chapter 6, Section 6.9 describes a workaround for this problem.

Most platforms have a utility that allows you to control one computer from another, provided some type of network connects them. If this is the case, and if it works for you, you might not need to use any of the fancy tools mentioned later in this chapter to handle your DAQ networking needs! Consider security and firewalls if you want to implement such a system. For Windows, my favorite tool is Symantec's pcAnywhere program, which you can read about at *www.symantec.com*.

10.1.1 Internet DAQ

LabVIEW can be used to build worldwide applications via the Internet. There are a number of ways to do this, and a whole book could be written on this topic. In fact, it has been written! If you're interested, get *LabVIEW Internet Applications,* by Jeffrey Travis, who (as I did) spent many years working at NI.

Since there's already that great book on the immense topic of LabVIEW Internet applications, I'll give only a quick summary of how you might use LabVIEW with the Internet for DAQ purposes.

DataSocket is a very nifty platform-independent way to exchange information over the Internet. Information and files can be exchanged among HTTP and FTP servers. You can specify DataSocket sources and targets (connections) using URLs (uniform resource locators) that adhere to the familiar URL model. DataSocket is essentially a subscription model in which you publish a data point (or arrays) to a DataSocket server, and other clients can read this data. Fortunately, many DataSocket examples ship with LabVIEW. All other major NI software products are using DataSocket.

NI has some interesting networking software for DAQ called RDA (Remote Device Access). RDA allows you to acquire data from DAQ boards in remote computers across the network, theoretically as if they were in your local computer.

LabVIEW still has an Internet Developers Toolkit (G Web Server), which allows you to do email and FTP, and to view your VIs across the Internet. Access control and security are part of this toolkit.

You can produce HTML (a common format for Web pages) with LabVIEW. First, you can print your VIs as HTML for documentation purposes. This ability is available from deep within the **File»Print...** menu item, after you hit the **Next** button quite a few times. Secondly, the Report Generation functions can output HTML.

Measurement Studio, another NI product, has an Internet component, which I will describe by stealing text directly from NI's product catalog (year 2000). Measurement Studio allows you to:

- View static snapshots of your VIs through standard Web browsers
- View animated VIs through standard Web browsers
- Specify update rates for animated displays
- Set up several client connections
- Control access to your VIs through the Web

Many items in the **Functions»Communication** palette can be used with the Internet. Each subpalette shown in Figure 10–1 has varying relevance to the Internet, described in Table 10.1.

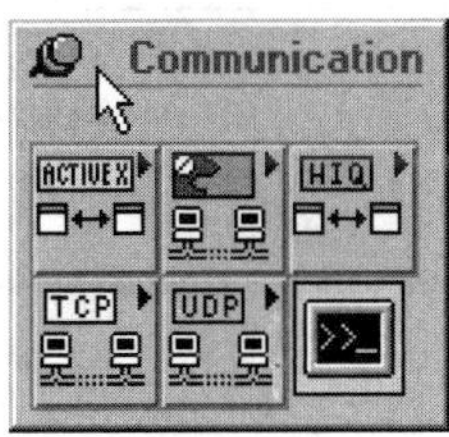

Figure 10–1
The **Functions»Communication** palette can be used with the Internet.

Table 10.1 The Subpalettes of the ***Functions»Communication*** *Palette*

Subpalette	Icon	Internet Relevance
ActiveX		(Windows only) ActiveX controls can be extremely useful. They can do almost anything that any program can do, so if you find one that has Internet capabilities, you can use it in LabVIEW.
DataSocket		DataSocket is a great way to stream data over the Internet.
HiQ		HiQ is not particularly relevant to the Internet.
TCP		TCP/IP (Transmission Control Protocol and the Internet Protocol) is a suite of communication protocols originally developed for military purposes, as was the Internet itself. Almost any kind of computer capable of networking can use these low-level protocols.
UDP		UDP (User Datagram Protocol) is another suite of communications protocols similar to TCP.
`System Exec.vi` (a VI, not a palette)		(Windows only) Performs text commands almost as if they were typed into the command line of a DOS box or entered via **Start»Run...** from the Windows task bar. These commands are not necessarily relevant to networks.

10.2 Intranet DAQ

Nearly everything mentioned in the Internet section is also relevant to intranets. There are a couple of differences, though. First, you don't need to worry about going through firewalls. Second, you can use the File I/O functions to fully manipulate files across different computers on your intranet, provided you've configured your network to allow such file sharing.

With regard to DAQ, by using file I/O across different computers, it is fairly easy for multiple computers to monitor and display data from one computer that is actually doing the DAQ work. With a little work, you can even control the DAQ computer from those multiple computers, though I recommend caution when trying this. If two computers are trying to control one computer or trying to write data to that computer at the same time, it can get confusing.

10.3 Locally Distributed DAQ Systems

We shift gears completely here, as we are no longer talking about traditional networks. Instead, we will talk about other ways to build a distributed system, by which I mean that DAQ devices might be connected to a computer tens, hundreds, or even thousands of feet away by a variety of electrical connections. Usually, such systems are found in industrial environments.

In order to transmit data over hundreds or thousands of feet of wire, you need a special electrical connection. A simple analog voltage is often not reliable, as noise will interfere too much. The two most common reliable connections are a current signal (usually either 0–20 mA or 4–20 mA) or an RS-485 or RS-422 (serial) signal. A current signal can usually provide good accuracy for thousands of feet or more; beyond that, wire resistance becomes a factor. An RS-232 connection is only good for up to 50 feet, whereas the RS-485 or RS-422 signal is good for up to 4,000 feet. In either case, using shielded wire will reduce the chance of any electrical noise affecting the signal.

If you want to lengthen just a few voltage signals, you might want to convert them to current for most of the signal length, then back to voltage on either end. If you want to lengthen just a few RS-232 signals, you might want to similarly convert them to RS-485 and back. But if you have many such signals, NI probably has cost-effective products for you. NI has an entire section in its product catalog on distributed DAQ systems.

Optical fiber is also effective for thousands of feet, and can be extended with repeater units. It is expensive, but immune to noise and electrical hazards.

One objective is to minimize your wiring. If you can take advantage of an existing network, and the networking techniques in the previous sections of this chapter will work, do it! Standard Ethernet networking, in conjunction with RDA or other techniques, can often do the job.

FieldPoint is a modular distributed I/O system for industrial DAQ applications. FieldPoint has 2-channel, 8-channel, and 16-channel I/O modules which are mounted on standard DIN rails (standard rack-mount rails), and can communicate via RS-232, RS-485, wireless hardware, Foundation Fieldbus (an all-digital communications network), or Ethernet. Figure 10–2 shows two different views of FieldPoint modules.

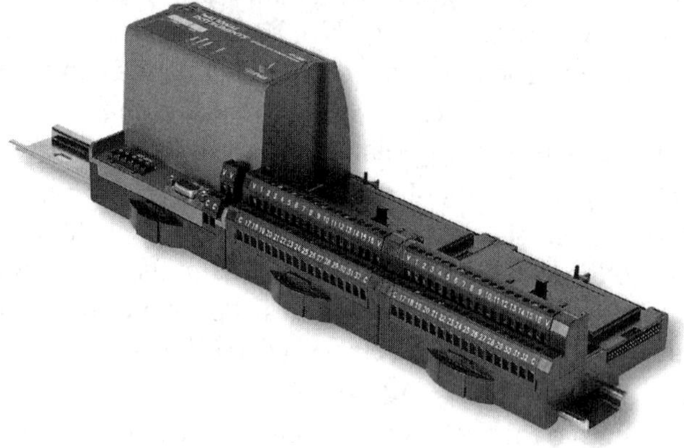

Figure 10–2
NI's FieldPoint hardware: in the field, then up close.

FieldPoint equipment is perfectly suited to industrial applications because it has a wide temperature range (-40 to 70 °C), isolation, programmable power-up states, and other industrial features.

Remote SCXI allows you to use standard SCXI products with a local digitizer in a remote chassis that communicates via RS-232 or RS-485 to a PC.

For widely distributed DAQ applications with few channels, the *6B Series* devices may be appropriate. They are single-channel digitizing DAQ modules for thermocouples, RTDs, low-level voltages, and current sources. These modules plug into a 1-slot, 4-slot, or 16-slot backplane that communicates to a PC via RS-232 or RS-485. Examples of these 6B Series devices are shown in Chapter 2, on the right in Figure 2–26.

Alternate Software for DAQ

11

Are you already familiar with another programming language, such as Microsoft Visual C++ or Microsoft Visual Basic? You may not want to use LabVIEW if this is the case. NI makes an excellent option to LabVIEW called LabWindows/CVI, which is based upon the C programming language. Measurement Studio is the name of NI's product that includes LabWindows/CVI, ComponentWorks (customized for Visual Basic), and ComponentWorks++ (customized for Visual C++). If you don't want to buy Measurement Studio, you can access NI-DAQ functionality directly from most C compilers, or from any environment that can access DLLs. See Section 11.4 for further details.

11.1 LabWindows/CVI

LabWindows/CVI, or just CVI hereafter, ships as a part of Measurement Studio. I just installed CVI 5.5 on my computer and ran the `ai_samp.prj` in CVI's `sample` folder, which is analogous to LabVIEW's `examples` folder.

ai_samp.prj collects and displays analog input data; I ran this CVI project with absolutely no modification. I connected analog input channel 0 to analog output channel 0 on my DAQ device, then free-ran LabWIEW's AO Update Channel.vi, as we've done earlier in this book (see Figure 11–1).

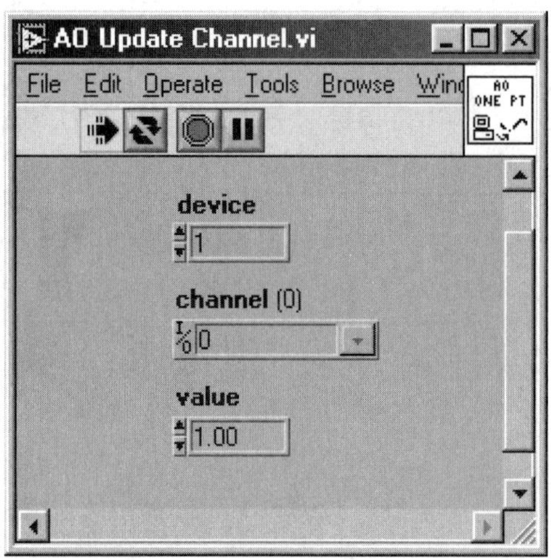

Figure 11–1
LabVIEW continuously updates analog output channel 0.

Clicking the up/down arrows on the LabVIEW screen shown in Figure 11–1, my CVI sample project displays the window shown in Figure 11–2.

Looks like LabVIEW, doesn't it? Sure it does, until you see the code that drives CVI. This code is not graphically oriented, like LabVIEW's, but text oriented. Following is the core of the actual code (with a bit of white space removed) used to update the CVI panel you see in Figure 11–2.

```
int CVICALLBACK TimerCallback (int panel, int control, int event,
        void *callbackData, int eventData1, int eventData2)
{
    short error;
    static int errorMsgActive = 0;
    static int inc = 0;
    static int needToRestartActualRateCalc = 1;
    static double startTime, elapsedTime;

    switch (event) {
        case EVENT_TIMER_TICK:
```

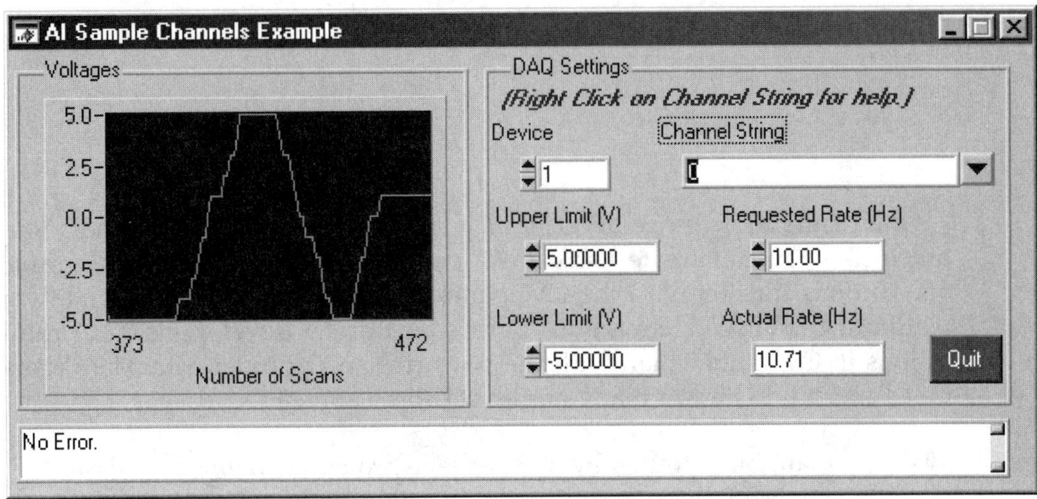

Figure 11–2
LabWindows/CVI continuously reads analog input channel 0.

```
            DisableBreakOnLibraryErrors ();
           error = AISampleChannels (device, channelString, upper,
lower, voltages);
            EnableBreakOnLibraryErrors ();
ErrorHandler (error);
if (numChannels)
                  PlotStripChart (panelHandle, PANEL_STRIPCHART,
voltages,
                          numChannels, 0, 0, VAL_DOUBLE);
if (needToRestartActualRateCalc)
            {
                startTime = Timer();
                needToRestartActualRateCalc = 0;
                inc = 0;
            }
if ((elapsedTime = (Timer() - startTime)) >= 1.0)
            {
                 SetCtrlVal (panelHandle, PANEL_ACTUAL_RATE, inc/
elapsedTime);
                needToRestartActualRateCalc = 1;
            }
```

```
inc++;
break;
    }
    return 0;
}
```

One of the nice things about CVI is that you can create its panels very much like you can create LabVIEW's panels. Just drop objects like graphs and buttons directly onto the CVI panels and arrange them much like you would in LabVIEW. Here is a list of reasons I prefer developing DAQ applications in CVI rather than with the powerful and popular Microsoft Visual C++ (which is better for many non-DAQ applications).

1. C++, although touted by most college professors *living in academia* as powerful, often turns out to be a time sink as compared to C when used by programmers *living in reality.*
2. The manner by which CVI's panels' objects are linked to the code is very elegant and simple, much more so than in Visual C++.
3. When you use CVI's standard debugging mode, CVI will automatically detect when you write to an "out of bounds" location of an allocated array, whether that array has been allocated dynamically or statically. Visual C++ is likely to crash, but it might let you slip on by while corrupting some other piece of your program. This CVI feature is a real time saver.
4. CVI is designed for use with DAQ, like LabVIEW, and has a huge library of example DAQ projects.

The one thing I like better about Visual C++ is that it can detect uninitialized local variables. These can result in crashes that are intermittent and not reproducible.

NI people, both CVI fans and LabVIEW fans, are likely to send me letter bombs for saying this, but it's true: CVI is better for more complex projects, while LabVIEW is better for simpler projects. Suppose you are good with C and good with LabVIEW, and you were to consider all possible DAQ projects, ranging from simple to complex. Figure 11–3 is an admittedly rough sketch of how you should choose which to use.

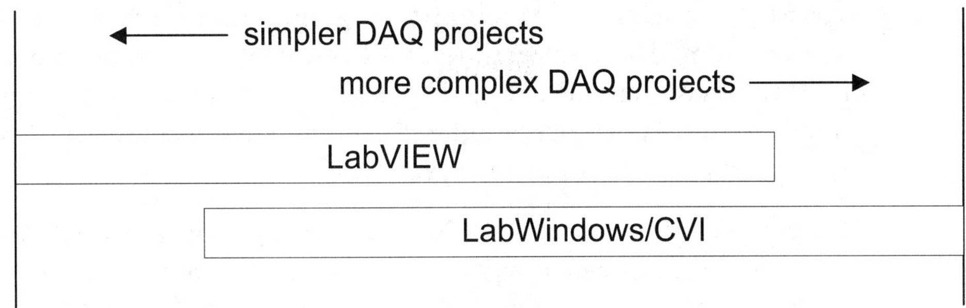

Figure 11–3
Deciding between LabWindows/CVI and LabVIEW. (**WARNING:** Controversy-generating diagram; lean towards CVI if you already know C.)

Hoping not to make any enemies at NI based on that last paragraph, I use some text directly from NI's Web site that describes many of CVI's features very well.

> **Design a Graphical User Interface.** Building an application in LabWindows/CVI begins with the user interface. You use an intuitive graphical user interface (GUI) editor to interactively design virtual instruments. Select from controls designed specifically for instrumentation, such as knobs, meters, gauges, dials, graphs, and strip charts, to build your GUI. As you place each control on the GUI, you double-click to customize its appearance and function to meet your needs.
>
> **Generate Program Code**. You can generate a C program using CodeBuilder. CodeBuilder automatically generates C source code to display and respond to the controls on your user interface. CodeBuilder creates code for you to respond to user events, such as mouse clicks, key presses, and menu selections.
>
> **Complete the Program with Function Panels.** Complete the application by inserting acquisition, analysis, and control code into the program. The LabWindows/CVI code generation tools, called function panels, help you use the built-in libraries and instrument drivers. A function panel, available for every function, is a graphical representation of a LabWindows/CVI function and its parameters. Simply set the values of each parameter to interactively build a function call. You can even execute the function from the function panel to test its operation, then automatically paste the function call into your source file. You save time,

bypassing the tedious process of typing and editing function calls in your program. The LabWindows/CVI development environment has an array of editing and debugging tools to streamline your programming, including a 32-bit compiler, linker, variable display, watch window, and full-function source editor.

Measurement Studio for CVI Multithreading. You can easily create and debug multithreaded applications in LabWindows/CVI. The LabWindows/CVI libraries are multithread-safe, and the LabWindows/CVI Utility Library contains a large set of functions to simplify multithreaded programs. The LabWindows/CVI development environment provides full multithreaded debugging capabilities, such as setting breakpoints that can be honored in any thread and viewing the state of each thread when the program is suspended.

Automatic Run-Time Checking When you run your LabWindows/CVI program in debug mode, LabWindows/CVI's patented User Protection feature automatically checks for program memory errors, such as writing beyond the end of an array or dereferencing an uninitialized pointer. If LabWindows/CVI encounters such an error, it stops the program and points to the offending line of code. LabWindows/CVI also checks the calls you make to its library functions for parameter values that might cause memory errors. For example, if you pass an array to one of the Analysis functions along with a count that indicates the array is larger than it really is, LabWindows/CVI stops your program and points to the function call. Additionally, you can configure the LabWindows/CVI debug mode to stop your program whenever a LabWindows/CVI library function returns an error. These User Protection features significantly speed up the development process.

Package Your Code for Delivery. When you finish your application, you can build an executable with a single mouse click. You can also build a dynamic link library (DLL) and incorporate your instrumentation code into external development tools or applications that work with DLLs, such as LabVIEW, Visual Basic, or other C/C++ development environments. You can package your code onto disks with the built-in LabWindows/CVI distribution kit builder and then download the code onto target computers.

11.2 Microsoft Visual Basic with DAQ (ComponentWorks)

ComponentWorks, like CVI, ships as a part of Measurement Studio. I defer yet again to the experts at NI, by snagging some wonderfully descriptive ComponentWorks text directly from their Web site.

> **User Interface Components for Visual Basic.** With these ActiveX controls in Measurement Studio, you can configure real-time 2D and 3D graphs, knobs, meters, gauges, dials, tanks, thermometers, binary switches, and LEDs to create professional instrument front panels in your computer-based measurement applications. For example, you can display waveforms acquired from DAQ boards or GPIB instruments. Or you can display slowly changing data, such as temperature, pressure, or strain, in a scrolling strip chart. The 2D graph displays multiple waveforms, X and Y axes, and interactive cursors.
>
> **Internet Components for Visual Basic.** Using the DataSocket ActiveX control, you can share live measurement data between applications via the Internet. DataSocket provides a simple mechanism for interacting with OPC, HTTP, FTP, and file servers from any ActiveX container. DataSocket delivers Plug and Play connectivity throughout your company by providing an easy interface for sharing live measurement data between applications separated by a network or the Internet. DataSocket delivers seamless access to live data, which empowers users throughout your company to improve productivity, reduce costs, and increase profitability. Use DataSocket to share data between Visual Basic, Visual C++, LabWindows/CVI, and LabVIEW applications.
>
> **Instrument Control Components for Visual Basic.** With Measurement Studio you can use GPIB, serial, and VISA I/O controls and intuitive property pages to configure communications with your instruments. You can easily send commands and receive response strings from instruments. To simplify the parsing of data strings, Measurement Studio features an interactive tool to define rules for parsing information out of instrument strings and keep only the data you need.
>
> **ActiveX Interchangeable Virtual Instrument (IVI) Controls for Visual Basic.** National Instruments Measurement Studio also includes ActiveX controls for communicating with two of the most popular IVI instrument classes, oscilloscopes and digital multimeters (DMMs). These

ActiveX controls increase your productivity when using IVI drivers as opposed to the DLL version of the IVI drivers. With the drivers, you make fewer software calls, thereby reducing configuration. They provide a less complicated programming interface to the hardware and enforce interchangeability by making sure you do not call instrument specific functions instead of the class functions. The controls also have a built-in user interface so you can access the functions of your instrument without programming. You can turn off this feature so that you can create your own user interface.

Plug-In DAQ Controls for Visual Basic. With the Measurement Studio DAQ controls, you can easily perform analog, digital, and timing I/O operations on all National Instruments DAQ boards. With these ActiveX controls, you configure your DAQ operations by setting properties in the intuitive property pages. There is no need for any low-level programming to set up your acquisition routines or to transfer buffers from your board to your computer—the ActiveX controls handle the details for you.

11.3 Microsoft Visual C++ with DAQ (ComponentWorks++)

ComponentWorks++, like CVI, ships as a part of Measurement Studio. I defer yet again to the experts at NI, by snagging some wonderfully descriptive CompentWorks++ text directly from their Web site.

User Interface Display Components for Visual C++. Because measurement applications often require real-time 2D and 3D graphs, knobs, meters, gauges, and more, Measurement Studio provides these flexible measurement-focused user interface components to simplify your development and save you time. Programmatically modify each property during the execution of the program to give you total programming flexibility.

Internet Components for Visual C++. Using the new DataSocket classes, you can easily communicate measurement data with multiple interfaces, such as OPC, HTTP, FTP, and DataSocket across any network, including the Internet. By using a client/server architecture, the data transfer across the network is optimized, making user interaction to live data across the Internet a reality. You will be able to view test information or control your system from anywhere in the world.

Analysis Components for Visual C++. Measurement Studio includes a powerful and comprehensive set of functions for analyzing data in Visual C++. You condition and transform your signal using smoothing windows, digital filters, frequency domain transforms, or measurement functions. With these powerful analysis routines, you can convert raw data into meaningful information and build robust virtual instruments.

Instrument Control Components for Visual C++. Interface your application to the outside world through Measurement Studio instrumentation classes. In addition to using the GPIB IEEE 488.2 library to send and receive commands to and from instruments, you can use VISA, an industry-standard I/O library, to communicate with your instruments. With VISA, you can control GPIB, VXI, or serial devices using the same set of components.

Application Wizard. To simplify the development of measurement applications in Visual C++, Measurement Studio features an application wizard. Built on the MFC AppWizard, you simply select the instrument drivers you want to include in your project and the type of measurement application you want to create, and the wizard automatically generates an MFC project with the necessary user interface, analysis, and instrumentation components.

Backward Compatibility with LabWindows/CVI. Use the LabWindows/CVI import wizard to preserve your legacy code. Any application developed in LabWindows/CVI can port to Visual C++ without changing a single line of code.

11.4 Other Programming Languages with DAQ

If you just need a simple interface to NI-DAQ from Visual C++, Visual Basic, or another programming environment, read this section.

For Windows users, you can access the free, low-level NI-DAQ functions directly without paying a dime for any of the Measurement Studio components, provided you have a development environment that can link to DLLs. Currently, NI-DAQ functionality ships in a DLL called `nidaq32.dll`, and its functions are documented in the *NI-DAQ Function Reference Manual*, which you can download from NI's Web site. Most C compilers and Microsoft Visual Basic can use this DLL directly.

If your application behaves as an ActiveX container, you can use the many powerful ActiveX Components of ComponentWorks, described in Section 11.2. For non-Windows users, there are similar mechanisms for accessing NI-DAQ functions in platform-dependent libraries from a variety of programming environments.

Finalizing Your LabVIEW Software

12

Once you've finally finished your project, there are certain steps you can take to lock your LabVIEW project, thus making sure it doesn't get changed or damaged. A non-LabVIEW programmer ultimately runs every single DAQ project I've ever done, hence this chapter. And all of these users are usually bright and/or curious people who will probably have the urge to modify the software. This is generally a bad thing, regardless of the user's intentions, for obvious reasons.

If you want to distribute your LabVIEW software to other computers, there are other steps you can take to make this distribution simple. These steps involve creating an `.exe` file by buying even more NI software, the LabVIEW Application Builder. Unlike other popular programming environments, LabVIEW hits you with a steep charge to distribute your application to other computers. This is one of the few disadvantages I see to using LabVIEW.

Section 12.1 describes how to protect your code from changes without building an `.exe` file, which may be useful if the code is still under development or on few machines. Section 12.2 describes how to protect your code by building an `.exe` file, which makes it even tougher to change the code, but requires the purchase of the LabVIEW application builder.

12.1 Finalizing Your LabVIEW Software on Your Development Computer

The objective here is to make it unlikely for your users to modify your code. You could create a standalone executable file, something like `My Tester.exe`, but in my opinion, that's not necessary on a machine that has LabVIEW installed. The subtle trickery shown in this section will make it very unlikely for anybody unfamiliar with LabVIEW programming to modify your VIs. Anybody who knows how to get around this soon-to-be-described trickery would also know how to build and overwrite your `.exe` file.

These steps are designed to allow the non-LabVIEW-knowledgeable user to fully run the VI, but to make it very unlikely that they can edit the VI.

1. You should save a backup copy of all of your code somewhere other than your on hard drive. See Appendix E, item 2, for more details.
2. Design your LabVIEW system so that one VI, called a top-level VI, controls every other VI.
   - Make your top-level VI run when opened (check the **File»VI Properties»Category»Execution»Run When Opened** box).
   - Make your top-level VI and all others whose front panels will be showing behave like a dialog box (click the **File»VI Properties»Category»Window Appearance»Dialog** button). Most important, no tool bars should be showing, hence no Abort Execution buttons and no built-in menu items will allow the user to reach any block diagrams, where they might find other Abort Execution buttons.
   - Make your top-level VI exit when stopped (use the **Functions»Application Control»Quit LabVIEW** function). Otherwise, the user will wonder why things have "stopped working" when the VI has stopped. If your tool bar is hidden, as per item b above, this can make it hard for you, the developer, to edit the VI, since you have no Abort Execution button. So, use any kind of secret trick you want to allow stopping without exiting. I always place a little secret dummy file called `testplat.txt` in my application's directory that causes the

top-level VI to not exit when stopped; my code checks to see if `testplat.txt` exists, and if not, it exits. `testplat.txt` should only exist while you're editing the LabVIEW code—rename it to `testplat2.txt` or anything else when you want to make it not exist. Finally, if your exit-prevention code does not work, and you cannot edit your VIs, simply rename any of your subVIs; then you can open your top-level VI in a broken state, fix the exit-prevention code, then properly name the renamed subVI.

3. To make it very easy for your user to run the VI, create a shortcut on the computer's desktop to said VI.

12.2 Moving Your LabVIEW Software to Other Computers

There are many fancy options you can use to build a distributable LabVIEW application, like creating a custom icon, but the following are the basic steps I recommend in order to move your LabVIEW executable code to other computers.

1. Reach for your wallet and buy the LabVIEW Application Builder from NI and install it (unless you have it already).
2. Follow steps 1 and 2 from the previous section, if you haven't already.
3. Open your top-level VI's front panel, and stop it without exiting LabVIEW. Select the **File»Save With Options...** menu item. Click the **Application Distribution** button, then click the **Save** button, saving an `.llb` file to the same folder as your top-level VI.
4. Close *all* VIs, then open a new one. You should have only one blank VI open at this point.
5. Using the **Tools»Build Application or Shared Library (DLL)...** menu item (this menu item exists only if you have installed the LabVIEW Application Builder), bring up the window shown in Figure 12–1.

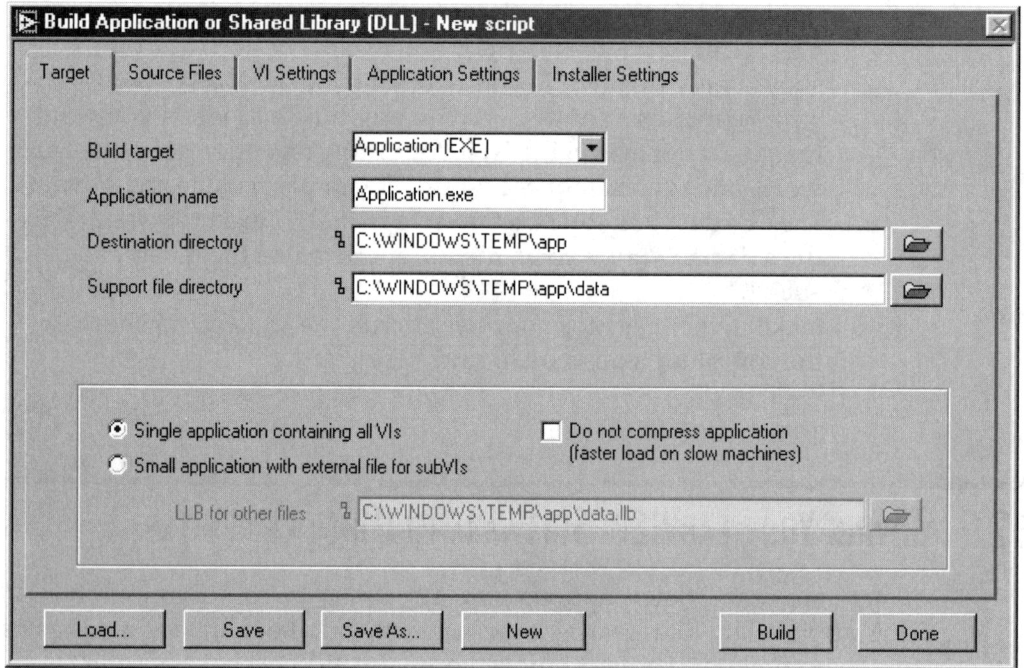

Figure 12–1
Main Window for building applications.

6. In the main window, do the following:

   - In the **Target** tab, set the **Application name** field to your desired `.exe` file name.
   - In the **Target** tab, set the **Destination directory** field the folder of your top-level VI.
   - In the **Source Files** tab, click the **Add Top-Level VI...** button and find the `.llb` file you created in step 3, double-click it, then add the top-level VI within this `.llb` file.
   - In the **Installer Settings** tab, check the **Create Installer** box, then set the various fields in this tab as you wish. I prefer installing all program files to my `C:\Program Files` folder.
   - Click the **Build** button, then watch your `.exe` file being built.

- Click the **Done** button, and create a script file (a `.bld` file) when asked so you can quickly rebuild later by reproducing these building steps with the **Load...** button, shown in Figure 12–1.

7. Optional: Create a shortcut on your desktop to your newly created `.exe` file on each computer on which your software is installed.

Appendix A

Fundamentals: Bits, Bytes, Files, and Data

The terms *bits, bytes,* and *files* will be explained in this appendix. Data formatting concepts will be discussed here as well.

General Data Concepts

Hexadecimal numbers (or *hex* numbers) are useful when describing bits and bytes, so understand hex first. Hex numbers are a base-16 counting system, whereas *binary numbers* are a base-2 counting system. You are probably more familiar with a base-10 counting system, since we humans have 10 fingers in general.

The 10 decimal digits are 0, 1, 2, 3, 4, 5, 6, 7, 8, and 9. The two binary digits are 0 and 1. The sixteen hex digits are 0, 1, 2, 3, 4, 5, 6, 7, 8, 9, A, B, C, D, E, and F.

First, have a look at Table A.1, then it will be explained.

Table A.1 A Comparison of Numeric Systems

Decimal Number	Hexadecimal Number	Binary Number	Binary Number with leading zeros
0	0	0	0000
1	1	1	0001
2	2	10	0010
3	3	11	0011
4	4	100	0100
5	5	101	0101
6	6	110	0110
7	7	111	0111
8	8	1000	1000
9	9	1001	1001
10	A	1010	1010
11	B	1011	1011
12	C	1100	1100
13	D	1101	1101
14	E	1110	1110
15	F	1111	1111
16	10	10000	00010000
17	11	10001	00010001

Each of the three counting systems above works exactly the same way: When the base number is reached on a particular digit, that digit is changed to zero and the next digit to the left is incremented. You can see this happen when counting with normal decimal numbers and you reach 10, which is the base number. The same thing happens when counting from 19 to 20, 29 to 30, 39 to 40, 99 to 100, and so on. With hexadecimal numbers, this happens in Table A.1 when counting from F to 10. With binary numbers, this happens very often—when counting from 1 to 10, 11 to 100, 101 to 110, 111 to 1000, 1001 to 1010, and so on.

In Table A.1, note the close correlation between the hex numbers and the binary numbers. For a multidigit hex number, each digit corresponds directly to a specific pattern of binary digits. For example, a hex 1234 is a binary 0001001000110100, where the hex 1 corresponds to binary 0001, the hex 2 corresponds to binary 0010, and so on. Because of this correlation, hex numbers allow you to more easily visualize the underlying binary representation of a number, while taking up less space than a binary number.

Because of the popular C programming language, hex numbers are often identified by a preceding 0x, like 0x10. Without this, a hex 10 might be confused with a decimal 10, when in reality, a hex 10 is a decimal 16.

Bits are the basis of all computer data. A bit is a unit of information that can take on only two values, often referred to as True and False, 1 and 0, or on and off. Imagine you have a penny lying flat on the floor; this could be considered a bit in that it is either True or False, depending on whether heads or tails is showing. According to *www.m-w.com*, a bit is "a unit of computer information equivalent to the result of a choice between two alternatives (as yes or no, on or off)."

A **byte** is an ordered group of eight bits, ordered from "most significant bit" (MSB) to "least significant bit" (LSB). For example, a byte could be viewed as a binary 00000101, where the MSB is 0 and the LSB is 1. This binary 00000101 is a decimal 5, or a hex 0x05. There are exactly 256 different combinations of eight bits, since $2^8 = 256$, meaning that one byte can take on exactly 256 different values. Imagine you have eight pennies lying flat on the floor, each with its own spot on the floor. You can flip them in exactly 256 different combinations. HHHHHHHT and THHHHHHH would be considered two different combinations, since each penny has its own spot on the floor.

Figure A–1 shows the first few and the last few of the 256 possible byte values, where the middle bytes are not shown to save space.

A byte can be interpreted in many different ways. In the context of human-readable, English text, a byte often has a one-to-one correspondence with letters. This is called ASCII (or *text*) data. Only the first 128 bytes are valid ASCII data, as illustrated by Table A.2.

Table A.2 ACII Data

Dec	Hex	Char	Dec	Hex	Char	Dec	Hex	Char	Dec	Hex	Char
0	0	NUL	32	20		64	40	@	96	60	`
1	1	SOH	33	21	!	65	41	A	97	61	a
2	2	STX	34	22	"	66	42	B	98	62	b
3	3	ETX	35	23	#	67	43	C	99	63	c
4	4	EOT	36	24	$	68	44	D	100	64	d
5	5	ENQ	37	25	%	69	45	E	101	65	e
6	6	ACK	38	26	&	70	46	F	102	66	f
7	7	BEL	39	27	'	71	47	G	103	67	g
8	8	BS	40	28	(	72	48	H	104	68	h
9	9	TAB	41	29	)	73	49	I	105	69	i
10	A	LF	42	2A	*	74	4A	J	106	6A	j
11	B	VT	43	2B	+	75	4B	K	107	6B	k
12	C	FF	44	2C	,	76	4C	L	108	6C	l
13	D	CR	45	2D	-	77	4D	M	109	6D	m
14	E	SO	46	2E	.	78	4E	N	110	6E	n
15	F	SI	47	2F	/	79	4F	O	111	6F	o
16	10	DLE	48	30	0	80	50	P	112	70	p
17	11	DC1	49	31	1	81	51	Q	113	71	q
18	12	DC2	50	32	2	82	52	R	114	72	r
19	13	DC3	51	33	3	83	53	S	115	73	s
20	14	DC4	52	34	4	84	54	T	116	74	t
21	15	NAK	53	35	5	85	55	U	117	75	u
22	16	SYN	54	36	6	86	56	V	118	76	v
23	17	ETB	55	37	7	87	57	W	119	77	w
24	18	CAN	56	38	8	88	58	X	120	78	x
25	19	EM	57	39	9	89	59	Y	121	79	y
26	1A	SUB	58	3A	:	90	5A	Z	122	7A	z
27	1B	ESC	59	3B	;	91	5B	[	123	7B	{
28	1C	FS	60	3C	<	92	5C	\	124	7C	|
29	1D	GS	61	3D	=	93	5D	]	125	7D	}
30	1E	RS	62	3E	>	94	5E	^	126	7E	~
31	1F	US	63	3F	?	95	5F	_	127	7F	DEL

Notice that many of the characters on your keyboard can be represented by a particular ASCII byte. The first 32 ASCII bytes are special characters, but there are a few worthy of mention. CR and LF are used together to form an *end of line* character on PCs, or they are used individually for the same purpose on other platforms. TAB is the tab character, which is similar to the space character (hex 20).

decimal	hexadecimal	binary
0	0x00	00000000
1	0x01	00000001
2	0x02	00000010
3	0x03	00000011
4	0x04	00000100
.	.	.
.	.	.
.	.	.
251	0xFB	11111011
252	0xFC	11111100
253	0xFD	11111101
254	0xFE	11111110
255	0xFF	11111111

Figure A–1
Comparing decimal, hex, and binary numbers.

Memory, or **RAM**, is what stores data in the form of bytes as a computer program is running. RAM is volatile, meaning that it loses its data when the computer's power is turned off.

A **file** stores nonvolatile digital data, meaning that data is retained when the computer's power is off. A file is an ordered group of bytes. Files are usually stored "to disk," since it is currently some kind of disk (rotating magnetic disk) that stores data. I suspect the term disk may not last long, as other technologies progress—but for now, we'll say disk. Disks take a relatively long time to access, compared to RAM, but their data hopefully stays put, regardless of whether the computer is on or off. Hard disks usually hold much more data than RAM. Floppy disks usually only hold about 1.4 megabytes (1,440,000 bytes) of data. If you had 8 × 1.44 million pennies lying on the floor (11.52 million pennies), you could arrange them in heads/tails patterns so that you had an exact copy of the information on a floppy disk! I do not recommend actually trying this experiment.

A file can even contain zero bytes, although that usually isn't very useful. Even if a file does contain zero bytes, every operating system retains a tiny bit of extra information associated with this file, such as when it was created and when it was last modified, but this other information should not be confused with the actual data inside the file.

If you opened up a text editor and typed in `Hello!`, then saved this information to disk in text-only format, you would have a file of six bytes. The first byte would represent the `H`, the second byte would represent the `e`, and so on. The binary (bit) representation of these bytes could be deduced from the ASCII table above:

01001000, 01100101, 01101100, 01101100, 01101111, 00100001

or in hex,

0x48, 0x65, 0x6C, 0x6C, 0x6F, 0x21

All software on your computer is composed of bytes, and is shipped to you in the form of files. When you launch a piece of software, or run a program, many bytes are pulled into RAM to facilitate speedy operation of your program. Any program you have is nothing but a collection of bytes, as is any file on your disks.

Numeric Processing

When any software manipulates data, it often needs to operate on numbers. Table A.3 lists two different types of numbers, as a computer sees them.

Table A.3 *Two Types of Numbers for Computers*

Numeric Category	Description
Integer	These numbers cannot have a fractional part. For example, 1, 2, 134, -43, and 0 are all integers, but 5.5 or -0.24 are not.
Floating-Point	These numbers can have a fractional part. For example, both 5 and 5.5 could be considered floating-point numbers.

Before discussing integers, we should cover some relevant terminology.

Signed vs. Unsigned: A *signed* integer can be positive, negative, or zero. An *unsigned* integer can be positive or zero, but not negative.

Range: Any type of integer has an inherent maximum and minimum value. One small type of integer is a signed 8-bit integer, which has a relatively small range of -128 to 127.

Bits: At the lowest level of the computer, all numbers are stored as a series of bits. Integers are often represented with 32, 16, or 8 bits. Bits affect the range of the integer.

Table A.4 lists the six basic integer types that computers currently use and their ranges.

Table A.4 Integer Types and Ranges

Integer Type	Range
Signed 32-bit	-2147483648 to 2147483647
Signed 16-bit	-32768 to 32767
Signed 8-bit	-128 to 127
Unsigned 32-bit	0 to 4294967295
Unsigned 16-bit	0 to 65535
Unsigned 8-bit	0 to 255

Notice in Table A.3 that the range for signed N-bit integers is $-2^{(N-1)}$ to $2^{(N-1)}-1$, whereas the range for unsigned N-bit integers is 0 to 2^N-1.

I prefer to use the signed 32-bit integer for most purposes. Suppose I wanted to represent the number 12345678 as a 32-bit integer; this number is 0x00BC614E in hex. On a Macintosh, if you stored this number in memory (or disk), it would occupy these four bytes (hex numbers shown),

0	BC	61	4E

which are composed of these 32 bits:

00000000 10111100 01100001 01001110.

On a "backwards" computer, such as a PC, numeric data has its bytes reversed, so the same number would occupy these four bytes (hex numbers shown),

4E	61	BC	0

which of course are composed of these 32 bits:

01001110 01100001 10111100 00000000.

A negative signed integer always has the most significant bit set to 1; non-negative signed integers have a 0 for the most significant bit.

Floating-point numbers have a different internal representation:

x = sign * (mantissa * 2 ^ exponent).

One bit is used for sign (negative or not), a few bits are used as the exponent, and the rest are used for the mantissa. IEEE has set a standard for 32-bit and 64-bit floating-point numbers called IEEE 754. While 32-bit floating-point numbers have this bit pattern,

```
s eeeeeeee mmmmmmmmmmmmmmmmmmmmmmm
```

64-bit floating-point numbers have this bit pattern:

```
s eeeeeeeeeee mmmmmmmmmmmmmmmmmmmmmmmmmmmmmmmmmmmmmmmmmmmmmmmmmmmm
```

The s's represent the sign bit. The e's represent bits in the exponent, and the m's represent bits in the mantissa. I could fill pages describing the technical details of floating-point numbers, but let's just skim their highlights here. Special bit patterns are reserved for Inf (infinity), -Inf (negative infinity), and NaN (not a number). Table A.5 lists the basic floating-point number types.

Table A.5 *Basic Floating-Point Number Types*

Floating-Point Type	Bits	Exponent Range	Mantissa Range
Single Precision	32	2^{-126} to 2^{127} roughly 1.2e-38 to 1.7e38	23 bits roughly 6 decimal places
Double Precision	64	2^{-1022} to 2^{1023} roughly 2.2e-308 to 9.0e307	52 bits roughly 15 decimal places
Extended Precision	>64	Varies with platform.	Varies with platform.

Appendix B

Top Ten DAQ Problems and Their Solutions

Problems

1. LabVIEW loops run at an irregular rate—some loop iterations take longer than others.
2. Your VI runs too slowly.
3. Delays are seen between analog input waveforms when trying to acquire one right after the other.
4. Digital outputs cannot drive relays.
5. Grounding issues not handled properly.
6. Electrical noise not taken into consideration.
7. Graphing during test slows down acquisition.
8. Third party board difficult to use with LabVIEW.
9. NI's DAQ device does not respond to anything in LabVIEW.
10. Program or computer crashes.

Solutions

1. Don't expect perfect LabVIEW loop timing unless you have a real-time operating system or are running LabVIEW RT with appropriate hardware (see Chapter 9).
2. There could be a number of problems, see Chapter 6, Section 6.6.
3. See Chapter 3, Section 3.3.1, *Continuous Analog Acquisition* (double buffering).
4. Make sure your digital outputs have enough current to drive your relays, plus reverse voltage protection if driving mechanical relays; NI makes many products with built-in relays.
5. See Chapter 2, Section 2.2.1, particularly the DAQ Connections Table (Figure 2–10).
6. See Chapter 2, Section 2.3.1.
7. See Chapter 6, Section 6.7.3.
8. Try to find a working LabVIEW driver for this board, or (a more common begrudging solution) buy an appropriate NI board, if available.
9. Many things could be wrong: (a) MAX is not set up correctly—make sure MAX's test panels work; (b) defective DAQ device—check its fuses, if any; (c) defective user—seek programming help (see Preface).
10. See Chapter 5, Section 5.3.

Please contact me (see Preface) with your top few DAQ problems so I can keep this list accurate for the next book revision.

Appendix C
Saving LabVIEW's VIs

You will often need to open a VI from LabVIEW's examples folder or a subVI from LabVIEW's functions palette. These VIs, and all VIs anywhere under LabVIEW's directory, belong to LabVIEW—you do not want to change these VIs and save them, or you may foul up some of LabVIEW's functionality! At some point, you may make a *minor* upgrade to a version of LabVIEW, in which all saved VIs will still be compatible with the previous version. In this case, it is okay to save these version-related changes to all VIs belonging to LabVIEW. For example, if you were to open LabVIEW after such a minor version upgrade, and you open any of LabVIEW's VIs, such as `Cont Acq&Chart (buffered).vi` from the `examples` folder, you may see a little asterisk like the one shown in the window's title bar in Figure C–1.

If you try to close the above VI, you will see a window like the one in Figure C–2.

You can hit the **Explain...** button to see what changes have been made. If the only kinds of changes are version-related, as shown in Figure C–3, and you're sure you haven't changed any of the subVIs, go ahead and save your VI with the **Yes to All** button, because you know you personally haven't fouled up any VIs.

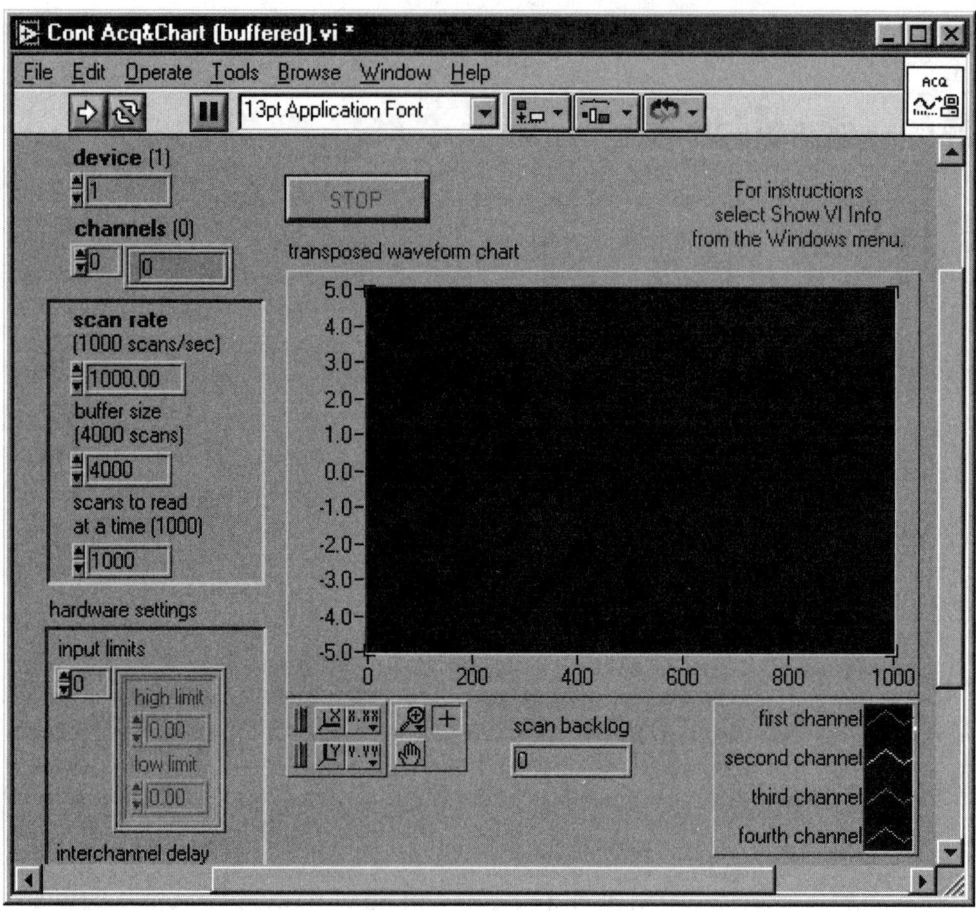

Figure C–1
The asterisk in the title bar indicates this VI has been changed.

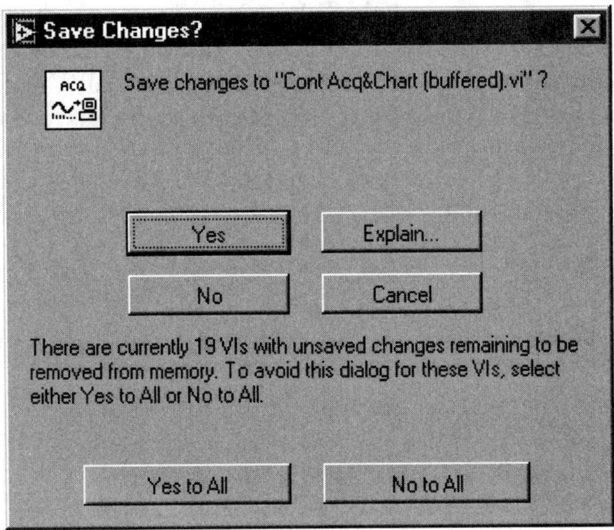

Figure C–2
Deciding whether to save your VIs.

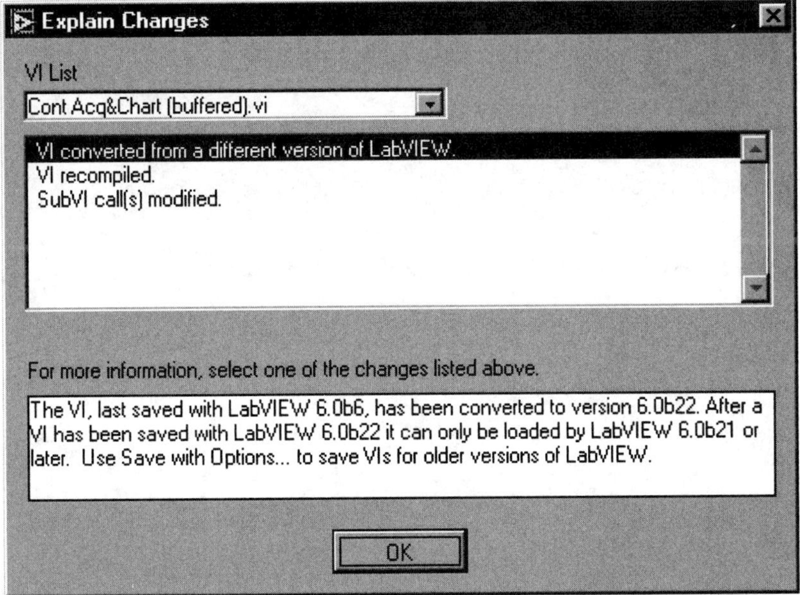

Figure C–3
Changes are only from a LabVIEW version change.

Whenever you upgrade LabVIEW, a nice trick for saving all of your VIs under LabVIEW's folder is to use the **Tools»Advanced»Mass Compile...** menu item to recompile (and automatically save) all VIs under LabVIEW's folder. This will take a long time, as there are many VIs; it took 25 minutes on my 400 MHz Pentium II.

Appendix D

Example Applications

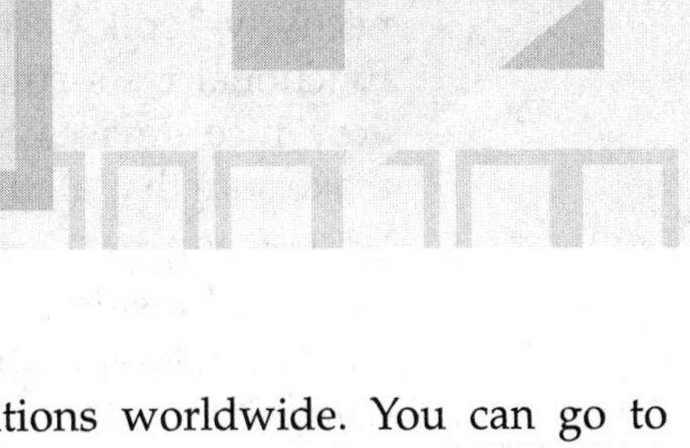

LabVIEW is used in thousands of situations worldwide. You can go to *www.ni.com* and see all the example applications you want through NI's Customer Applications link.

Following are a few LabVIEW projects I've done personally, and I'm saying just enough not to violate any NDAs.

1. LabVIEW is used to measure the force on materials as they are being pulled apart on a five station stress/strain machine. Force and position is recorded at 100 Hz during the tests, which may last hours. The data is displayed graphically during the test, and logged to disk throughout.
2. LabVIEW is used in a semiconductor-manufacturing environment to log data from gas purity detection equipment, as well as MFCs (mass flow controllers) throughout the plant. The MFCs detect the flow rate of gasses and are placed hundreds of feet apart, so 4-20 mA signals are sent to SCXI equipment. Inside three computers are eight-port RS-232 boards, each of which is connected to very high-dollar purity devices, some of which can detect particles, others moisture, and others oxygen. Three com-

puters are placed hundreds of feet apart and communicate all information to a series of other computers via Ethernet, using simple file I/O, nothing TCP/IP-specific. The system runs 24/7, and all data is 1 Hz except particle data, which takes about a minute to measure due to the equipment. This was not an overnight project!

3. LabVIEW collects data from multiple gamma radiation detectors, using memory-mapped data rather than NI DAQ devices, at 100 Hz, recording and displaying continuously how strong the radiation is at a variety of frequencies. I wish I could say why—it is a neat idea.
4. LabVIEW simultaneously tests 32 network-monitoring devices that have both serial and Ethernet connections. The Internet Connectivity Toolkit was needed for this one. A series of about 20 functional tests runs, which takes about 20 minutes. These are very nice devices with very diverse functionality, so we must make sure that all the parts work.
5. LabVIEW tests 20 pressure transducers at a time, while they are all placed inside a temperature-controlled chamber and taken to all sorts of temperatures, from freezing to hotter than Texas, while data is being recorded and displayed.
6. Same as item 5, but for batteries and motors that go inside airplanes, thus the test had a much wider temperature range. Airplanes get really cold up there, like –40°.
7. Monitor and control an Instron stress/strain machine. Instron's machines currently use a GPIB interface.
8. Very large objects are placed on a rocket sled, accelerated to insane speed, then crashed into a wall. Data is being recorded and displayed throughout this process. I am dying to go into detail, but I would be violating an NDA, which could possibly lead to my being used as a test object in this project.

Appendix E
LabVIEW/DAQ Tips and Tricks

1. Avoid redundant code. If you come back later and change one piece of code, you might forget to change the other. SubVIs are helpful in this regard.
2. Back up your work often (**File»Save All** is helpful). Hard drives often die after a few years, so back up *all* your important work frequently, not just LabVIEW code. Backing it up at a physically different site gives you automatic fire and theft protection, and coincidentally, free storage sites on the Internet abound. If you create three backup disks daily, and store them in your office, then the building burns down, that's no good. If you choose to back up via the Internet, zipping your files and password-protecting them is a good idea.
3. Did you make a mistake while building your VI? Use the **Edit»Undo...** feature <Ctrl-Z> when needed, or the **File»Revert...** menu item to return to the last saved version of your VI.
4. Simplify your block diagrams as much as possible—use subVIs wires where needed (to avoid redundant code and to reduce

block diagram clutter), and avoid crossing wires. Make big block diagrams tall or wide, but never both. See item 7.

5. If doing DAQ, buy NI's DAQ devices.
6. Use the keyboard, rather than your mouse, whenever possible.
7. Use a big monitor with a minimum resolution of 800 × 600 pixels. Make your front panel and especially your block diagram no larger than 800 × 600 pixels, if there is any likelihood that the VI will ever be used by someone with a smaller monitor. If you cannot do this on the block diagram, then expand either in the horizontal direction or vertical direction, but NEVER both, as it would become very difficult to navigate your block diagram. For development, I recommend using the largest monitor possible for LabVIEW in terms of *screen resolution*, not necessarily physical size.
8. Use a good mouse. Optical mice are nice, except for their price. They do not "gunk up" and require cleaning, like even the best mechanical mice. Speaking of good mechanical mice, I've had the best results from Logitech and Microsoft mice. Use a fabric-topped mouse pad for good traction; a slick plastic top on a mouse pad is worthless to the user once the mouse becomes the slightest bit sticky, but it saves the manufacturer a few cents. If your mouse starts sticking, clean the gunk from the internal rollers with audiocassette cleaning solution and a Q-Tip (or "cotton bud"), or use whatever it takes if the gunk is too thick for this to work quickly. I've had to scrape it off with a knife before!
9. <Ctrl-F> helps you find LabVIEW objects and/or text.
10. <Ctrl-T> tiles your windows, which is useful if your window is off screen.
11. Like most Microsoft applications, you can single-, double-, and triple-click text for different effects. Single-clicking places a cursor between characters, double-clicking selects an entire word, and triple-clicking selects an entire paragraph.
12. Have plenty of fuses laying around for your DAQ device.
13. Don't electrocute yourself when working with high voltages.
14. Don't electrocute your DAQ device when working with any voltage level—watch out for static electricity when handling your DAQ device.

15. If LabVIEW requires a login, you may want to disable this through the **Tools»Options...»Revision History** menu item's window. Setting the radio buttons therein to **Login automatically with the LabVIEW Registration name**, provided you have such a registration name, should eliminate this login box.
16. If using Microsoft Windows, you can capture the active window, such as a LabVIEW front panel or block diagram, to the Windows clipboard via the <Alt-Print Screen> key, then paste it to Paint or any other drawing program for printing or for use in other documents. This "quick and dirty" technique can be done, given similar trickery, on non-Windows LabVIEW platforms as well. There exist shareware packages that allow you to print the foremost window directly, without the need for an external drawing program.
17. For faster file I/O, use SCSI drives.
18. Become familiar with the Speed of Execution tips in Chapter 6, Section 6.6.
19. Always select **This Connection Is»Required** for input terminals in the Connector Pane, unless the input really must be optional.
20. To make room on the block diagram, <Ctrl-drag> the Positioning Tool. This is particularly useful within structures, like loops and the Case Structure.
21. Do not use bright colors as the background for your front panels. Keep them gray or pastel; otherwise, it looks unprofessional. Bright colors may be used sparingly for very small objects, or when you are flashing a warning light.
22. Most DAQ applications, even the very complex ones, can be controlled with just a few buttons on the front panel. But if you need so many buttons that your front panel looks cluttered, you may add your own custom run-time menu through the **Edit»Run-Time Menu...** menu item. LabVIEW's documentation will show you how to use such menu items, which can be found in the **Functions»Application Control»Menu** palette.

Index

A

B

C

D

E

F

G

H

I

L

M

N

O

P

R

S

T

U

V

W

About the Author

Bruce Mihura is owner of LC Technology, an National Instruments certified consultancy based in Austin, Texas specializing in writing custom data acquisition and control software. He holds a B.S. in electrical engineering and an M.S. in electrical engineering and computer science from the Massachusetts Institute of Technology. From 1989–1996 he worked at National Instruments as a LabVIEW software engineer and as the sole programmer of DAQ Designer, a data acquisition configuration utility. Since 1996, Bruce has been a software consultant using LabVIEW in about 50% of his work.

PH01

LICENSE AGREEMENT AND LIMITED WARRANTY

READ THE FOLLOWING TERMS AND CONDITIONS CAREFULLY BEFORE OPENING THIS SOFTWARE MEDIA PACKAGE. THIS LEGAL DOCUMENT IS AN AGREEMENT BETWEEN YOU AND PRENTICE-HALL, INC. (THE "COMPANY"). BY OPENING THIS SEALED SOFTWARE MEDIA PACKAGE, YOU ARE AGREEING TO BE BOUND BY THESE TERMS AND CONDITIONS. IF YOU DO NOT AGREE WITH THESE TERMS AND CONDITIONS, DO NOT OPEN THE SOFTWARE MEDIA PACKAGE. PROMPTLY RETURN THE UNOPENED SOFTWARE MEDIA PACKAGE AND ALL ACCOMPANYING ITEMS TO THE PLACE YOU OBTAINED THEM FOR A FULL REFUND OF ANY SUMS YOU HAVE PAID.

1. **GRANT OF LICENSE:** In consideration of your payment of the license fee, which is part of the price you paid for this product, and your agreement to abide by the terms and conditions of this Agreement, the Company grants to you a nonexclusive right to use and display the copy of the enclosed software program (hereinafter the "SOFTWARE") on a single computer (i.e., with a single CPU) at a single location so long as you comply with the terms of this Agreement. The Company reserves all rights not expressly granted to you under this Agreement.

2. **OWNERSHIP OF SOFTWARE:** You own only the magnetic or physical media (the enclosed software media) on which the SOFTWARE is recorded or fixed, but the Company retains all the rights, title, and ownership to the SOFTWARE recorded on the original software media copy(ies) and all subsequent copies of the SOFTWARE, regardless of the form or media on which the original or other copies may exist. This license is not a sale of the original SOFTWARE or any copy to you.

3. **COPY RESTRICTIONS:** This SOFTWARE and the accompanying printed materials and user manual (the "Documentation") are the subject of copyright. You may not copy the Documentation or the SOFTWARE, except that you may make a single copy of the SOFTWARE for backup or archival purposes only. You may be held legally responsible for any copying or copyright infringement which is caused or encouraged by your failure to abide by the terms of this restriction.

4. **USE RESTRICTIONS:** You may not network the SOFTWARE or otherwise use it on more than one computer or computer terminal at the same time. You may physically transfer the SOFTWARE from one computer to another provided that the SOFTWARE is used on only one computer at a time. You may not distribute copies of the SOFTWARE or Documentation to others. You may not reverse engineer, disassemble, decompile, modify, adapt, translate, or create derivative works based on the SOFTWARE or the Documentation without the prior written consent of the Company.

5. **TRANSFER RESTRICTIONS:** The enclosed SOFTWARE is licensed only to you and may not be transferred to any one else without the prior written consent of the Company. Any unauthorized transfer of the SOFTWARE shall result in the immediate termination of this Agreement.

6. **TERMINATION:** This license is effective until terminated. This license will terminate automatically without notice from the Company and become null and void if you fail to comply with any provisions or limitations of this license. Upon termination, you shall destroy the Documentation and all copies of the SOFTWARE. All provisions of this Agreement as to warranties, limitation of liability, remedies or damages, and our ownership rights shall survive termination.

7. **MISCELLANEOUS:** This Agreement shall be construed in accordance with the laws of the United States of America and the State of New York and shall benefit the Company, its affiliates, and assignees.

8. **LIMITED WARRANTY AND DISCLAIMER OF WARRANTY:** The Company warrants that the SOFTWARE, when properly used in accordance with the Documentation, will operate in substantial conformity with the description of the SOFTWARE set forth in the Documentation. The Company does not

warrant that the SOFTWARE will meet your requirements or that the operation of the SOFTWARE will be uninterrupted or error-free. The Company warrants that the media on which the SOFTWARE is delivered shall be free from defects in materials and workmanship under normal use for a period of thirty (30) days from the date of your purchase. Your only remedy and the Company's only obligation under these limited warranties is, at the Company's option, return of the warranted item for a refund of any amounts paid by you or replacement of the item. Any replacement of SOFTWARE or media under the warranties shall not extend the original warranty period. The limited warranty set forth above shall not apply to any SOFTWARE which the Company determines in good faith has been subject to misuse, neglect, improper installation, repair, alteration, or damage by you. EXCEPT FOR THE EXPRESSED WARRANTIES SET FORTH ABOVE, THE COMPANY DISCLAIMS ALL WARRANTIES, EXPRESS OR IMPLIED, INCLUDING WITHOUT LIMITATION, THE IMPLIED WARRANTIES OF MERCHANTABILITY AND FITNESS FOR A PARTICULAR PURPOSE. EXCEPT FOR THE EXPRESS WARRANTY SET FORTH ABOVE, THE COMPANY DOES NOT WARRANT, GUARANTEE, OR MAKE ANY REPRESENTATION REGARDING THE USE OR THE RESULTS OF THE USE OF THE SOFTWARE IN TERMS OF ITS CORRECTNESS, ACCURACY, RELIABILITY, CURRENTNESS, OR OTHERWISE.

IN NO EVENT, SHALL THE COMPANY OR ITS EMPLOYEES, AGENTS, SUPPLIERS, OR CONTRACTORS BE LIABLE FOR ANY INCIDENTAL, INDIRECT, SPECIAL, OR CONSEQUENTIAL DAMAGES ARISING OUT OF OR IN CONNECTION WITH THE LICENSE GRANTED UNDER THIS AGREEMENT, OR FOR LOSS OF USE, LOSS OF DATA, LOSS OF INCOME OR PROFIT, OR OTHER LOSSES, SUSTAINED AS A RESULT OF INJURY TO ANY PERSON, OR LOSS OF OR DAMAGE TO PROPERTY, OR CLAIMS OF THIRD PARTIES, EVEN IF THE COMPANY OR AN AUTHORIZED REPRESENTATIVE OF THE COMPANY HAS BEEN ADVISED OF THE POSSIBILITY OF SUCH DAMAGES. IN NO EVENT SHALL LIABILITY OF THE COMPANY FOR DAMAGES WITH RESPECT TO THE SOFTWARE EXCEED THE AMOUNTS ACTUALLY PAID BY YOU, IF ANY, FOR THE SOFTWARE.

SOME JURISDICTIONS DO NOT ALLOW THE LIMITATION OF IMPLIED WARRANTIES OR LIABILITY FOR INCIDENTAL, INDIRECT, SPECIAL, OR CONSEQUENTIAL DAMAGES, SO THE ABOVE LIMITATIONS MAY NOT ALWAYS APPLY. THE WARRANTIES IN THIS AGREEMENT GIVE YOU SPECIFIC LEGAL RIGHTS AND YOU MAY ALSO HAVE OTHER RIGHTS WHICH VARY IN ACCORDANCE WITH LOCAL LAW.

ACKNOWLEDGMENT

YOU ACKNOWLEDGE THAT YOU HAVE READ THIS AGREEMENT, UNDERSTAND IT, AND AGREE TO BE BOUND BY ITS TERMS AND CONDITIONS. YOU ALSO AGREE THAT THIS AGREEMENT IS THE COMPLETE AND EXCLUSIVE STATEMENT OF THE AGREEMENT BETWEEN YOU AND THE COMPANY AND SUPERSEDES ALL PROPOSALS OR PRIOR AGREEMENTS, ORAL, OR WRITTEN, AND ANY OTHER COMMUNICATIONS BETWEEN YOU AND THE COMPANY OR ANY REPRESENTATIVE OF THE COMPANY RELATING TO THE SUBJECT MATTER OF THIS AGREEMENT.

Should you have any questions concerning this Agreement or if you wish to contact the Company for any reason, please contact in writing at the address below.

Robin Short
Prentice Hall PTR
One Lake Street
Upper Saddle River, New Jersey 07458

About the CD

CONTENTS

This CD contains the evaluation version of LabVIEW 6*i* (for Microsoft Windows only). The CD also contains software built by the user in the book *LabVIEW for Data Acquisition* by Bruce Mihura. Updated versions of the software on this CD can be dowloaded at *www.LCtechnology.com/booksoftware.htm.*

This evaluation version of LabVIEW 6*i* must be installed on your hard drive. The installation program should automatically run when the CD is inserted—if not, run Autorun.exe in the root directory of the CD. This evaluation version of LabVIEW 6*i* is good for 30 days and will allow you to run your software for up to five minutes.

Additional software relevant to the book is in the "VIs For DAQ Book" folder on the CD. It contains most of the relevant software built and saved from the book, and should only be referenced if you're having trouble building the software.

SYSTEM REQUIREMENTS

You can use LabVIEW with the operating system of your choice without sacrificing portability or reuseability. You can share LabVIEW VIs on Microsoft Windows, Apple Mac OS, Linux, Sun Solaris, and Hewlett Packard HP-UX. Full system details can be found in README.txt on the CD.

TECHNICAL SUPPORT

Prentice Hall does not offer technical support for this CD. However, if there is problem with the media, you may obtain a replacement copy by emailing us at *disc_exchange@prenhall.com.*